Lecture Notes in Mathematics

Edited by A. Dold, F. Takens and B. Teissier

Editorial Policy
for the publication of monographs

1. Lecture Notes aim to report new developments in all areas of mathematics –
quickly, informally and at a high level. Monograph manuscripts should be reason-
ably-self-contained and rounded off. Thus they may, and often will, presentnot only
results of the author but also related work by other people. They may bebased on
specialized lecture courses. Furthermore, the manuscripts should provide sufficient
motivation, examples and applications. This clearly distinguishes Lecture Notes
from journal articles or technical reports which normally are very concise. Articles
intended for a journal but too long to be accepted by most journals, usually do not
have this "lecture notes" character. For similar reasons it is unusual for doctoral
theses to be accepted for the Lecture Notes series.

2. Manuscripts should be submitted (preferably in duplicate) either to one of the
series editors or to Springer-Verlag, Heidelberg. In general, manuscripts will be
sent out to 2 external referees for evaluation. If a decision cannot yet be reached on
the basis of the first 2 reports, further referees may be contacted: The author will be
informed of this. A final decision to publish can be made only on the basis of
the complete manuscript, however a refereeing process leading to a preliminary
decision can be based on a pre-final or incomplete manuscript. The strict minimum
amount of material that will be considered should include a detailed outline
describing the planned contents of each chapter, a bibliography and several sample
chapters.
Authors should be aware that incomplete or insufficiently close to final manuscripts almost always result in longer refereeing times and nevertheless unclear
referees' recommendations, making further refereeing of a final draft necessary.
Authors should also be aware that parallel submission of their manuscript to
another publisher while under consideration for LNM will in general lead to
immediate rejection.

3. Manuscripts should in general be submitted in English.
Final manuscripts should contain at least 100 pages of mathematical text and
should include
– a table of contents;
– an informative introduction, with adequate motivation and perhaps some
  historical remarks: it should be accessible to a reader not intimately familiar
  with the topic treated;
– a subject index: as a rule this is genuinely helpful for the reader.

Lecture Notes in Mathematics

1703

Editors:
A. Dold, Heidelberg
F. Takens, Groningen
B. Teissier, Paris

Springer
Berlin
Heidelberg
New York
Barcelona
Budapest
Hong Kong
London
Milan
Paris
Singapore
Tokyo

Richard M. Dudley   Rimas Norvaiša

Differentiability of Six Operators on Nonsmooth Functions and p-Variation

With the collaboration
of Jinghua Qian

 Springer

Authors

Richard M. Dudley
Department of Mathematics
Massachusetts Institute of Technology
Room 2-245
Cambridge, MA 02139, USA
e-mail: rmd@math.mit.edu

Rimas Norvaiša
Institute of Mathematics
and Informatics
Akademijos 4
Vilnius 2600, Lithuania
e-mail: norvaisa@ktl.mii.lt

With the collaboration of
Jinghua Qian
Department of Mathematics
Bucknell University
Lewisburg, PA 17837, USA
e-mail: jqian@bucknell.edu

Cataloging-in-Publication Data applied for
Die Deutsche Bibliothek - CIP-Einheitsaufnahme

Dudley, Richard M.:
Differentiability of six operators on nonsmooth functions and p-variation / Richard M. Dudley ; Rimas Norvaisa. With the collab. of Jinghua Qian. - Berlin ; Heidelberg ; New York ; Barcelona ; Hong Kong ; London ; Milan ; Paris ; Singapore ; Tokyo : Springer, 1999
   (Lecture notes in mathematics ; Vol. 1703)
   ISBN 3-540-65975-7

Mathematics Subject Classification (1991): Primary: 46G05, 58C20, 26A45, 26A42, 26E15; Secondary; 26A15, 45A05, 46G20, 60F15, 62G30

ISSN 0075-8434
ISBN 3-540-65975-7 Springer-Verlag Berlin Heidelberg New York

Typesetting: Camera-ready TeX output by the authors
SPIN: 10650182   41/3143-543210 - Printed on acid-free paper

Preface

This volume consists of four parts. Part I is a survey on differentiability of the six operators in view, namely, the two-function composition operator $(F, G) \mapsto F \circ G$, the two-function extended Stieltjes integral operator $(F, G) \mapsto \int F dG$, the convolution operator $(F, G) \mapsto F * G$, the multiplication operator $(F, G) \mapsto F \cdot G$, the inverse or quantile operator $F \mapsto F^{\leftarrow}$, and the product integral operator, which takes the coefficient matrix of a system of first-order linear differential equations with variable coefficients to a solution matrix-valued function, whose columns are vector-valued solutions. All the operators are defined on spaces of possibly discontinuous functions of a real variable. Part I also compares types of differentiability with a view toward differentiable statistical functionals and presents some facts needed for probability and statistical applications of the rest of the work. Part II treats the product integral with respect to Banach algebra valued functions and Fréchet differentiability of the induced operator with respect to p-variation norms, $1 \leq p < 2$, and quasinorms for $0 < p < 1$. It includes a treatment of the three-function operator $(f, g, h) \mapsto \int g \, dh \, f$ for functions with values in a possibly non-commutative and infinite-dimensional Banach algebra. None of the functions f, g, h need be of bounded variation, and they may have jumps at the same points, so ordinary Lebesgue-Stieltjes or Riemann-Stieltjes integrals are not sufficient. Rather, we expand on integrals defined by L. C. and W. H. Young. Part III treats differentiability of the composition and quantile operators uniformly over large sets of functions. Part IV is a bibliography including all items we could find on p-variation and the related notion of ϕ-variation. Parts I, II and III each have their own reference lists of items cited. Part IV, co-authored by Jinghua Qian, includes listings of references not cited in the other Parts.

We thank Dr. Qian very much for her contributions. We also thank Richard Gill, Aad van der Vaart, and Jon Wellner very much for stimulating and informative discussions. We especially thank Richard Gill for comments leading to Appendix E of Part I and for supplying code for a product integral symbol.

MIT, Cambridge, Massachusetts   Richard M. Dudley[1]
November 1998   Rimas Norvaiša[2]

[1] R. M. Dudley was partially supported by National Science Foundation Grants.

[2] R. Norvaiša was partially supported by a Fulbright Grant at Massachusetts Institute of Technology, by an NSERC Canada Collaborative Grant at Carleton University, Ottawa, Canada, and by U.S. National Science Foundation Grants.

CONTENTS

Part IV. **Bibliographies on *p*-variation and *φ*-variation**
by R. M. Dudley, R. Norvaiša and Jinghua Qian

PART I

A SURVEY ON DIFFERENTIABILITY OF SIX OPERATORS IN RELATION TO PROBABILITY AND STATISTICS

BY R. M. DUDLEY AND R. NORVAIŠA

CONTENTS

1. Introduction; kinds of differentiability. We consider six operators:

1. two-function composition $(F, G) \mapsto FoG$, where $(FoG)(x) := F(G(x))$;
2. two-function integration $(F, G) \mapsto I(F, G)$, where $I(F, G)(x) := \int_a^x F \, dG$ for functions with real values or values in a Banach algebra $\mathbb{B}$, and the three-function extension $(F, G, H) \mapsto \int_a^{(\cdot)} F \, dG \, H$ when $\mathbb{B}$ is noncommutative;
3. convolution $(F, G) \mapsto F * G$, where $(F * G)(x) := \int_{-\infty}^{\infty} F(x - y) dG(y)$;
4. multiplication $(F, G) \mapsto F \cdot G$ for functions;
5. the inverse or quantile operator $F \mapsto F^{\leftarrow}$, where $F^{\leftarrow}(x) := \inf\{y : F(y) \geq x\}$, for x and y in some intervals, and F is real-valued;
6. product integration $F \mapsto \mathcal{P}_a F$ for functions F with values in a Banach algebra $\mathbb{B}$ having identity $\mathbb{1}$, where $(\mathcal{P}_a F)(x) := \prod_a^x (\mathbb{1} + dF)$ is the limit of products $\Pi_{i=1}^n (\mathbb{1} + F(x_i) - F(x_{i-1}))$ as partitions $a = x_0 < t_1 < \cdots < x_n = x$ are refined, if the limit exists.

The integration operator I is an indefinite extended Riemann-Stieltjes integral. The exact definition of the central Young, or (CY), form of integral we mainly use is given in Section 4 below, where it is also shown how this integral extends other integrals of Riemann-Stieltjes type and works well for our purposes. The definition and sufficient conditions for the existence of the (CY) integral go back to the work of L. C. Young (1936). One great advantage of the (CY) integral is that $(CY) \int_a^b F \, dG$ can be defined when F and G have jumps on the same side of the same point. For G of bounded variation and F regulated (having only jump discontinuities), the (CY) integral also agrees with the Lebesgue-Stieltjes integral (Theorem 4.2 below). Thus, it seems natural to define the convolution operator in terms of the (CY) integral and we will do so.

Given a matrix or operator-valued function F on $[a, b]$, the indefinite product integral $G(t) := (\mathcal{P}_a F)(t)$, $t \in [a, b]$, under some conditions, solves the integral equation $G(t) = \mathbb{1} + \int_a^t dF \, G$ (cf. Part II, after (1.6)). If F is (continuously) differentiable we have the differential equation $G'(t) = F'(t)G(t)$ with $G(a) = \mathbb{1}$. Gill and Johansen (1990), see also Gill (1994), gave several applications of the product integral in statistics and probability. It takes a hazard function into a survival function (for real-valued functions), and matrix-valued functions arise for Markov chains in continuous time with non-stationary transitions.

The six operators are natural ones and analysts have studied them on various domains. Statisticians have been interested in cases where the directions along which one differentiates are themselves non-differentiable and indeed discontinuous functions. Specifically, here and in Section 6 below we treat the case where F and G are probability distribution functions, having corresponding empirical distribution functions F_n and G_m, which are random step functions converging to F and G respectively as $m, n \to \infty$. If T is a nonlinear operator $F \mapsto T(F)$, such as the quantile or product integral operator, one then looks for a representation

$$T(F_n) = T(F) + L(F_n - F) + R_n, \qquad (1.1)$$

where L is a linear operator, the derivative of T at F, and R_n is a remainder of smaller order of magnitude in probability as $n \to \infty$ than the derivative term. For an operator acting on two functions, say $T(F, G)$, we would have a similar expansion with two linear partial derivative terms.

We then have three main steps to pursue:

(1) Find the facts about differentiability of the six operators. To find in what directions the operators are differentiable and with what degree of uniformity will be of interest, we trust, not only to statisticians. It turns out that p-variation norms $\|\cdot\|_{[p]}$, to be defined below after an Example, work well in giving differentiability with bounds on the norms of the remainders. The differentiability of the six operators is treated more generally (not only for distribution functions) in Sections 1-5, and for non-Fréchet modes of differentiability in Section 7.

(2) Find the behavior of $\|F_n - F\|_{[p]}$ as $n \to \infty$, where F_n is the empirical distribution function corresponding to F;

(3) Find how (1) and (2) fit together to give good bounds on the size of R_n in (1.1). Steps (2) and (3) are treated in Section 6.

Although the six operators are, in a sense, well known, there are possibly surprising aspects of their differentiability:

• The chain rule gives the derivative of a function $x \mapsto (F \circ G)(x) \equiv F(G(x))$ for suitable F and G, but not that of the composition operator $(F, G) \mapsto F \circ G$. In fact, taking $(F + f) \circ (G + g)$ as $f, g \to 0$ for fixed F and G, to differentiate the composition operator, *only* F, not f, G nor g, needs to have a derivative. Indeed, as many analysts knew (e.g. Appell and Zabrejko, 1990), G and g can be defined on an abstract measure space or probability space without any differentiable structure. Statisticians had noticed that f may be discontinuous. See Section 2.

• The quantile operator $f \mapsto (F + f)^{\leftarrow}$ is differentiable at $f = 0$ for functions f such that $F + f$ can be neither one-to-one nor onto any non-degenerate interval, while f can be discontinuous again as for composition, as some statisticians had also noticed. Again, F does need to be differentiable. See Section 3.

• The multiplication operator $(F, G) \mapsto F \cdot G$, the integration operator $(F, G) \mapsto I(F, G)$ and the convolution operator $(F, G) \mapsto F * G$ are bilinear but not linear. For F and G neither continuous nor of bounded variation, the (CY) integral we use, although far from new (L. C. Young, 1936) is also far from being as well known as we think it should be. See Section 4.

• The product integral operator $\mathcal{P}_a$, while operating on possibly discontinuous, nowhere differentiable functions, turns out to be not only differentiable but analytic (holomorphic). See Section 5.

• Statisticians had worked on differentiability of such operators and noted they were not Fréchet differentiable for the supremum norm but, with some modifications, were compactly differentiable for that norm. Thus, compact differentiability gained some preference among statisticians. But, a family of norms - the p-variation norms, originating with Wiener (1924) and developed by Young (1936) - turns out to yield, for suitable ranges of p depending on the operator, not only Fréchet differentiability but remainder bounds of optimal order for all six operators, as will be seen.

The composition operator for fixed F and $f \equiv 0$, namely $g \mapsto F \circ (G + g)$, is a special case of the so-called Nemytskii or superposition operator which has been extensively studied for g nonsmooth, e.g. in L^p, cf. Appell (1988), Appell and Zabrejko (1990). For the multiplication, integration and convolution operators, which are bilinear, differentiability and indeed analyticity follow from joint continuity in suitable norms, which holds for p-variation norms as shown respectively by Krabbe (1961a, 1961b); Young (1936); and Young (1937) and Gehring (1954), see Theorem 1.1 below.

Statisticians contributed to the exploration of differentiability, for integration but also notably for other operators. Reeds (1976), see also Fernholz (1983), to our knowledge, first showed differentiability of the two-function composition operator in the direction of discontinuous f, with g also varying, and similarly for the quantile operator (itself of particular interest to probabilists and statisticians). For product integrals, there was some early work on differentiability along curves (see Part II, after (1.5)). Gill and Johansen (1990), see also Gill (1994), made a substantial advance in showing infinite-dimensional differentiability with uniformity and including discontinuous directions f.

A form of differentiability for operators on infinite-dimensional spaces, which includes the other best-known forms, is C-differentiability, defined as follows.

Definition. Let X be a normed space with norm $\|\cdot\|$ and $x \in A \subset X$. Let C be a collection of bounded subsets of X containing all finite sets. Let T be a mapping from A into another normed space Y with norm also written $\|\cdot\|$. Then T will be called *C-differentiable* at x on A if there is a bounded linear operator $(DT)_x \equiv L$ from X into Y such that for each $C \in C$, as $t \to 0$,

$$\sup_{u \in C,\ x+tu \in A} \|T(x+tu) - T(x) - tL(u)\| = o(|t|).$$

If C is the class of all finite sets, T is *Gateaux* differentiable at x on A; if C is the class of all norm-compact subsets of X, T is *compactly* differentiable at x on A; if C is the class of all norm-bounded sets in X, or equivalently if it consists just of the unit ball of the norm, T is *Fréchet* differentiable at x on A.

On C-differentiability more generally see Section 7 and Part III.

If A is an open set V, then the definition simplifies a little: for t small, since C is bounded, $x + tu \in V$ automatically. Non-open sets A have been found to be needed for compact differentiability of some of our six operators for the supremum norm, e.g. Gill (1989), Gill and Johansen (1990), Gill (1994) and Tables 3 and 4 below. Such sets A also occur in other ways:

Example. A function will be said to be C_b^k if its partial derivatives through order k all exist and are continuous and bounded. A C_b^k function will be said to be a C_b^k diffeomorphism if it is 1-1 and its inverse is also C_b^k. The set $A = \mathrm{Diff}^k(M)$ of C_b^k diffeomorphisms onto itself of a compact manifold M of dimension m can be viewed as a subset of the set $C_b^k(M, \mathbb{R}^N)$ of C_b^k functions from M into a Euclidean space of dimension $N > m$, e.g. for the circle $M := \{(x,y) : x^2 + y^2 = 1\}$ where $m = 1 < N = 2$. There is a natural inverse operator $f \mapsto f^{-1}$ of A onto itself whose differentiability has been studied, see the end of Section 3. Here A is neither convex nor open in $C_b^k(M, \mathbb{R}^N)$.

Let T be a mapping between normed spaces X, Y and let U be an open subset of X. Some authors define the *Gateaux variation* of T at $x \in U$ as

$$\delta T(x; h) := \lim_{t \to 0} [T(x+th) - T(x)]/t$$

if the limit exists for all $h \in X$. Then $\delta T(x; h)$ is homogeneous in h of degree one, but need not be linear in general. If $\delta T(x; \cdot)$ is a bounded linear operator then it is denoted by DT_x and is called the *Gateaux derivative*. Necessary and sufficient conditions for linearity of the Gateaux variation are given in Vainberg (1964, Theorem 3.2).

Bibliographical remarks: In the note Gateaux (1913), often cited as the first place his differentiability was defined, Gateaux gives, in the same sentence, a sufficient condition for linearity. Gateaux was killed in World War I, in September 1914. At the suggestion of J. Hadamard, Paul Lévy edited some posthumous papers of Gateaux (1919, 1922). Lévy (1922, pp. 51-52) further emphasized linearity. Some authors refer to the Gateaux derivative as the Gateaux-Lévy derivative (e.g., Averbukh and Smolyanov, 1968).

Let Z also be a normed space. For differentiability of an operator $(F, G) \mapsto T(F, G)$ from the product $X \times Z$ into Y, the derivative will be a bounded linear operator D from $X \times Z$ into Y, which can be written $D = (B, C)$, where B, C are bounded linear operators from X, Z respectively into Y. Here B and C can be viewed as partial derivatives of T along X and Z respectively.

Reeds, Fernholz, Gill, Johansen, and others working on differentiable statistical functionals noted that for the composition, quantile, integration and product integral operators, Fréchet differentiability fails for the supremum norm (on f for composition) but, for suitable choices of A, compact differentiability holds. The same holds, in fact, for convolution. Compact differentiability also suffices for some probability limit theorems useful in statistics, and so seemed to provide satisfactory and complete answers. It did not, however, give good bounds for the remainder $o(|t|)$ such as $O(|t|^\gamma)$, $\gamma > 1$. Other norms will give Fréchet differentiability with such bounds. We next need the following

Definitions. Let $J \subset \mathbb{R}$ be an interval, possibly unbounded. If f is a function from J into a normed space X with norm $\|\cdot\|$ and $0 < p < \infty$ then the *p-variation* $v_p(f) = v_p(f; J)$ of f is

$$v_p(f; J) := \sup \left\{ \sum_{i=1}^{n} \|f(x_i) - f(x_{i-1})\|^p : x_0 \in J, \, x_0 < x_1 < \cdots < x_n \in J \right\}.$$

Let $\|f\|_{(p)} := V_p(f) := V_p(f; J) := v_p(f; J)^{1/p}$ and $\|f\|_\infty := \sup\{|f(x)| : x \in J\}$. If $v_p(f) < \infty$ and $1 \le p < \infty$ then the *p-variation norm* of f is

$$\|f\|_{[p]} := \|f\|_{(p)} + \|f\|_\infty.$$

Let $\mathcal{W}_p(J; X) := \{f : J \mapsto X, \, v_p(f) < \infty\}$ if $p < \infty$. For $p = \infty$, let $\|f\|_{(\infty)} := \sup\{\|f(x) - f(y)\| : x, y \in J\}$ and let $\|f\|_{[\infty]} := \|f\|_{(\infty)} + \|f\|_\infty$.

For a function $\phi : [0, \infty) \mapsto [0, \infty)$, the ϕ-variation $v_\phi(f; [a, b])$ of a function f is defined as the p-variation except that the p-th powers $\|f(x_i) - f(x_{i-1})\|^p$ are replaced by $\phi(\|f(x_i) - f(x_{i-1})\|)$. Usually, ϕ is a strictly increasing continuous function from $[0, \infty)$ onto itself.

For each J and X as in the definition and $p \ge 1$, $(\mathcal{W}_p(J; X), \|\cdot\|_{[p]})$ is a normed space. Let $\mathcal{W}_p := \mathcal{W}_p(J; \mathbb{R})$ unless otherwise specified. In the definition of p-variation the supremum is over all partitions. The mesh $\max_i(x_i - x_{i-1})$ of partitions $\{x_i : i = 0, \ldots, n\}$ may not approach 0. Thus, for example, for $p = 2$, the 2-variation may be infinite while the "quadratic variation" as defined by probabilists, for sequences of partitions with mesh going to 0 at some rate, or nested partitions, is finite.

The spaces $\mathcal{W}_p$ and norms $\|\cdot\|_{[p]}$ are invariant under the transformation $f \mapsto f \circ g$ for any strictly increasing, continuous function g from the interval J onto itself. This invariance property, which may be called nonparametric invariance in statistical terms, is shared with the supremum norm. Note that on ℓ_∞ the norms $\|\cdot\|_\infty$ and $\|\cdot\|_{[\infty]}$ are equivalent, with $\|\cdot\|_\infty \leq \|\cdot\|_{[\infty]} \leq 3\|\cdot\|_\infty$.

As will be seen in more detail below, for all six operators T, Fréchet differentiability holds for $(f,g) \mapsto T(F+f, G+g)$ or $f \mapsto T(F+f)$ with respect to $\|f\|_{[p]}$ and $\|g\|_{[q]}$ for suitable p, q and, for the quantile and composition operators, at suitable F (and G), and except that, for composition, an L^s norm $\|g\|_s$ is used. For the meaning of "suitable" F and G for composition, necessary and sufficient conditions are given in Theorem 2.2 and sufficient conditions with remainder bounds in Theorem 2.10. For the quantile operator, Theorem 3.1 gives a sufficient condition on F for a remainder bound, and necessary conditions for differentiability at F are mentioned near the beginning of Section 3 and in Appendix C. Restriction to a subset A is not needed for any of the six operators, in these cases, although it is if one wants to have the sup norm on the range space of the composition operator (Table 4 below). For the quantile operator with the sup norm on the range, we have not stated any new result.

The p-variation Fréchet differentiability properties are in different senses optimal among all possible norms, giving remainder bounds of best possible order for the composition and quantile operators (Table 2 and Sections 2, 3) and analyticity of the product integral operator (Section 5). Regarding norms on the ranges of the operators see Table 4.

Recall that the integration, convolution and multiplication operators are all bilinear and so analytic if they are jointly continuous for given norms. The multiplication operator $(F, G) \mapsto F \cdot G$ is the only one of the six operators to be Fréchet differentiable for the supremum norm. The joint continuity and analyticity still hold, from $\mathcal{W}_p \times \mathcal{W}_p$ into $\mathcal{W}_p$, for the stronger p-variation norms by the inequality

$$\|FG\|_{[p]} \leq \|F\|_{[p]}\|G\|_{[p]}, \quad 1 \leq p \leq \infty$$

(Krabbe, 1961a, 1961b). So, we will not need to consider the multiplication operator further.

A function f on an interval J is said to be *regulated*, or $f \in \mathcal{R} := \mathcal{R}(J)$, if it has right and left limits at each point in the interior of J, a right limit at the left endpoint and a left limit at the right endpoint. For integration, Theorem 4.4 below recalls the Love-Young inequality: for some finite constants $C_{p,q}$,

$$\left| (CY) \int_a^b F\, dG \right| \leq C_{p,q}\|F\|_{[p]}\|G\|_{(q)} \quad \text{for} \quad \frac{1}{p} + \frac{1}{q} > 1, \quad p, q > 0, \qquad (1.2)$$

where L. C. Young's (CY) integral is defined in Section 4 below. Although this integral has not been widely known, the existence of $(CY) \int F\, dG$ for F, G regulated follows from existence of the Riemann-Stieltjes integral, its extension the Moore-Pollard-Stieltjes integral (where partitions are refined rather than just letting the mesh approach 0), or the Lebesgue-Stieltjes integral, and in each case the values of the integrals then coincide: Theorems 4.2 and 4.3 below. Another common extension of Riemann-Stieltjes and Lebesgue-Stieltjes integrals is the gauge integral, treated in Appendix F.

For the indefinite integral operator I, there are finite constants $D_{p,q}$ such that

$$\|I(F,G)\|_{[q]} \leq D_{p,q}\|F\|_{[p]}\|G\|_{(q)} \text{ for } \frac{1}{p} + \frac{1}{q} > 1 \text{ and } p,q > 0$$

(Theorem 4.7 below; Part II, Proposition 3.32).

Let F and G be real-valued regulated functions of a real variable. Define their convolution $F*G$ on $\mathbb{R}$ in terms of Young integrals by

$$(F*G)(x) := \lim_{a \to -\infty,\, b \to \infty} (CY) \int_a^b F(x-y)\,dG(y)$$

if the (CY) integrals over $[a,b]$ exist for each x, a, b and if the limit exists. The following fact, a slight extension of a statement of Gehring (1954, Theorem 4.1.4), is proved after Corollary A.2 in Appendix A.

1.1 Theorem. Let $1 \leq p,q < \infty$ and $p^{-1} + q^{-1} > 1$. Let ν be given by $\nu^{-1} = p^{-1} + q^{-1} - 1$. There exists a finite constant $K = K(p,q)$ such that for all $F \in \mathcal{W}_p$ and $G \in \mathcal{W}_q$, we have $F*G \in \mathcal{W}_\nu$ and $\|F*G\|_{[\nu]} \leq K\|F\|_{[p]}\|G\|_{(q)}$.

Thus $(F,G) \mapsto F*G$ is a bounded bilinear operator from $\mathcal{W}_q \times \mathcal{W}_p$ into $\mathcal{W}_\nu$ if $1 \leq p,q,\nu < \infty$, and $\frac{1}{p}+\frac{1}{q}-\frac{1}{\nu} = 1$. There is a corresponding theorem in Zygmund (1959, (1.26) on p. 39) for convolution $(F*G)(x) := (2\pi)^{-1}\int_0^{2\pi} F(x-y)\,dG(y)$ defined in the Riemann-Stieltjes sense for each x such that the functions $F(x-\cdot)$ and G have no discontinuities in common, when $F(x+2\pi)-F(x)$ is constant for $-\infty < x < +\infty$, F has bounded variation in $[0,2\pi]$ and G satisfies similar conditions. Young (1937, pp. 458-459) gave a proof (of a more general Theorem A.1) when $\frac{1}{p}+\frac{1}{q}-\frac{1}{\nu} > 1$ under certain periodicity and continuity assumptions. The conclusion fails for $\frac{1}{p}+\frac{1}{q}-\frac{1}{\nu} < 1$. For the supremum norm of $F*G$ we have $\|F*G\|_\infty \leq C_{p,q}\|F\|_{[p]}\|G\|_{(q)}$ for $p \geq 1$, $q \geq 1$, and $\frac{1}{p}+\frac{1}{q} > 1$. Young (1970, Theorem 7.8) generalized Theorem 1.1 to functions with bounded ϕ-variation using a different extended Riemann-Stieltjes integral.

Some tables will summarize various aspects of differentiability for the five operators: integration, convolution, composition, quantile and product integral. For composition, the tables give information on differentiability with respect to f in $(f,g) \mapsto (F+f)\circ(G+g)$ into L^r, where $g \in L^s$ for $s > r$, as in Theorems 2.1 and 2.2 below. Also, $F \mapsto F^\leftarrow$ is considered as a map into L^r. Both situations are as treated by Reeds (1976) and Fernholz (1983, pp. 66, 77), where for compact differentiability one can take $s = r$, see Appendix D below. For composition and $F \mapsto F^\leftarrow$, hypotheses are needed on the F and G at which one differentiates (for any form of differentiability). Theorems 2.5, 2.10 and 3.1 give such hypotheses.

Table 1. *Does Fréchet differentiability hold for the given norm(s)?*

$(F,G) \mapsto:$ (or $F \mapsto:$)	$I(F,G)$	$F*G$	$F \circ G$	$F^\leftarrow$	$\mathcal{P}_a F$
sup norm	No	No	No	No	No
$\|\cdot\|_{[p]}, \|\cdot\|_{[q]}$	Yes	Yes	Yes	Yes	Yes
For what p (and q)?	$p^{-1}+q^{-1}>1$		$p \leq r$	$p \leq r$	$p < 2,$

where the entries in the last line hold by Theorems 4.7, 1.1, 2.10, 3.1 and 5.7 respectively.

So, these failures of Fréchet differentiability for the supremum norm can be viewed as failures of the norm, rather than of the Fréchet aspect. It will be seen in Section 6 that the given values of p work well.

In the statistical literature the supremum norm has often been taken on spaces D of right-continuous functions with left limits. Functions $F^{\leftarrow}$ are left-continuous with right limits for any right-continuous F. Thus the space $\mathcal{R}$ of regulated functions contains both functions. The p-variation spaces are strictly included in $\mathcal{R}$. In each case the spaces are defined on some interval, which may be the whole line.

For normed spaces X, Y and for $F, f \in X$, an operator $T\colon X \mapsto Y$ will be said to have a *remainder bound of order* $\gamma > 1$ *at* F *along* f, or $T \in RB_\gamma(F,f)$, if for some $L(f)$, as $t \to 0$,

$$\|T(F + tf) - T(F) - tL(f)\| = O(|t|^\gamma). \tag{1.3}$$

Here $T \in RB_2(F,f)$ if T has two or more derivatives at F in the direction f. A value γ will be called *best for* T, F *and* f, or $\gamma = B(T,F,f)$, if $T \in RB_\gamma(F,f)$ but for any $\beta > \gamma$, $T \notin RB_\beta(F,f)$. For T as above, Fréchet differentiable at $F \in X = (X, \|\cdot\|_X)$, we say that T has a *(Fréchet) remainder bound of order* $\gamma > 1$ *at* F, or $T \in FRB_\gamma(F;X) := FRB_\gamma(F;X, \|\cdot\|_X)$, if $L(\cdot)$ is a bounded linear operator from X into Y and

$$\|T(F + f) - T(F) - L(f)\| = O(\|f\|_X^\gamma) \tag{1.4}$$

as $\|f\|_X \to 0$. A value γ will be called *best for* T, F *and* $X = (X, \|\cdot\|_X)$ if $T \in FRB_\gamma(F;X)$ and $T \notin FRB_\beta(F;X)$ for any $\beta > \gamma$. If $T \in FRB_\gamma(F;X)$ then $T \in RB_\gamma(F,f)$ for all $f \in X$. If also $\gamma = B(T,F,f)$ for some f, in other words γ is best for some f in (1.3), then γ is also best for T, F and $X = (X, \|\cdot\|_X)$. The norm $\|\cdot\|_X$ is itself optimal in the sense that if also $f \in Z$ where $(Z, \|\cdot\|_Z)$ is a normed space such that $T(F + g)$ is defined for $g \in Z$ with $\|g\|_Z$ small enough and $T \in FRB_\beta(F;Z)$ then $\beta \leq \gamma$.

We will write $\gamma = 2(\infty)$ if T is analytic in some given norms, i.e., $T(F + f) = \sum_{k=0}^\infty a_k(f)$, where $a_0 = T(F)$, $a_k(f) = a_k(F,f)$ is a homogeneous kth order functional polynomial in f for fixed F and $k \geq 1$ (Part II, Section 5.3) and the series converges uniformly for $\|f\| < r$ for some $r > 0$. For normed spaces X, Y, Z and $(F,G) \in X \times Z$, an operator $T\colon X \times Z \mapsto Y$, Fréchet differentiable at (F,G), will be said to have a *remainder bound of order* $\gamma > 1$ *at* (F,G), or $T \in FRB_\gamma(F,G)$, if

$$\|T(F + f, G + g) - T(F,G) - D_1T(F,G)(f) - D_2T(F,G)(g)\| = O(\|f\|^\gamma + \|g\|^\gamma)$$

as $\|f\| \to 0$ and $\|g\| \to 0$, where each first partial derivative operator $D_iT(F,G)(\cdot)$, $i = 1, 2$, is a bounded linear operator between the appropriate spaces. The bilinear operators $I(F,G)$ and $F * G$ have finite and so convergent power series expansions. Thus whenever $\gamma = 2$ holds for them, $\gamma = 2(\infty)$ will also hold, by definition.

In the following tables, compact(A) will mean compact (Hadamard) differentiability for the supremum norm when restricted to a suitable set A. Also, $\|\cdot\|_{[p]}, \|\cdot\|_{[q]}$ refers to Fréchet differentiability with respect to $\|\cdot\|_{[p]}$ (and $\|\cdot\|_{[q]}$) on the domain, for the ranges of p and q given in Table 1 for each operator, except that for composition we use $\|g\|_s$ instead of $\|g\|_{[q]}$, and take $(f,g) \mapsto (F+f)\circ(G+g)\colon \mathcal{W}_p \times L^s \mapsto L^p$ for $1 \le p < s$ and suitable F, G as in Theorem 2.10 below. For $\gamma = \gamma(p,s) := 1 + s/(p(1+s))$, the remainder is $O(\|f\|_{[p]}\|g\|_s^{\gamma-1} + \|g\|_s^\zeta)$ as $\|f\|_{[p]}, \|g\|_s \to 0$, and in the term with f, the exponents 1 on $\|f\|_{[p]}$ and $\gamma(p,s) - 1 = s/(p(1+s))$ on $\|g\|_s$ are separately optimal for all norms (Theorem 2.5 below); the exponent $\zeta = \zeta(p,s)$ varies with different hypotheses in Theorem 2.10. The operator $f \mapsto (F+f)^{\leftarrow}$ is taken from $\mathcal{W}_p$ of an interval to L^p of another, at suitable F (Theorem 3.1 below).

Table 2. *Does the given differentiability hold? Does it give the best γ?*

$(F,G) \mapsto:$ (or $F \mapsto:$)	$I(F,G)$	$F * G$	$F \circ G$	$F^{\leftarrow}$	$\mathcal{P}_a F$
Differentiability:					
compact (A) holds?	Yes	Yes	Yes	Yes	Yes
best γ?	No	No	No	No	No
$\|\cdot\|_{[p]}, \|\cdot\|_{[q]}$ holds?	Yes	Yes	Yes	Yes	Yes
best γ?	Yes	Yes	Yes	Yes	Yes
$\gamma =$	$2(\infty)$	$2(\infty)$	$\gamma(p,s)$	$(1+p)/p$	$2(\infty)$.

So p-variation gives remainder bounds of best order in these cases, including analyticity when it holds, while compact differentiability does not.

Suppose T is Fréchet differentiable at F and has a remainder bound of order $\gamma > 1$ at F. Then for a sequence $f_n \to 0$ such that $L(f_n) \to 0$ faster than $\|f_n\| \to 0$, also $T(F + f_n) \to T(F)$ faster. This does occur in some probability limit theorems (Section 6 below). Thus, limit theorems of a certain order, specifically central limit theorems, order $O(n^{-1/2})$, can be proved for $T \in FRB_\gamma(F)$ with $\gamma > 1$ via p-variation and limit theorems for L, while they may not hold for $\|\cdot\|_{[p]}$, e.g. for $p < 2$.

Some operators are not defined on the whole space $\mathcal{R}$ of regulated functions, so restrictions are needed. Table 3 reviews what sets A have been used as domains in published work. Here $\mathcal{W}_{1,M}$ will denote a bounded set in $\mathcal{W}_1$, in other words a set of functions of uniformly bounded total variation. In Table 3, p and q are as in Table 1, for each operator. Although apparently convolution had not been treated explicitly by way of compact differentiability, it seems natural to assign it the same sets A as for $I(F,G)$. For compact differentiability the set A is as in Gill (1989, Lemma 3 p. 110) for $I(F,G)$ at $F, G \in \mathcal{W}_1$, and thus for $F*G$; Reeds (1976) and Fernholz (1983) for $F\circ G$ and $F^{\leftarrow}$; and Gill (1994, p. 119) for $\mathcal{P}_a F$. See also Andersen, Borgan, Gill and Keiding (1993, Section II.8). Smaller sets A may be needed to get stronger norms on range spaces (Table 4 below).

Table 3. *On what sets A have the operators been defined in proving differentiability?*

$(F,G) \mapsto:$ (or $F \mapsto:$)	$I(F,G)$	$F * G$	$F \circ G$	$F^{\leftarrow}$	$\mathcal{P}_a F$
for $\|\cdot\|_{[p]}, \|\cdot\|_{[q]}:$	$\mathcal{W}_p \times \mathcal{W}_q$	$\mathcal{W}_p \times \mathcal{W}_q$	$\mathcal{W}_p \times L^s$	$\mathcal{W}_p$	$\mathcal{W}_p$
for compact(A)	$\mathcal{R} \times \mathcal{W}_{1,M}$	$\mathcal{R} \times \mathcal{W}_{1,M}$	$\mathcal{R} \times L^s$	$\mathcal{R}$	$\mathcal{W}_{1,M}$

For differentiability with respect to norms $\|\cdot\|_{[p]}$ (and $\|\cdot\|_{[q]}$) no further restrictions to sets A are needed here beyond the obvious ones (whole spaces $\mathcal{W}_p, \mathcal{W}_q$), although they are for the sup norm on the range of the composition operator. For the product integral operator $\mathcal{P}_a$ we have only treated higher order derivatives and analyticity so far for right-continuous functions (Part II, Theorem 5.17). For compact differentiability, however, for three of the five operators a subset A is needed, and the sets chosen (for values of $G + tg$) have been bounded in $\mathcal{W}_1$. The sets $\mathcal{W}_{1,M}$ are applicable since all differences $H - G$, e.g. $G_m - G$, where H and G are probability distribution functions, belong to them. On the other hand, $\|F_n - F\|_{[p]} \to 0$ as $n \to \infty$ only for $p > 1$. Moreover, for the product integral operator $\mathcal{P}_a$ to be defined and not 0, for a real-valued function f, it's necessary that $f \in \mathcal{W}_2$ (Part II, Theorem 4.3), and sufficient for analyticity, even in the Banach algebra-valued case, that $f \in \mathcal{W}_p$ for some $p < 2$ and f be right-continuous (Part II, Theorem 5.17). Thus, for at least some operators, the use of p-variation and spaces $\mathcal{W}_p$ with $p < \infty$ in defining and differentiating the operators has not in fact been avoided by use of compact differentiability or (in defining $\mathcal{P}_a$) cannot be avoided.

Another question is, to what range spaces do we have differentiability.

In the Reeds-Fernholz results on the composition and quantile operators, L^p norms are taken on g and on the ranges of the operators. It's desirable if possible to get stronger norms, such as the supremum norm, on the range, provided that $g \to 0$ also, say, in the supremum norm. Andersen et al. (1993, Proposition II.8.8) did so for compact differentiability tangential to the space of continuous functions (see the definition following (8.5)), and when the closure of the range of G is included in an open interval on which F is continuously differentiable. The larger open interval can be avoided by using a set A of functions g such that $G + g$ takes values in the same closed interval that G does, e.g. $[0, 1]$ for probability distribution functions, if F is a diffeomorphism of that interval.

Two-function composition is not jointly differentiable, with supremum norm on its range, in the direction f if f is not continuous. Thus, the "tangential to C" condition cannot be deleted. Neither can the compactness of a set of functions h such that $f = th$ with $t \to 0$.

Esty, Gillette, Hamilton and Taylor (1985) gave a compact differentiability property of $f \mapsto (F + f)^{\leftarrow}$ with the supremum norm on the range, where F and f are continuous. Andersen et al. (1993, Prop. II.8.5) extended this to compact differentiability on $D[0, 1]$ tangential to $C[0, 1]$, under further conditions. Compact and/or tangential differentiability is sufficient for several statistical purposes: see Appendix E.

When compact differentiability holds for an operator T and a norm $|\cdot|$, then Fréchet differentiability clearly also holds for any norm $\|\cdot\|$ whose closed unit ball $\{x : \|x\| \leq 1\}$ is compact for $|\cdot|$. For some purposes the Fréchet differentiability for $\|\cdot\|$ can suffice. Specifically, for any normed space $(X, \|\cdot\|)$ and set S let $\ell^{\infty}(S, X)$ be the Banach space of all bounded functions f from S into X, with the supremum norm $\|f\|_{\infty} := \sup_{s \in S} \|f(s)\|$. For any metric space (T, d) and $0 < \alpha \leq 1$ Let $H(\alpha) := H_\alpha(T, X)$ be the Banach space of all functions g from T into X such that

$$\|g\|_{H(\alpha)} := \|g\|_{\infty} + \sup_{s \neq t} \|g(s) - g(t)\|/d(s, t)^\alpha < \infty.$$

Thus $H(\alpha)$ is the space of bounded, α-Hölder functions from T into X. For any bounded interval $[a, b] \subset \mathbb{R}$ and $0 < \alpha \leq 1$, the closed unit ball $\{f \in H_\alpha([a, b], \mathbb{R}) : \|f\|_{H(\alpha)} \leq 1\}$ is compact for the sup norm by the Arzelà-Ascoli theorem. Also, if $f \in H_\alpha([a, b], \mathbb{R})$ for a finite interval $[a, b]$, then f has bounded $1/\alpha$-variation with $\|f\|_{(1/\alpha)} \leq (b - a)^{1/\alpha}\|f\|_{H(\alpha)}$ for $0 < \alpha \leq 1$. We then have:

1.2 Proposition. Let S be a set and $(X, \|\cdot\|)$ a normed space. Let $B \subset X$ and let G be a function from S into B. Let $(Y, |\cdot|)$ be a normed space, $0 < \alpha \leq 1$, $H(\alpha) := H_\alpha(B, Y)$, and let A be the set of all functions h from S into B such that $h - G \in \ell^\infty(S, X)$. Then for any $f \in H(\alpha)$ and $g \in \ell^\infty(S, X)$ such that $G + g \in A$, we have

$$\|f \circ (G + g) - f \circ G\|_\infty \leq \|f\|_{H(\alpha)} \|g\|_\infty^\alpha \leq (\|f\|_{H(\alpha)}^{1+\alpha} + \alpha \|g\|_\infty^{1+\alpha})/(1 + \alpha).$$

Proof. The first inequality follows directly from the definitions and assumptions and the second from an inequality of W. H. Young used in the proof of Hölder's, e.g. Dudley (1993, Lemma 5.1.3).   □

Note that $R_1 := f \circ (G + g) - f \circ G$ is the only part of the remainder involving f in differentiating $(f, g) \mapsto (F + f) \circ (G + g)$ at $f = g = 0$, see Section 2. Proposition 1.2 will be applied near the end of Section 6. There, a remainder bound is based on a probability theorem, and tangential differentiability is not needed.

Table 4 displays results on range spaces and whether the norms there are stronger than the supremum norm ("Yes, >"), equal to it ("Yes, =") or weaker ("No"). As always, F and G for the composition operator and F for the quantile operator must be suitable.

Table 4. *Differentiability of $(f, g) \mapsto T(f, g)$ or $f \mapsto T(f)$ to what range spaces?*

$T(f,g)$ or $T(f)$	domain	range	norm $\geq$ sup norm?	Fréchet or?	what $p, q, \cdots$?	reference
$f \cdot g$	$\mathcal{W}_p \times \mathcal{W}_p \mapsto$	$\mathcal{W}_p$	Yes, >	Fréchet	$p \geq 1$	Krabbe
$\int^{(\cdot)} f \, dg$	$\mathcal{W}_p \times \mathcal{W}_q \mapsto$	$\mathcal{W}_q$	Yes, >	Fréchet	$\frac{1}{p} + \frac{1}{q} > 1$	Th. 4.7
$f * g$	$\mathcal{W}_p \times \mathcal{W}_q \mapsto$	$\mathcal{W}_\nu$	Yes, >	Fréchet	$\frac{1}{p} + \frac{1}{q} - \frac{1}{\nu} \geq 1$	Th. 1.1
$\mathcal{P}_a f$	$\mathcal{W}_p \mapsto$	$\mathcal{W}_p$	Yes, >	Fréchet	$1 \leq p < 2$	Th. 5.7

$(F + f) \circ (G + g)$ as $f, g \to 0$, $(F + f, G + g) \in$ domain:

	$D \times L_{[0,1]}^p \mapsto L^p$		No	compact	$1 \leq p < \infty$	Reeds
	$\mathcal{W}_p \times L^s \mapsto L^p$		No	Fréchet	$1 \leq p < s$	Th. 2.10
	$E \times E_{[0,1]} \mapsto D$		Yes, =	CTF		ABGK
	$H(\alpha) \times \mathcal{G} \mapsto \ell^\infty$		Yes, =	Fréchet	$0 < \alpha \leq 1$	Pr. 2.7

$(F + f)^\leftarrow$ as $f \to 0$, $F + f \in$ domain:

	$D \mapsto$	$L_{[0,1]}^p$	No	compact	$1 \leq p < \infty$	Reeds
	$C \mapsto$	D	Yes, =	compact		EGHT
	$\mathcal{W}_p \mapsto$	L^p	No	Fréchet	$1 \leq p < \infty$	Th. 3.1
	$D_0 \mapsto$	D	Yes, =	CTF		ABGK

Legend: $D = D[0, 1] =$ space of right-continuous functions with left limits on $[0, 1]$, $C = C[0, 1] =$ space of continuous functions $[0, 1] \mapsto \mathbb{R}$, both with supremum norm; $E =$ set of nondecreasing functions in D; $E_{[0,1]} = \{h \in E : 0 \leq h \leq 1\}$; $D_0 =$ a subspace of D; $\mathcal{W}_p = \mathcal{W}_p(J, X)$, $J = \mathbb{R}$ or $[a, b]$ or $[0, 1]$, $X = \mathbb{R}$ or $\mathbb{B}$ as the case may be; $L^p = L^p[0, 1]$; $L_{[0,1]}^p =$ set of measurable functions from $[0, 1]$ into itself, with L^p norm; $H(\alpha) = H_\alpha([0, 1], \mathbb{R})$; $\ell^\infty =$ bounded functions $S \mapsto \mathbb{R}$ for a set S;

$\mathcal{G} = \{h \in \ell^\infty(S) : 0 \le h \le 1\}$; CTF = compact differentiability tangentially to $f \in C$ (and $g \in D$) with further conditions; ABGK = Andersen, Borgan, Gill and Keiding (1993), Section II.8; EGHT = Esty, Gillette, Hamilton and Taylor (1985); Th. m.n = Theorem m.n of this Part; Pr. 2.7 = Proposition 2.7 below.

Note: the space $L^p_{[0,1]}$ is not explicit in Reeds (1976) or Fernholz (1983) but gives some (not the only) results which follow from their proofs.

So, Fréchet differentiability for p-variation norms holds with norms strictly stronger than the supremum norm on the range for the multiplication, integration, convolution and product integral operators - a further benefit, for these operators.

2. The two-function composition operator $(F, G) \mapsto F{\circ}G$. Recall that $(F{\circ}G)(x) := F(G(x))$. The operator $(F, G) \mapsto F{\circ}G$ is linear in F for fixed G. Thus in differentiating the operator, letting $f, g \to 0$ for fixed F and G in $(F + f){\circ}(G + g)$, the partial derivative with respect to f is $f \mapsto f{\circ}G$. The linear operators $f \mapsto f{\circ}G$ for various G and linear spaces X of functions f have been extensively studied: the book of Singh and Manhas (1993) treats operators $f \mapsto \pi{\cdot}f{\circ}G$ (composition followed by multiplication) and lists 415 references, with an emphasis on the case that X is an L^p space. On cases where f and G are analytic, the book of Shapiro (1993) cites 147 references by 122 authors, of whom less than a third are also cited by Singh and Manhas (1993). Cf. also the review of both books by Rosenthal (1995). These works on linear composition operators relate, however, very little to the present work, where we are concerned with the two-function composition operator and the size of the remainder in its differentiation, measuring the extent to which it is nonlinear.

If on the other hand we let $f \equiv 0$, then the operator $g \mapsto F{\circ}(G + g)$ is a special case of the so-called superposition or Nemytskii operator $g \mapsto (x \mapsto H(x, g(x)))$ for a function H of two variables. In our case $H(x, y) \equiv F(G(x) + y)$. There has been much work done on such operators, surveyed by Appell (1988) and Appell and Zabrejko (1990), which does provide information useful in differentiating the two-function composition operator.

For G and g in $\mathcal{L}^s(S, \mathcal{S}, \mu)$, where $(S, \mathcal{S}, \mu)$ is a finite measure space, we consider under what conditions $(f, g) \mapsto (F + f){\circ}(G + g)$ is differentiable at $f = g = 0$ into $L^p(S, \mathcal{S}, \mu)$ where $1 \le p < s$. For $p = s$ and $f \equiv 0$, Fréchet differentiability of $g \mapsto F{\circ}(G + g)$ only holds for F affine, while compact differentiability holds for suitable F (Appendix D); for $p > s$, F must be constant (Krasnosel'skiĭ, Zabreĭko, Pustyl'nik and Sobolevskiĭ, 1966, Theorem 20.1).

Appell and Zabrejko (1990, Chapter 3) treat superposition operators from L^s to L^p. Their Notes to Chapter 3 cite some 61 of the book's 395 references. The following necessary condition for differentiability is a corollary of results they give:

2.1 Theorem. If F is everywhere continuous, $G \in \mathcal{L}^s(S, \mathcal{S}, \mu)$ and $g \mapsto F{\circ}(G + g)$ is Fréchet differentiable at $g = 0$ from L^s to L^p for $p < s$, then the derivative is $g \mapsto (F'{\circ}G){\cdot}g$, where $F'(y)$ exists for $(\mu{\circ}G^{-1})$-almost all y and $F'{\circ}G \in \mathcal{L}^{sp/(s-p)}$.

Proof. Since F is continuous, $x \mapsto F(G(x) + u)$ is measurable for all $u \in \mathbb{R}$ and $u \mapsto F(G(x) + u)$ is continuous for all $x \in S$. Thus $(x, u) \mapsto F(G(x) + u)$ is a Caratheodory function and we can apply Appell and Zabrejko (1990, Theorems 3.12 and 3.13, and the sentence after (2.56)). $\qquad\square$

Wang Sheng-Wang (1963) and Krasnosel'skiĭ et al. (1966, Section 20) gave some necessary conditions and some sufficient conditions for differentiability of $g \mapsto F{\circ}(G + g)$: $L^s \mapsto L^p$ for $1 \le p < s$. Appell (1983), see also Appell and Zabrejko (1990, Theorem 3.13), gave criteria (necessary and sufficient conditions) for the differentiability. Sufficient conditions for Fréchet differentiability of $g \mapsto F{\circ}(G + g)$ at $g = 0$ from L^s to L^p for all p and s with $1 \le p < s$ are that F is Lipschitz and the measure $\mu{\circ}G^{-1}$ is absolutely continuous with respect to Lebesgue measure on $\mathbb{R}$.

We define remainders $R_0(f,g) := R_1(f,g) + R_2(f,g)$, where $R_1(f,g) := f{\circ}(G + g) - f{\circ}G$, and $R_2(f,g) := R_2(g) := F{\circ}(G + g) - F{\circ}G - (F'{\circ}G){\cdot}g$. Here are basic conditions for joint differentiability of the composition operator:

2.2 Theorem. Let $(Y, \|{\cdot}\|)$ be a normed space of functions from $\mathbb{R}$ into $\mathbb{R}$ and $(S, \mathcal{S}, \mu)$ a finite measure space. The map $(f,g) \mapsto (F + f){\circ}(G + g)$ is jointly Fréchet differentiable at $f = g = 0$ from $(Y, \|{\cdot}\|) \times (L^s, \|{\cdot}\|_s)$ into $(L^p, \|{\cdot}\|_p)$ for $1 \le p < s$ if and only if the following three conditions all hold:
(a) $f \mapsto f \circ G$ is defined and a bounded operator from Y into L^p;
(b) $F' \circ G$ is defined μ-almost everywhere and in $\mathcal{L}^{(sp)/(s-p)}(S, \mathcal{S}, \mu)$;
(c) We have, for $i = 0$,

$$\|R_i(f,g)\|_p = o(\|f\| + \|g\|_s) \quad \text{as} \quad \|f\| + \|g\|_s \to 0. \tag{2.1}$$

Proof. First suppose joint Fréchet differentiability holds. Then taking $g \equiv 0$, (a) must hold. Or, taking $f \equiv 0$, Theorem 2.1 implies that (b) holds. Then (c) holds by definition of joint Fréchet differentiability.

Conversely, suppose (a), (b) and (c) hold. Then the two given partial derivative operators are indeed bounded linear operators between the appropriate spaces, and (c) implies joint Fréchet differentiability. □

2.3 Corollary. If F is continuous on $\mathbb{R}$, then joint Fréchet differentiability as in Theorem 2.1 holds if and only if both
(d) $g \mapsto F{\circ}(G + g)$ is Fréchet differentiable at $g = 0$ from L^s into L^p and
(e) (a) holds and (2.1) holds for $i = 1$.
Whenever the joint differentiability holds, the partial derivative with respect to f is $f \mapsto f{\circ}G$ and that with respect to g is $g \mapsto (F'{\circ}G){\cdot}g$.

Proof. If (c) holds, i.e. (2.1) holds for $i = 0$, then it also holds for $f \equiv 0$ as $\|g\|_s \to 0$, i.e. it holds for $i = 2$. Thus (c) holds if and only if (2.1) holds for both $i = 1$ and $i = 2$. Now, by Theorem 2.1, (d) holds if and only if both (b) holds and (2.1) holds for $i = 2$. It then follows easily that (a), (b) and (c) all hold if and only if (d) and (e) both hold, and that the partial derivatives are as stated. □

Note that (e) does not involve F.

In the cases we have found in the literature outside of statistics where the two-function composition operator $(f,g) \mapsto (F + f){\circ}(G + g)$ has been considered, f was assumed differentiable at least once: Brokate and Colonius (1990), Gray (1975), Hartung and Turi (1997); regarding C^k spaces for $k \ge 1$ and Sobolev spaces see Ebin and Marsden (1970, p. 108), who give earlier references, and on C^k spaces, also Garay (1996), whom we thank for pointing out his paper and some other references. On the C^∞ case, which apparently has been much studied in connection with infinite-dimensional Lie groups, cf. Milnor (1984). On the composition operator for holomorphic functions, cf. Stevenson (1974, 1977); or for linear operators Dieudonné (1960, (8.3.1) p. 148). Thus it was a striking innovation by Reeds (1976) to take f non-differentiable and indeed discontinuous, in the space $D[0, 1]$ of right-continuous functions with left limits on $[0, 1]$. Reeds and, following his lead, Fernholz (1983) proved compact differentiability facts for the composition operator. For p-variation norms we have:

2.4 Proposition. Statement (a) in Theorem 2.2 holds when $\|\cdot\|$ is the p-variation norm for any p, $1 \leq p < \infty$, and G is any μ-completion measurable function.

Proof. We have $\|\cdot\|_{[p]} \geq \|\cdot\|_\infty$ and a function f in $\mathcal{W}_p$, being regulated (and so continuous except for at most countably many jumps), is Borel measurable, so $f \circ G \in L^\infty \subset L^p$ for any μ-measurable function G. The proposition follows. $\square$

For (2.1) when $i = 1$ we have the following:

2.5 Theorem. Let $1 \leq p < \infty$ and let $(S, \mathcal{S}, \mu)$ be $[0, 1]$ with Lebesgue measure. Let G be an increasing function such that, for some $\beta > 0$, $G(y) - G(x) \geq \beta(y - x)$ for $0 \leq x \leq y \leq 1$. Then there is a constant $K < \infty$ depending only on β such that

$$\|f \circ (G + g) - f \circ G\|_p \leq K\|g\|_s^{\gamma-1}\|f\|_{[p]} \leq K\big(\|f\|_{[p]}^\gamma + \|g\|_s^\gamma\big)$$

for any $f \in \mathcal{W}_p(\mathbb{R})$ and $g \in L^s[0, 1]$, $1 \leq s \leq \infty$, where $\gamma := 1 + s/(p(1 + s))$ if $s < \infty$, or $1 + 1/p$ if $s = \infty$.

Conversely, for any normed space $(Y, \|\cdot\|)$ of functions on $\mathbb{R}$ with $h \in Y$ for $h(x) := 1_{x>0}$ and $U(x) \equiv x$ for $x \in [0, 1]$, if for some $K < \infty$

$$\|f \circ (U + g) - f\|_p \leq K\|g\|_s^\alpha\|f\|^\lambda$$

for all $f \in Y$ with $\|f\| \leq 1$ and $g \in L^s[0, 1]$ with $1 \leq s < \infty$, then $\alpha \leq s/(p(1 + s))$ and $\lambda \leq 1$, and the above γ is best.

Proof. This is essentially Theorem 2.2 of Dudley (1994). We supply its easier proof for $s = \infty$ since it is not given there. Let $\delta := \|g\|_\infty$. We can assume $\delta < 1$, or take $K = 2$. Let m be a positive integer such that $1/(2m) \leq \delta \leq 1/m$. We can assume that $\beta \leq 1$. Let J be the smallest integer greater than or equal to $1/\beta$ and $B := [J/m, 1 - J/m]$. Then

$$\int_{[0,1]\setminus B} |f(G + g) - f(G)|^p(x)\, dx \leq 2^p \frac{2J}{m}\|f\|_\infty^p$$

and, as in the proof of Theorem 2.2 of Dudley (1994), we have

$$\int_B |f(G + g) - f(G)|^p(x)\, dx \leq (2J + 1)2\delta\|f\|_{(p)}^p.$$

It follows that $\int_{[0,1]} |f(G + g) - f(G)|^p(x)\, dx \leq C_p\|g\|_\infty\|f\|_{[p]}^p$, where

$$C_p := C_{p,\beta} := 4J + 2 + 2^p 4J \leq (6 + 2^p 4) \cdot \left(1 + \frac{1}{\beta}\right) \leq 10 \cdot 2^p \cdot \left(1 + \frac{1}{\beta}\right).$$

So the first inequality in the statement with $s = \infty$ holds for $K := 20(1 + 1/\beta)$.

To show that $\gamma = 1 + s/(p(1 + s))$ is best, in the proof of Theorem 2.2 of Dudley (1994) ("converse direction"), suppose we have for some $C < \infty$ and $\eta > 0$, $t\delta^{1/p} \leq C(t^\eta + \delta^{[1+(1/s)]\eta})$ as $t \downarrow 0$ and $\delta \downarrow 0$. Letting $t = \delta^{1+(1/s)}$ we get $\eta \leq 1 + s/(p(1+s))$ as stated. $\qquad\square$

The last theorem shows that $(\mathcal{W}_p(\mathbb{R}), \|\cdot\|_{[p]})$ provides a normed space $(Y, \|\cdot\|)$ for joint Fréchet differentiability of the two-function composition operator $(f, g) \mapsto (F + f) \circ (G + g)$ from $Y \times L^s$ to L^p of a finite measure space, giving a remainder bound of optimal order by maximizing both exponents α and λ.

If one allows larger remainders (with smaller exponent α or λ), then naturally there are larger normed spaces $(Y, \|\cdot\|)$ for which joint Fréchet differentiability holds. Under some conditions one can characterize the 1-dimensional spaces Y for which it holds:

2.6 Proposition. Let $(S, \mathcal{S})$ be a Borel set in a complete separable metric space with Borel σ-algebra and μ a finite measure on $(S, \mathcal{S})$. Let $G \in \mathcal{L}^s(S, \mathcal{S}, \mu)$ and $1 \leq p < s$. Let f be a function from $\mathbb{R}$ into $\mathbb{R}$. Then

$$\|t[f \circ (G + g) - f \circ G]\|_p = o(|t| + \|g\|_s) \quad \text{as} \quad t \to 0 \quad \text{and} \quad \|g\|_s \to 0$$

for $g \in L^s(S, \mathcal{S}, \mu)$, if and only if the following three conditions all hold:
(f) f is continuous almost everywhere for Lebesgue measure on $\mathbb{R}$;
(g) For some $C < \infty$, and all real x, $|f(x)| \leq C(1 + |x|^{s/p})$;
(h) f is universally measurable, i.e. measurable for the completion of every finite Borel measure on $\mathbb{R}$.

Proof. This follows from Part III, Theorem 5.1 for "if" and Theorem 5.3 for "only if." $\qquad\square$

Next, recall the definitions as in Proposition 1.2.

2.7 Proposition. Let F be a continuous function from an interval $[a, b]$ into $\mathbb{R}$ such that for some $\beta > 0$ and $C < \infty$, the derivative $F'(x)$ exists for $a < x < b$ and satisfies $|F'(x) - F'(y)| \leq C(y - x)^\beta$ for $a \leq x \leq y \leq b$. Let S be a set and let G be a function from S into $[a, b]$. Let E be the set of all real-valued functions g on S such that $G + g$ has values in $[a, b]$. Let $H(\alpha) := H_\alpha([a, b], \mathbb{R})$. Then $(f, g) \mapsto (F + f) \circ (G + g)$ is Fréchet differentiable on $H(\alpha) \times E$ with norm $\|(f, g)\| := \|f\|_{H(\alpha)} + \|g\|_\infty$ and with remainder bounds $\|R_1\|_\infty = O(\|f\|_{H(\alpha)}\|g\|_\infty^\alpha)$ and $\|R_2\|_\infty = O(\|g\|_\infty^{1+\beta})$ as $\|f\|_{H(\alpha)}, \|g\|_\infty \to 0$.

Proof. The bound on $\|R_1\|_\infty$ is as in Proposition 1.2, and that on $\|R_2\|_\infty$ follows from the assumption on F' and the mean value theorem of calculus. $\qquad\square$

For the supremum norm on the range of the composition operator one has the straightforward bound $\|f \circ G\|_\infty \leq \|f\|_\infty$. Whitt (1980, Section 3) treated composition for cadlag functions, i.e. right-continuous functions f and G with left limits. If G is monotone then $\|f \circ G\|_{[p]} \leq \|f\|_{[p]}$ for any $f \in \mathcal{W}_p(\mathbb{R})$. Whitt noted that for G nondecreasing, $f \circ G$ is cadlag, but otherwise $f \circ G$ may not have left limits. Whitt's counterexample extends as follows, and shows that for $f \in \mathcal{W}_p$ for all $p > 0$ but not continuous and G very smooth but not monotone, $f \circ G$ can be quite wild.

2.8 Proposition. There exists a C^∞ function G from $[0,1]$ into $[-1,1]$ such that for $f := 1_{[0,1]}$, $f \circ G$ has no right limit at 0, and so is not regulated and is not of bounded p-variation for any $p < \infty$.

Proof. Let $G(x) := e^{-1/x} \sin(1/x)$ for $x > 0$ and $G(x) := 0$ for $x \leq 0$. Then G is C^∞, with all derivatives 0 at 0, and the other conclusions clearly hold. $\square$

We still need to supply concrete sufficient conditions for differentiability of a Nemytskii operator $g \mapsto F \circ (G + g) : L^s \mapsto L^p$ as in Corollary 2.3(d), with remainder bounds. Let $k = 1, 2, \cdots$. A function F on a closed bounded interval $J := [a, b]$ is called C^k if it has continuous derivatives through order k on (a, b) which extend to continuous functions on $[a, b]$. F will be differentiable of order between 1 and 2 on J in terms of Hölder conditions on F'.

2.9 Theorem. Let F be a function from $\mathbb{R}$ into $\mathbb{R}$ whose restriction to a closed bounded interval $J := [a, b]$ is C^1. Let $(X, \mathcal{A}, \mu)$ be a finite measure space and G a measurable function from X into J. For $1 \leq p < s$ and $g \in L^s(X, \mu)$ let

$$R(g) := F \circ (G + g) - F \circ G - (F' \circ G)g.$$

Let F be $C^1 : J \mapsto \mathbb{R}$ and $\alpha := \frac{s}{p} - 1$. If $0 < \alpha \leq 1$, and either
  (a) F is C^1 on $\mathbb{R}$ with $F' \in H_\alpha(\mathbb{R}, \mathbb{R})$, or
  (b) $F' \in H_\alpha(J, \mathbb{R})$ and $G + g$ has values in J,
then $\|R(g)\|_p \leq C\|g\|_s^{s/p}$ where $C := \|F'\|_{H(\alpha)}$.
  Or, if
  (c) $F' \in H_\beta(J, \mathbb{R})$ where

$$\beta := \frac{s(p+1)}{(s+1)p} - 1 = \frac{s-p}{(s+1)p},$$

$F(x) = F(a)$ for all $x \leq a$, $F(x) = F(b)$ for all $x \geq b$, and for some $K < \infty$,

$$(\mu \circ G^{-1})([a, a + x] \cup [b - x, b]) \leq Kx \qquad (2.2)$$

for all $x \geq 0$, then as $\|g\|_s \to 0$, $\|R(g)\|_p = O(\|g\|_s^{1+\beta})$.

Remarks. The condition that F be constant on $(-\infty, a]$ and on $[b, \infty)$ applies to distribution functions of probability measures concentrated on $[a, b]$, cf. Section 6.

If $s > 1$ and $0 < \alpha \leq \min(1, s-1)$ or $0 < \beta \leq (s-1)/(s+1)$ are given, then we can define p by $p = s/(1+\alpha)$ or $p = s/(1 + \beta + s\beta)$ respectively, with $1 \leq p < s$ in each case. Also, since $H(\alpha) \subset H(\delta)$ for $0 < \delta < \alpha$, if the given conditions on α or β do not hold, we can replace them by α' or β' with $0 < \alpha' < \alpha$ or $0 < \beta' < \beta$ respectively, for which the conditions do hold.

Proof. If (a) or (b) holds then by the mean value theorem of calculus

$$|R(g)| = |F' \circ (G + \lambda g) - F' \circ G| \cdot |g| \leq C|g|^{1+\alpha} = C|g|^{s/p}$$

for a variable λ with $0 \leq \lambda \leq 1$ and the conclusion follows.

Under (c), note that $0 < \beta < 1$. We can assume $a = 0$ and $b = 1$, possibly changing K. Let $A := \{G+g \leq 0\}$, $B := \{0 < G+g < 1\}$, and $C := \{G+g \geq 1\}$.

Let $g_1 := 1_A \cdot (G + g)$, $g_3 := 1_C \cdot (G + g - 1)$, and $g_2 := g - g_1 - g_3$. Then $G + g_2 = \max(0, \min(G + g, 1))$ so (b) holds for g_2 with β in place of α. We also have

$$R(g) = R(g_2) - (F' \circ G) \cdot (g_1 + g_3)$$

and $|F' \circ G| \leq M$ for some $M < \infty$. Now by Hölder's inequality,

$$\int |g_1|^p d\mu = \int_A (-G - g)^p d\mu \leq \|g\|_s^p \mu(A)^{1-p/s}. \qquad (2.3)$$

For any $\delta > 0$, the maximum of $\mu(A)$ for $\|g\|_s \leq \delta$ is attained when $g = -G$ on A, since replacing g by $-G$ on A, where $0 \geq -G \geq g$, does not increase $\|g\|_s$ or change A. Then, $\mu(A)$ is maximized under the constraint $\int_A G^s d\mu \leq \delta^s$ when A is a set such that for some $t \geq 0$, $\{G < t\} \subset A \subset \{G \leq t\}$, by the Neyman-Pearson Lemma (normalizing μ and $G^s\mu$ to be probability measures), e.g. Lehmann (1991, Theorem 3.1 p. 74). Let $\mu_A(E) := \mu(A \cap E)$ for all $E \in \mathcal{A}$, $\nu := \mu_A \circ G^{-1}$ and $L(x) := \nu([0, x])$. Then $\int_A G^s d\mu = \int_{[0,t]} x^s dL(x)$, and $L(t) = \mu(A)$. We can assume t is the unique smallest number for which the latter holds. We have $L(x) \leq Kx$ for all $x \geq 0$ by (2.2). Integrating by parts, we see that for fixed $\mu(A)$, $\int_A G^s d\mu$ is minimized, also as μ varies, when $L(x) = Kx$ for $0 \leq x \leq u := \min(t, \mu(A)/K)$, where, if $u < t$, then $L(x) = \mu(A)$ for $u \leq x \leq t$. But then by choice of t, we must have $t = u$ and $Kt = \mu(A)$. Thus $\|g\|_s^s \geq K(\int_0^t x^s dx) = Kt^{s+1}/(s + 1)$, so $\mu(A) \leq T\|g\|_s^{s/(s+1)}$ for a constant $T = T(s, K)$. It follows by (2.3) that $\|g_1\|_p \leq V\|g\|_s^{1+\beta}$ for a constant $V = V(K, p, s)$. The function g_3 and set C can be treated symmetrically, interchanging G with $1 - G$ and g with $-g$, so Theorem 2.9 is proved.   □

The constants α and β are best possible under the given conditions, as shown in Dudley (1997, Section 13.4) even under the stronger condition $F \in C^2$ on J.

2.10 Theorem. For $1 \leq p < s$ and F, G satisfying any of the hypotheses (a), (b) or (c) of Theorem 2.9, the map $(f, g) \mapsto (F + f) \circ (G + g)$ from $\mathcal{W}_p \times L^s$ into L^p is Fréchet differentiable as $f \to 0$ and $g \to 0$ with remainder bound $O(\|f\|_{[p]}^\gamma + \|g\|_s^\zeta)$ where $\gamma := 1 + s/[p(1 + s)]$ and under (a) or (b), $\zeta = \min(s/p, \gamma)$ while under (c), $\zeta = s(p + 1)/[(s + 1)p] = 1 + (s - p)/[p(s + 1)] < \gamma$.

Proof. We apply Theorems 2.5 and 2.9. Theorem 2.5 gives exponents γ on both norms. Under (a) or (b), Theorem 2.9 gives an exponent s/p on $\|g\|_s$ and under (c), the given $\zeta < \gamma$. The conclusions follow.   □

In this survey the composition operator $(f, g) \mapsto (F + f) \circ (G + g)$ is treated mainly for G and g real-valued. Part III, Section 6 treats G and g Banach-valued, as did Gray (1975).

3. The quantile (inverse) operator. For a function H from an interval J into $\mathbb{R}$, let $H^\leftarrow(y) := H_J^\leftarrow(y) := \inf\{x \in J : H(x) \geq y\}$, or the right endpoint of J (which may be $+\infty$) if there is no such x. The notation $H^\leftarrow$ has appeared in Bingham, Goldie and Teugels (1987) and Beirlant and Deheuvels (1990). It is used when H is not necessarily 1-1, while H^{-1} is reserved for the inverse image of sets, or a point function in case H is 1-1. The operator $f \mapsto (F + f)^\leftarrow$ will be studied when F (but not necessarily $F + f$) is strictly increasing and continuous and so has an inverse $F^{-1} \equiv F^\leftarrow$ on the range interval $F[J] := \{F(x) : x \in J\}$. Then, as is known from elementary calculus, for any x at which F has a derivative $F'(x) > 0$, F^{-1} has a derivative at $y = F(x)$ given by

$$(F^{-1})'(y) \; = \; 1/(F'(F^{-1}(y))).$$

Appendix C shows that in defining $F^{\leftarrow}$ or $(F+f)^{\leftarrow}$, F can be replaced by its upper semicontinuous envelope $\overline{F}$, and for differentiability of $t \mapsto (\overline{F} + t)^{\leftarrow}$ into L^p for $p > 1$ at $t = 0$ along constant functions, $\overline{F}$ must be strictly increasing.

In the cases we consider, the derivative of the operator $f \mapsto (F+f)^{\leftarrow}$ at $f = 0$ will be given by $f \mapsto -(f \circ F^{-1})/(F' \circ F^{-1})$ with a remainder R_f defined by

$$R_f \; := \; (F+f)^{\leftarrow} - F^{\leftarrow} + (f \circ F^{-1})/(F' \circ F^{-1}). \qquad (3.1)$$

Specifically, Theorems 3.1 and 3.3 below include the conclusion that the derivative must be $f \mapsto -(f \circ F^{-1})/(F' \circ F^{-1})$ under their hypotheses.

Whenever F' is bounded below by some $c > 0$, the operator derivative is a bounded linear operator with respect to the supremum norm, both on the domain and range spaces, and so with respect to a stronger norm on the domain, such as $\| \cdot \|_{[p]}$, and a weaker norm on the range space, such as an L^p norm $\|\cdot\|_p$. On the other hand if F' is not bounded below, then in the operator derivative, $f \circ F^{-1}$ is multiplied by an unbounded function, giving an unbounded operator between some function spaces, e.g. from $\mathcal{W}_p$ to ℓ_∞ for $1 \leq p < \infty$. Thus it is reasonable to assume that F' is bounded below. Also, F can be so chosen in some applications to probability and statistics, see Theorem 6.3 below.

Definition. A function F will be called an *increasing diffeomorphism* of an interval $[a, b]$ onto an interval $[c, d]$ if F is a continuous, increasing function from $[a, b]$ onto $[c, d]$ having a derivative F' everywhere on the open interval (a, b) which extends to a continuous, strictly positive function on $[a, b]$.

For the remainder R_f defined by (3.1) we have the following bound:

3.1 Theorem. Let $1 \leq p < \infty$ and let F be an increasing diffeomorphism from an interval $[a, b]$ onto an interval $[c, d]$. Suppose that F' satisfies the Hölder condition $\sup_{a \leq x < u \leq b} |F'(u) - F'(x)|/(u-x)^{1/p} \; < \; \infty$. Then there is a constant $K_p < \infty$ such that

$$\|R_f\|_p \leq K_p \|f\|_{[p]}^{(p+1)/p} \qquad \forall \, f \in \mathcal{W}_p([a,b]) \text{ with } \|f\|_{[p]} \leq 1. \qquad (3.2)$$

Proof. Dudley (1994), Corollary 2.4, applies since the extension from $[0, 1]$ to $[a, b]$ and $[c, d]$ is straightforward. $\qquad\qquad\qquad\qquad\qquad\qquad\qquad\qquad\qquad\qquad\qquad\qquad\quad \square$

The remainder bound $O(\|f\|_{[p]}^{(p+1)/p})$ in (3.2) cannot be replaced by $o(\|f\|^{(p+1)/p})$, for the p-variation norm or any other norm $\|\cdot\|$, as the following shows:

3.2 Proposition. Let $1 \leq p < \infty$, $[a, b] = [c, d] = [0, 1]$ and $F(x) = x$, $0 \leq x \leq 1$. Let $h = 1_{[u,v]}$ for any fixed $0 < u < v < 1$. Then there is a constant $C_p > 0$ such that for $0 < t < \min(v - u, 1 - v)$, we have $\|R_{th}\|_p = C_p t^{(p+1)/p}$.

Proof. This is Dudley (1994), Proposition 2.5 (in which y should be replaced by Y). $\qquad\qquad\qquad\qquad\qquad\qquad\qquad\qquad\qquad\qquad\qquad\qquad\qquad\qquad\qquad\qquad\quad \square$

Thus (3.2) gives the best value of γ, namely $(1+p)/p$, for the operator $F \mapsto F^{\leftarrow}$ Fréchet differentiable at $U \in \mathcal{W}_p$, in the sense defined in Section 1. For differentiability in specific directions we also have the following:

3.3 Theorem. Let $1 \leq p < \infty$ and let F be an increasing diffeomorphism of an interval $[a, b]$ onto an interval $[c, d]$. Let h be a bounded real-valued function on $[a, b]$, continuous almost everywhere for Lebesgue measure. Then $\|R_{th}\|_p = o(|t|)$ as $|t| \to 0$.

Proof. This follows from Part III, Theorem 4.5. $\qquad\qquad\qquad\qquad\qquad\square$

Theorem 4.5 in Part III actually gives $\mathcal{C}$-differentiability for a large class $\mathcal{C}$ of sets of such functions h. See also Section 7 below. The conditions on h in Theorem 3.3 are nearly necessary, as follows:

3.4 Proposition. Let $1 \leq p < \infty$ and $F(x) = x$, $0 \leq x \leq 1$. Let $f : [0, 1] \mapsto \mathbb{R}$ and $g(x) := f(1 - x)$ for $0 \leq x \leq 1$. If $\|R_{th}\|_p = o(|t|)$ as $t \to 0$ for $h = f$ and for $h = g$ then f is bounded and continuous almost everywhere for Lebesgue measure.

Proof. This is Corollary 4.15 of Part III. $\qquad\qquad\qquad\qquad\qquad\square$

Here is a fact about combining the composition and quantile operators.

3.5 Proposition. (a) For any two functions F, G from $\mathbb{R}$ into $\mathbb{R}$, and any $y \in \mathbb{R}$, $(F \circ G)^{\leftarrow}(y) \geq (G^{\leftarrow} \circ F^{\leftarrow})(y)$.
(b) If F is non-decreasing and right-continuous, then for any G,
$$(F \circ G)^{\leftarrow} \equiv G^{\leftarrow} \circ F^{\leftarrow}.$$

Proof. For (a), if $F(G(x)) \geq y$, then $G(x) \geq F^{\leftarrow}(y)$, so $x \geq G^{\leftarrow}(F^{\leftarrow}(y))$ and (a) follows.

For (b), we have $F(x_n) \geq y$ for some $x_n \downarrow F^{\leftarrow}(y)$, so since F is right-continuous, $F(F^{\leftarrow}(y)) \geq y$. If $G(x) \geq F^{\leftarrow}(y)$, then since F is non-decreasing, $F(G(x)) \geq y$ and with (a), now (b) follows. $\qquad\qquad\qquad\qquad\qquad\square$

Examples. If $F(x) \equiv G(x) \equiv -x$ then $(F \circ G)(x) \equiv x$ so $(F \circ G)^{\leftarrow}(y) = y$ for all y, but $F^{\leftarrow}(y) = G^{\leftarrow}(y) = -\infty$ for all y.

Let $F := 1_{(0,\infty)}$ and $G := 1_{[1,\infty)}$. Then F and G are non-decreasing but only G is right-continuous. For $0 < y \leq 1$, we have $(F \circ G)^{\leftarrow}(y) = 1$ while $G^{\leftarrow}(F^{\leftarrow}(y)) = G^{\leftarrow}(0) = -\infty$. Thus the right-continuity of F in Proposition 3.5 cannot be dispensed with.

We also have the following right and left inverse properties, whose proofs are straightforward; (a) is stated e.g. by M. Csörgő (1983, p. 1).

3.6 Proposition. (a) Let F be a non-decreasing, continuous function on a possibliy unbounded interval $J \subset \mathbb{R}$. Then for all y in the range of F, $F(F_J^{\leftarrow}(y)) = y$ where if u is the left endpoint of J and $-\infty \leq u \notin J$, we let $F(u) := \lim_{v \downarrow u} F(v)$.
(b) If F is strictly increasing: $J \mapsto \mathbb{R}$ then $F^{\leftarrow}(F(x)) = x$ for all $x \in J$.

Inverse operators $f \mapsto f^{-1}$ have been considered in the literature on spaces of invertible functions. The following remarks are based on Garay (1996). Let M be a compact m-dimensional C^{∞} manifold. Then M can be C^{∞} embedded in a Euclidean space $\mathbb{R}^N$ with $N > m$. M is covered by finitely many open sets U_i, $i = 1, \ldots, I$, such that for each i, there is a C^{∞} homeomorphism α_i of a neighborhood of the closure $\overline{U}_i$ onto an open set W_i in $\mathbb{R}^m$. Let α_i take U_i onto $V_i \subset W_i$. A function f on M is C^k if $f \circ \alpha_i^{-1}$ is C^k on V_i for each i. If f takes values in M, we can view it as a map: $M \mapsto \mathbb{R}^N$. Let $C^k(M, \mathbb{R}^N)$ be the set of all C^k functions from M into $\mathbb{R}^N$ and $C^k(M, M)$ the subset with values in M. On $C^k(M, \mathbb{R}^N)$ a norm is defined by $\|f\|_{k,N} := \sup\{|D^p f_j(\alpha_i^{-1}(x))|: j = 1, \ldots, N, i = 1, \ldots, I, x \in V_i, |p| := p_1 + \cdots + p_m \leq k, p_r \in \mathbb{N}, r = 1, \ldots, m\}$, where $D^p := \partial^{|p|}/\partial x_1^{p_1} \cdots \partial x_m^{p_m}$, $f = (f_1, \ldots, f_N)$. Let $\operatorname{Diff}^k(M)$ be the set of all one-to-one functions f in $C^k(M, M)$ taking M onto itself with $f^{-1} \in C^k(M, M)$. Thus $\operatorname{Diff}^k(M)$ is the set of all C^k diffeomorphisms of M. The norm $\| \cdot \|_{k,N}$ defines a metric on $C^k(M, M)$, Lipschitz equivalent to the metric defined for a different choice of U_i and α_i. The map $f \mapsto f^{-1}$ of $\operatorname{Diff}^k(M)$ onto itself is a homeomorphism. According to Garay (1996): $f \mapsto f^{-1}$ is nowhere Fréchet differentiable on $A := \operatorname{Diff}^1(M)$ into $C^1(M, M)$; it is compactly differentiable on A at each point of $\operatorname{Diff}^2(M)$ into $\operatorname{Diff}^1(M)$, and it is Fréchet C^q from $\operatorname{Diff}^r(M)$ into $\operatorname{Diff}^{r-q}(M)$ for $q = 0, 1, \ldots, r$.

4. The integration operator and Young integrals. By *integral functional* we mean the real-valued bilinear functional $(F, G) \mapsto \int_a^b F \, dG$. The indefinite integral gives rise to the *integration* or *indefinite integral operator* I defined by $I(F, G) := \int_a^{(\cdot)} F \, dG$. As will be seen, a good way to define the integral for this purpose is the (CY) integral of L. C. Young (1936), defined below. The domain set of pairs (F, G) for which $(CY) \int_a^b F \, dG$ is defined overlaps with, but is different from, the domains of the better-known Riemann-Stieltjes and Lebesgue-Stieltjes integrals, as follows.

Recall that a real-valued function f on $[a, b]$ is called *regulated* if it has a right limit

$$f_+^{(b)}(x) := f_+(x) := f(x+) := \lim_{y \downarrow x} f(y) \tag{4.1}$$

at each $x \in [a, b)$, with $f_+^{(b)}(b) := f(b)$, and a left limit

$$f_-^{(a)}(x) := f_-(x) := f(x-) := \lim_{y \uparrow x} f(y) \tag{4.2}$$

at each $x \in (a, b]$, with $f_-^{(a)}(a) := f(a)$. The set of all such functions is called $\mathcal{R} = \mathcal{R}[a, b]$. Functions in $\mathcal{W}_p$ for any $p < \infty$ are regulated. The same definition applies to functions with values in normed spaces.

Detailed definitions of various integrals are given in Part II, Section 3, also for functions with values in a Banach algebra, some other integrals are defined in Appendix F. Here we note the following points.

The Riemann-Stieltjes integral $(RS) \int_a^b F \, dG$ is defined as the mesh of partitions goes to 0, where a *partition* of $[a, b]$ is a finite set $\{a = x_0 < x_1 < \cdots < x_n = b\}$ with *mesh* $\max_i(x_i - x_{i-1})$. Its extension, the *Moore-Pollard-Stieltjes integral* $(MPS) \int_a^b F \, dG$ is defined by way of refinements of partitions. The Riemann-Stieltjes integral is not defined if F and G have a common discontinuity, and the (MPS) integral is not defined if F and G have a jump on the same side of the same point, e.g. if $F(x+) \neq F(x)$ and $G(x+) \neq G(x)$ for some x with $a \leq x < b$.

The Lebesgue-Stieltjes integral $(LS) \int_a^b F \, dG$ is defined only for G of bounded variation, and G is usually taken to be right-continuous, $G_+^{(b)} \equiv G$. For G of bounded variation, the Lebesgue-Stieltjes integral is defined for much more general F than are the other integrals being treated here. The Lebesgue-Stieltjes integral has sometimes been defined by a formal integration by parts if F is of bounded variation and G is rather general, but on such a definition see Theorem 4.8(d) below and the example after its proof. At any rate the Lebesgue-Stieltjes integral is not defined if neither F nor G is of bounded variation.

One form of L. C. Young integral (see p. 263 in Young, 1936 and p. 1975 in Dudley, 1992) is defined by

$$(Y_1) \int_a^b F \, dG := (MPS) \int_a^b F_+^{(b)} \, dG_-^{(a)} - [(\Delta^+ F)(\Delta^+ G)](a)$$

$$+ [F(\Delta^- G)](b) - \sum_{a < x < b} [(\Delta^+ F)(G_+ - G_-)](x)$$

if the (MPS) integral exists and the sum converges unconditionally, so that, for real-valued functions, it converges absolutely. Here and below for any regulated function h on an interval $[a, b]$ let $(\Delta^+ h)(x) := h(x+) - h(x)$ for $a \le x < b$ and $(\Delta^- h)(x) := h(x) - h(x-)$ for $a < x \le b$. In this definition, G must be regulated and F must have right limits. We will actually assume that F is also regulated. Another form of L. C. Young integral, $(Y_2) \int_a^b F \, dG$, is defined as $(MPS) \int_a^b F_-^{(a)} \, dG_+^{(b)}$ plus a corresponding sum (Definition 3.4 in Part II; Dudley, 1992). The definitions of the two forms apply to functions with values in a Banach algebra and are justified by the next three theorems.

4.1 Theorem. If F and G are regulated functions from an interval $[a, b]$ into a Banach algebra $\mathbb{B}$ and $(Y_1) \int_a^b F \, dG$ or $(Y_2) \int_a^b F \, dG$ exists then the other also exists and the two are equal.

Proof. If one integral exists, then Part II, Theorem 3.7 gives that the other does. If (Y_1) and (Y_2) integrals both exist then they are equal in the real-valued case by Theorem 3.2 of Dudley (1992) and in the $\mathbb{B}$-valued case by the same proof. $\square$

The common value of $(Y_i) \int_a^b F \, dG$ for $i = 1, 2$ if either exists, for F and G regulated, will be called the *central Young integral* $(CY) \int_a^b F \, dG$.

Let G be any function of bounded variation (i.e., finite 1-variation) on $[a, b]$. Then there is a unique finite signed measure $\nu = \nu^G$ on $[a, b]$ such that $G(x+) - G(a) = \nu([a, x])$ for $a \le x < b$ and $G(b) - G(a) = \nu([a, b])$. For that ν, also $G(x-) - G(a) = \nu([a, x))$ for $a < x \le b$. The Lebesgue-Stieltjes integral $(LS) \int_a^b F \, dG$ is defined as $\int_{[a,b]} F \, d\nu^G$ (cf. p. 407 in Phillips, 1971).

The preceding definition of the Lebesgue-Stieltjes integral differs in endpoint terms from some other definitions (Saks, 1937, pp. 64–66, 99; Hildebrandt, 1963, p. 242). Let G be a function of bounded variation on $\mathbb{R}$. An interval function μ^G is defined by $\mu^G((a, b)) := G(b-) - G(a+)$ for $a < b$ and $\mu^G([a, b]) := G(b+) - G(a-)$ for $a \le b$. Then μ^G extends uniquely to a countably additive, finite signed measure on the Borel sets of $\mathbb{R}$. The integral $\int_a^b F \, dG$ had been defined as $\int_J F \, d\mu^G$ for an interval J with endpoints a, b. If $J = [a, b]$, the additivity

$$\int_a^b F \, dG = \int_a^c F \, dG + \int_c^b F \, dG, \quad a < c < b,$$

can fail. Hildebrandt (1963, p. 242) suggests taking $J = (a, b]$, which gives additivity, but does not allow a contribution from an atom at a, and still may depend on values of G outside of $[a,b]$ via $G(b+)$. Such values, and also $G(a-)$ in Saks's definitions, can cause integrals to differ which are equal except for such endpoint terms (e. g. Saks, 1937, p. 208). The definition of $(LS) \int_a^b F\, dG$ as above, gives additivity, only depends on the values of F, G on $[a,b]$, and allows for atoms at a and/or b.

4.2 Theorem. For real-valued functions, if G is of bounded variation and F is any regulated function on $[a,b]$, then the central Young and Lebesgue-Stieltjes integrals $\int_a^b F\, dG$ both exist and are equal.

Proof. Since each function of bounded variation is a difference of two nondecreasing functions, it is enough to prove the theorem when G is nondecreasing. By Theorem 5.32 in Hildebrandt (1938), the integral $(MPS) \int_a^b F_-^{(a)}\, dG_+^{(b)}$ exists. This and boundedness of the variation of G imply the existence of the integral $(Y_2) \int_a^b F\, dG$, defined in Part II, Section 3.2. Since F is bounded and ν-measurable, $(LS) \int_a^b F\, dG$ exists.

To prove that the (Y_2) and (LS) integrals are equal we will show that

$$\int_{[a,b]} F_-^{(a)}\, d\nu^G = (MPS) \int_a^b F_-^{(a)}\, dG_+^{(b)} + F(a)\Delta^+ G(a) \tag{4.3}$$

and

$$\int_{[a,b]} (F - F_-^{(a)})\, d\nu^G = \int_{[a,b]} (F - F_-^{(a)})\, d\nu_{at}^G = \sum_{(a,b]} \Delta^- F \Delta^- G_+^{(b)}, \tag{4.4}$$

where ν_{at}^G denotes the purely atomic part of ν^G.

To prove (4.3), let $\epsilon > 0$. Then there is a partition λ of $[a,b]$ with the following property. For every refinement $\kappa := \{a = x_0 < \cdots < x_n = b\}$ of λ, i.e. $\lambda \subset \kappa$, and for all $y_i \in [x_{i-1}, x_i]$, $i = 1, \ldots, n$,

$$\Big| \sum_{i=1}^n F_-^{(a)}(y_i)\nu^G(T_i) - (MPS) \int_a^b F_-^{(a)}\, dG_+^{(b)} \Big| \le \epsilon, \tag{4.5}$$

where $T_0 := \{a\}$ and $T_i := (x_{i-1}, x_i]$ for $i = 1, \ldots, n$. Then $\{T_0, \ldots, T_n\}$ is a decomposition of $[a,b]$ into mutually disjoint measurable sets. Let $m_i := \inf\{F_-^{(a)}(y): y \in T_i\}$ and let $M_i := \sup\{F_-^{(a)}(y): y \in T_i\}$ for $i = 1, \ldots, n$. Letting $F_-^{(a)}(y_i) \downarrow m_i$ and $F_-^{(a)}(y_i) \uparrow M_i$ in (4.3) for each $i = 1, \ldots, n$, we get that (4.5) holds with all $F_-^{(a)}(y_i)$ replaced by m_i or all $F_-^{(a)}(y_i)$ replaced by M_i. Therefore it follows that

$$-\epsilon \le \sum_{i=1}^n m_i\, \nu^G(T_i) - (MPS) \int_a^b F_-^{(a)}\, dG_+^{(b)}$$

$$\le \int_{[a,b]} F_-^{(a)}\, d\nu^G - (MPS) \int_a^b F_-^{(a)}\, dG_+^{(b)} - F(a)\Delta^+ G(a)$$

$$\le \sum_{i=1}^n M_i\, \nu^G(T_i) - (MPS) \int_a^b F_-^{(a)}\, dG_+^{(b)} \le \epsilon.$$

Since ϵ is arbitrary (4.3) holds.

To prove the first equality in (4.4) it is enough to use the fact that $|F(x) - F_-^{(a)}(x)| > \epsilon$ holds only for at most finitely many $x \in [a, b]$. The second equality follows similarly by choosing a finite sum which is close to the right side of (4.4). The proof of Theorem 4.2 is complete. □

A related fact, where G is right-continuous and ν has no point mass at a, is Lemma 3.3 of Dudley (1992).

Thus, the sums in the definitions of (Y_1) integral (above) and the (Y_2) integral (Part II, Section 3.2) are exactly what they should be for the integrals to agree with each other and with the Lebesgue-Stieltjes integral under the conditions of Theorem 4.2.

4.3 Theorem. If F and G are two regulated real-valued functions on $[a, b]$ and the integral $(MPS) \int_a^b F \, dG$ is defined, then $(CY) \int_a^b F \, dG$ is also defined and has the same value.

Proof. This is Corollary 3.18 of Part II. □

Thus, for regulated functions, the (CY) integral is a common extension of the Riemann-Stieltjes, Moore-Pollard-Stieltjes and Lebesgue-Stieltjes integrals. It also extends the refinement-Young-Stieltjes integral $(RYS) \int_a^b F \, dG$ used by L. C. Young (1938b) and Gehring (1954). The (RYS) integral is defined as the Moore-Pollard-Stieltjes integral except that instead of the Riemann-Stieltjes interval function $F(y)[G(d) - G(c)]$, $c \le y \le d$, one takes the interval function (R. C. Young, 1929)

$$\Phi(y; [c, d]) := F(c)(\Delta^+ G)(c) + F(y)(G(d-) - G(c+)) + F(d)(\Delta^- G)(d), \quad (4.6)$$

where y varies with $c < y < d$, defined by W. H. Young (1914) with $y = c+$ for step functions. Hildebrandt discussed several properties of the (RYS) integral in his survey paper (1938, p. 275) and in his book (1963, pp. 88-95).

For regulated functions existence of the (MPS) integral implies that of the (RYS) integral which implies that of the (CY) integral (Part II, Propositions 3.15 and 3.17), and the values of the integrals agree. There exist g, h which are in $\mathcal{W}_r[0, 1]$ for all $r > 2$ such that $(CY) \int_0^1 g \, dh = 0$ but $(RYS) \int_0^1 g \, dh$ does not exist (Part II, Proposition 3.16). Thus the (CY) integral strictly includes the other integrals of Riemann-Stieltjes type. The central Young (CY) integral also has a variety of other properties one would like an integral to have (bilinearity, additivity over adjoining intervals, etc.), as shown in Part II, Section 3.4.

Recall that $V_p(f) := v_p(f)^{1/p}$ and $V_{p,\infty}(f) := V_p(f) + \sup_{a \le x \le b} |f(x)|$. The following will imply among other things that the (CY) integral is defined in many cases when the (MPS) and Lebesgue-Stieltjes integrals are not. It was proved by E. R. Love and L. C. Young, see L. C. Young (1936, pp. 251, 256) and Love (1993).

4.4 Theorem (Love-Young inequality). For any $p > 0$ and $q > 0$ with $1/p + 1/q > 1$ there is a constant $C_{p,q} < \infty$ such that for all $F \in \mathcal{W}_p([a, b])$ and $G \in \mathcal{W}_q([a, b])$, $(CY) \int_a^b F \, dG$ and $(RYS) \int_a^b F \, dG$ exist and are equal, and

$$\left| (CY) \int_a^b F \, dG \right| \le C_{p,q} V_{p,\infty}(F) V_q(G).$$

Proof. This follows from Part II, Theorem 3.27, (3.94) and Corollary 3.20.   □

The statement of Theorem 4.4 also holds for the Ward–Perron–Stieltjes integral $(WPS) \int_a^b F\, dG$ and the equivalent Henstock-Kurzweil integral, defined in Appendix F, because under the hypotheses of Theorem 4.4, both equal $(RYS) \int_a^b F\, dG$, by Theorem F.2 and the fact that $\mathcal{W}_p \subset \mathcal{W}_r^*$ for $1 \le p < r < \infty$ (e.g. Lemma 2.14 in Part II). The agreement of the integrals under the given conditions (although not in general) confirms the values of the integrals and provides alternate characterizations that can be useful in different situations: the refinement limit for the (RYS) integral, the explicit (MPS) integrals plus series in the definitions of (Y_1) and (Y_2) integrals, and gauge functions for the (WPS) and (HK) integrals.

L. C. Young (1938b, Theorem 5.1) extended the preceding theorem to functions of bounded ϕ-variation as follows.

4.5 Theorem. Let ϕ and ψ be strictly increasing continuous function from $[0, \infty)$ onto itself and such that

$$\sum_{n=1}^{\infty} \phi^{-1}\left(\frac{1}{n}\right) \psi^{-1}\left(\frac{1}{n}\right) < \infty. \qquad (4.7)$$

If F and G are real-valued functions on $[a, b]$ of bounded ϕ-variation and of bounded ψ-variation, respectively, then $(RYS) \int_a^b F\, dG$ exists and satisfies the inequality

$$\left| (RYS) \int_a^b F\, dG - F(x)[G(b) - G(a)] \right| \le K \sum_{n=1}^{\infty} \phi^{-1}\left(\frac{v_\phi(F)}{n}\right) \psi^{-1}\left(\frac{v_\psi(G)}{n}\right)$$

for some constant K and any $x \in [a, b]$.

Essentially the same theorem was proved in a different way by Leśniewicz and Orlicz (1973, Theorem 4.11) when either F or G is continuous, using the Riemann-Stieltjes integral. They also proved that the series condition (4.6) is best possible for convex functions ϕ and ψ satisfying the Δ_2 and ∇_2 growth conditions: for some constants $c, d > 1$ and $u_0 > 0$, $\phi(2u) \le c\phi(u)$ and $2\phi(u) \le d^{-1}\phi(du)$ for $0 \le u \le u_0$, respectively, and the same for ψ instead of ϕ. The functions $\phi(u) = u^p$ and $\psi(u) = u^q$ for $p > 1$ and $q > 1$ do satisfy both the Δ_2 and ∇_2 growth conditions. For such ϕ and ψ, let

$$S_\phi(t) := \sum_{k=1}^{\infty} \phi^{-1}\left(\frac{1}{m^k}\right) \sin(2\pi m^k t) \ \text{ and } \ C_\psi(t) := \sum_{k=1}^{\infty} \psi^{-1}\left(\frac{1}{m^k}\right) \cos(2\pi m^k t)$$

for some integer $m \ge 4$ and all $t \in [0,1]$. Leśniewicz and Orlicz (1973, Theorem 4.21) then showed that $v_\phi(S_\phi) < \infty$, $v_\psi(C_\psi) < \infty$ and

$$\left| \sum_{i=1}^{m^n} S_\psi\left(\frac{i}{m^n}\right) \left[C_\psi\left(\frac{i}{m^n}\right) - C_\psi\left(\frac{i-1}{m^n}\right) \right] \right| \ge 2 \sum_{k=1}^{n-1} m^k \phi^{-1}\left(\frac{1}{m^k}\right) \psi^{-1}\left(\frac{1}{m^k}\right).$$

By their Theorem 1.12, the right side of the preceding inequality increases to infinity as $n \to \infty$ whenever the series (4.7) is divergent. Therefore the integral

$(RS) \int_0^1 S_\phi \, dC_\psi$ does not exist in this case. Since S_ϕ and C_ψ are continuous, by Theorem II.10.9 in Hildebrandt (1963), the more general (MPS), (RYS) and (CY) integrals also do not exist in this case. For $\phi(u) = u^p$ and $\psi(u) = u^q$, the series (4.7) converges if and only if $1/p + 1/q > 1$. Therefore the condition $1/p + 1/q > 1$ in Theorem 4.4 cannot be replaced by the condition $1/p + 1/q = 1$ if $1 < p, q < \infty$. It will be seen in Appendix F after Proposition F.3 that the same is true for the Ward–Perron–Stieltjes or Henstock–Kurzweil integral.

Using Theorem 4.4, the integral functional is now easy to analyze as follows.

4.6 Corollary. For any interval J, any $p \geq 1$ and $q \geq 1$ with $1/p + 1/q > 1$, and any $F \in W_p(J)$ and $G \in W_q(J)$, the map $(f,g) \mapsto (CY) \int_J (F+f) \, d(G+g)$ is jointly Fréchet differentiable for $f \in W_p(J)$ and $g \in W_q(J)$ at $f = g = 0$, with partial derivatives $f \mapsto (CY) \int_J f \, dG$ and $g \mapsto (CY) \int_J F \, dg$.

Proof. The central Young integral is bilinear (Part II, Proposition 3.21), so

$$(CY) \int_J (F+f) \, d(G+g) \;=\; (CY) \int_J F \, dG + (CY) \int_J f \, dG + (CY) \int_J F \, dg$$
$$+ (CY) \int_J f \, dg. \tag{4.8}$$

By Theorem 4.4 three times, $f \mapsto (CY) \int_J f \, dG$ is a bounded linear functional on $W_p(J)$, $g \mapsto (CY) \int_J F \, dg$ is a bounded linear functional on W_q, and for the remainder term,

$$\left| (CY) \int_J f \, dg \right| \;\leq\; C_{p,q} \|f\|_{[p]} \|g\|_{(q)} \;=\; o(\|f\|_{[p]} + \|g\|_{(q)})$$

as $\|f\|_{[p]} \to 0$ and $\|g\|_{(q)} \to 0$. Since $\|f\|_{(p)} = V_p(f)$ is the p-variation seminorm and $\|f\|_{[p]} = V_{p,\infty}(f)$ is the p-variation norm, both for $1 \leq p < \infty$, the conclusion follows. $\square$

In fact, the integral functional $(F, G) \mapsto (CY) \int_J F \, dG$, being bilinear and a quadratic polynomial, is analytic on $W_p(J) \times W_q(J)$ at any F, G, with the four-term power series expansion given by (4.8).

We have been treating the real-valued bilinear functional $(F, G) \mapsto (CY) \int_a^b F \, dG$. The map $(F, G) \mapsto I(F, G)$, where $I(F, G)(x) := (CY) \int_a^x F \, dG$ for $a \leq x \leq b$, when defined, takes two functions into a function.

4.7 Theorem. If $1 \leq p < \infty$, $1 \leq q < \infty$ and $1/p + 1/q > 1$, then $I(\cdot, \cdot)$ is a bounded bilinear operator from $W_p([a,b]) \times W_q([a,b])$ into $W_q([a,b])$.

Proof. This follows from Part II, Proposition 3.32. $\square$

We next recall some facts and give some others about integration by parts for integrals treated in this section.

4.8 Theorem (integration by parts). Let F and G be real-valued functions on an interval $[a, b]$. Then
(a) For the (MPS) or (RS) integral, if $\int_a^b F \, dG$ exists then so does $\int_a^b G \, dF$ and

$$\int_a^b F \, dG + \int_a^b G \, dF \;=\; F(b)G(b) - F(a)G(a). \tag{4.9}$$

(b) Let $F, G \in \mathcal{W}_1([a,b])$. Then the Lebesgue-Stieltjes integrals $(LS)\int_a^b F_-^{(a)}\,dG$ and $(LS)\int_a^b G_+^{(b)}\,dF$ both exist and

$$(LS)\int_a^b F_-^{(a)}\,dG + (LS)\int_a^b G_+^{(b)}\,dF = F(b)G(b) - F(a)G(a).$$

(c) Suppose that either $(CY)\int_a^b F\,dG$ or $(CY)\int_a^b G\,dF$ exists. Then $(CY)\int_a^b \widehat{G}\,dF$ and $(CY)\int_a^b \widehat{F}\,dG$ exist and

$$(CY)\int_a^b \widehat{G}\,dF + (CY)\int_a^b \widehat{F}\,dG = F(b)G(b) - F(a)G(a),$$

where $\widehat{G} := [G_-^{(a)} + G_+^{(b)}]/2$ and $\widehat{F} := [F_-^{(a)} + F_+^{(b)}]/2$.

(d) Let $F, G \in \mathcal{W}_1([a,b])$. Then for the (LS), (RYS) and (CY) integrals, which all exist and are equal for $\int_a^b F\,dG$ and for $\int_a^b G\,dF$,

$$\int_a^b F\,dG + \int_a^b G\,dF = F(b)G(b) - F(a)G(a) + \sum_{a < x \le b} [\Delta^- F \Delta^- G](x)$$

$$- \sum_{a \le x < b} [\Delta^+ F \Delta^+ G](x). \tag{4.10}$$

Remarks. A proof of statement (b) can be found in Hewitt and Stromberg (1969, p. 419). Hewitt (1960) proved the formula in (c) under nearly the same conditions. According to Hildebrandt (1938, p. 276), statement (d) for the (RYS) integral is proved by de Finetti and Jacob (1935). Statement (d) for the Ward-Perron-Stieltjes integral defined in Appendix F was proved by Kurzweil (1958). Furthermore, Tvrdý (1989, Theorem 2.15) proved the same formula when only $F \in \mathcal{W}_1([a,b])$ and $G \in \mathcal{R}([a,b])$. Norvaiša (1997b) extends (d) to $F, G \in \mathcal{W}_p([a,b])$ for $1 < p < 2$ for the (RYS) and (CY) integrals.

Proof. See Hildebrandt (1963), 11.7 p. 53 for statement (a).

Since $F_-^{(a)}$ and $G_+^{(b)}$ are bounded and Borel measurable the Lebesgue-Stieltjes integrals in statement (b) both exist. By Theorems 4.1 and 4.2 and the definition of (CY) integral, $(MPS)\int_a^b F_-^{(a)}\,dG_+^{(b)}$ and $(MPS)\int_a^b G_+^{(b)}\,dF_-^{(a)}$ exist. The first (MPS) integral is related to $(LS)\int_a^b F_-^{(a)}\,dG$ via (4.3). Similarly one can show that

$$\int_{[a,b]} G_+^{(b)}\,d\nu^F = (MPS)\int_a^b G_+^{(b)}\,dF_-^{(a)} + G(b)\Delta^- F(b).$$

Then part (b) follows from (a) for $F_-^{(a)}$ and $G_+^{(b)}$.

To prove (c) suppose $(CY)\int_a^b F\,dG$ exists. Then by Theorem 4.1, $(Y_1)\int_a^b F\,dG$ and $(Y_2)\int_a^b F\,dG$ exist. Hence $(Y_1)\int_a^b F_+^{(b)}\,dG$ and $(Y_2)\int_a^b F_-^{(a)}\,dG$ exist. By linearity (Proposition 3.21 of Part II), $(CY)\int_a^b \widehat{F}\,dG$ exists and

$$2(CY)\int_a^b \widehat{F}\,dG = (Y_2)\int_a^b F_-^{(a)}\,dG + (Y_1)\int_a^b F_+^{(b)}\,dG$$

$$= (MPS) \int_a^b F_-^{(a)} \, dG_+^{(b)} + [F\Delta^+ G](a) + (MPS) \int_a^b F_+^{(b)} \, dG_-^{(a)} + [F\Delta^- G](b). \quad (4.11)$$

By statement (a), the integrals $(MPS) \int_a^b G_+^{(b)} \, dF_-^{(a)}$ and $(MPS) \int_a^b G_-^{(a)} \, dF_+^{(b)}$ exist. Hence $(Y_1) \int_a^b G_+^{(b)} \, dF$ and $(Y_2) \int_a^b G_-^{(a)} \, dF$ exist. Also, by linearity, $(CY) \int_a^b \widehat{G} \, dF$ exists and the right side of (4.11) is equal to

$$F(b-)G(b) - (MPS) \int_a^b G_+^{(b)} \, dF_-^{(a)} - F(a)G(a) - F(a+)G(a) - (MPS) \int_a^b G_-^{(a)} \, dF_+^{(b)}$$

$$+ F(b)G(b) = 2F(b)G(b) - (Y_1) \int_a^b G_+^{(b)} \, dF - 2F(a)G(a) - (Y_2) \int_a^b G_-^{(a)} \, dF$$

$$= 2[F(b)G(b) - F(a)G(a)] - 2(CY) \int_a^b \widehat{G} \, dF.$$

This proves statement (c) when $(CY) \int_a^b F \, dG$ exists. When $(CY) \int_a^b G \, dF$ exists the proof is symmetric and, hence, we omit it.

Statement (d) for (RYS) integrals is proved in Hildebrandt (1963, 19.3.13 p. 93). The (CY) and (LS) integrals exist and are equal by Proposition 3.17 of Part II and Theorem 4.2 above. □

For $F = 1_{[1/2,1]}$, the classical integration by parts (4.9) fails for $\int_0^1 F \, dF$, which has value 1 for the Lebesgue-Stieltjes, (RYS) and (CY) integrals. The (MPS) and (RS) integrals are not defined. There is a non-zero term in the summations in (4.10), namely $(\Delta^- F)(1/2)^2 = 1$.

If F is left-continuous and G is right-continuous or vice versa, and both are regulated, then the sums in (4.10) vanish, so if (4.10) holds, it becomes (4.9).

5. The product integral. Here, we consider functions with values in a Banach algebra $\mathbb{B}$ with identity $\mathbb{1}$. The product integral $\prod$, the p-variation $v_p(f)$ and the space $\mathcal{W}_p([a,b]; \mathbb{B})$ were already defined in Section 1 for a function f. For $f \in \mathcal{W}_p([a,b]; \mathbb{B})$, as before let $V_p(f) := v_p(f)^{1/p}$ and $V_{p,\infty}(f) := V_p(f) + \sup_{x \in J} \|f(x)\|$. Freedman (1983) showed that the product integral $\prod_a^b (\mathbb{1} + df)$ exists for continuous f with $v_p(f) < \infty$ and $p < 2$. We extend Freedman's result to possibly discontinuous f in Part II.

When $\mathbb{B}$ is noncommutative, it turns out to be useful to consider the three-function integral $(CY) \int_a^b G \, dH \, F$ with two integrands G and F, defined very similarly to $(CY) \int_a^b G \, dH$, so that wherever values $G(x)$ appear to the left of dH or an increment of H, a factor $F(x)$ is inserted to the right (see Part II, Section 3.2). The Love-Young inequality (Theorem 4.4 above) extends as follows to a form useful for $\mathbb{B}$ noncommutative.

5.1 Theorem. Let $\mathbb{B}$ be a Banach algebra with norm $\| \cdot \|$. For any $p > 0$ and $q > 0$ with $1/p + 1/q > 1$ there is a constant $B_{p,q} < \infty$ such that

$$\left\| (CY) \int_a^b G \, dH \, F \right\| \le B_{p,q} V_{p,\infty}(F) V_{p,\infty}(G) V_q(H)$$

for all G and $F \in \mathcal{W}_p([a,b]; \mathbb{B})$ and $H \in \mathcal{W}_q([a,b]; \mathbb{B})$.

Proof. This follows from Part II, Theorem 3.27, (3.94).   □

In differentiating the product integral, another form of integral arises.

Definition. If G, H, F are regulated functions from an interval $[a, b]$ into a Banach algebra $\mathbb{B}$, the *left Young integral* is defined by

$$(LY)\int_a^b G\, dH\ F := (CY)\int_a^b G^{(a+)}\, dH\ F_-^{(a)}$$
$$+ \sum_{a < x < b} [G^+(\Delta^+ H)F - G(\Delta^+ H)F_-](x)$$

if the (CY) integral exists and the sum converges unconditionally in $\mathbb{B}$, where $G^{(a+)}(x) := G(x)$ for $a < x \leq b$ and $G^{(a+)}(a) := G(a+)$.

A symmetrical definition of (RY) integral is given in Part II, Definition 3.11. Theorem 3.12 in Part II shows that the two integrals are equal whenever both are defined.

5.2 Theorem. Theorem 5.1 also holds for the (LY) or (RY) integral in place of the (CY) integral, with a possibly larger constant in place of $B_{p,q}$.

Proof. This is Part II, Corollary 3.28.   □

One relation between p-variation and product integrals, proved for F continuous in Theorem 5.1 of Freedman (1983), is:

5.3 Theorem. For any Banach algebra $\mathbb{B}$ with identity $\mathbb{1}$, if $F \in \mathcal{W}_p([a, b], \mathbb{B})$ for some p with $0 < p < 2$, then the product integral $\prod_a^b (\mathbb{1} + dF)$ exists in $\mathbb{B}$.

Proof. This is Part II, Theorem 4.23.   □

The product integral $\prod_a^b (\mathbb{1} + dF)$ may exist even for F unbounded, for at least two kinds of reasons. First, the product integral may exist and be 0 if there are increments $F(x_i) - F(x_{i-1}) = -\mathbb{1}$ for all partitions which are refinements of some given partition. This may occur on a short interval while F is arbitrary outside the interval. So we define: let $\mathcal{P}([a, b]; \mathbb{B})$ be the set of all functions F from $[a, b]$ into $\mathbb{B}$ such that $\prod_a^b (\mathbb{1} + dF)$ exists and is invertible.

Secondly, many Banach algebras contain elements $A \neq 0$ with $A^2 = 0$, e.g. the 2×2 matrix

$$A = \begin{pmatrix} 0 & 1 \\ 0 & 0 \end{pmatrix}.$$

For such cases we have:

5.4 Proposition. For any Banach algebra $\mathbb{B}$ with identity $\mathbb{1}$ containing an $A \neq 0$ with $A^2 = 0$, and for an arbitrary function h from $(0, 1)$ into $[0, \infty)$, there is an $F \in \mathcal{P}([0, 1]; \mathbb{B})$ with $\|F(x)\| = h(x)$ for $0 < x < 1$.

Proof. This follows from Part II, Proposition 4.15.   □

On the other hand, for real-valued functions we have:

5.5 Theorem. If $F \in \mathcal{P}([a, b]; \mathbb{R})$ then $F \in \mathcal{W}_2([a, b]; \mathbb{R})$.

Proof. This is Part II, Theorem 4.3. ☐

Since $F \in \mathcal{W}_p([a,b]; \mathbb{B})$ for some p, $0 < p < 2$, is sufficient for existence of the product integral (Theorem 5.3 above), we see that at least for $\mathbb{B} = \mathbb{R}$, existence of a non-zero product integral is closely linked to p-variation properties. In fact, $\mathcal{P}([a,b]; \mathbb{R})$ is characterized as the set of $F \in \mathcal{W}_2([a,b]; \mathbb{R})$ not having jumps -1 and for which the local 2-variation comes only from jumps: Part II, Theorem 4.4.

Now we return to general $\mathbb{B}$. For $0 < p < 2$ and $F \in \mathcal{W}_p([a,b]; \mathbb{B})$, let $(\mathcal{P}_a F)(x) := \prod_a^x (\mathbb{1} + dF)$, $x \in [a,b]$, which always exists in $\mathbb{B}$ by Theorem 5.3.

5.6 Theorem. For any $a < b$, Banach algebra $\mathbb{B}$ with identity and $0 < p < 2$, the product integral operator $\mathcal{P}_a$ takes $\mathcal{W}_p([a,b]; \mathbb{B})$ into itself.

Proof. This follows from Part II, Theorem 5.16. ☐

Let $\mathcal{P}^b(F)(y) := \prod_y^b (\mathbb{1} + dF)$ for $y \in [a,b]$. Then $\mathcal{P}^b$ is another indefinite product integral operator with properties symmetric to those of $\mathcal{P}_a$.

Gill and Johansen (1990) and Gill (1994, pp. 133–135) proved compact differentiability of the product integral operator in the supremum norm on sets A bounded in 1-variation. We now have:

5.7 Theorem. For any $1 \le p < 2$, any Banach algebra $\mathbb{B}$ with identity $\mathbb{1}$ and any $a < b$, the product integral operator $\mathcal{P}_a$ is Fréchet differentiable from $\mathcal{W}_p([a,b]; \mathbb{B})$ into itself at any $F \in \mathcal{W}_p([a,b]; \mathbb{B})$. Its derivative at F is the bounded linear operator

$$f \mapsto D\mathcal{P}_a(F)(f) := (LY) \int_a^{(\cdot)} \mathcal{P}^{(\cdot)}(F)\, df\, \mathcal{P}_a(F).$$

Proof. This follows from Part II, Theorem 5.16. ☐

The above formula for the derivative of $\mathcal{P}_a$ was one main reason for introducing the (LY) integral. The (RY) integral is used to represent the derivative of $\mathcal{P}^b$.

For $0 < p < 1$, $\mathcal{W}_p([a,b]; \mathbb{B})$ consists of pure jump functions (Part II, Theorem 2.11, Proposition 2.12). With the topology of the quasinorm $V_{p,\infty}$, it is not locally convex (Part II, Proposition 2.17). Still, the definition of Fréchet differentiability extends to such quasinormed spaces (Part II, Sections 2.3 and 5.3) and then Theorem 5.7 above holds also for $0 < p < 1$ (Part II, Theorem 5.16).

The indefinite product integral operator $\mathcal{P}_a$ is analytic, at least under a restriction. Let $\mathcal{W}_p^r([a,b]; \mathbb{B})$ be the set of all right-continuous functions in $\mathcal{W}_p([a,b]; \mathbb{B})$.

5.8 Theorem. For $0 < p < 2$, any $a < b$ and Banach algebra $\mathbb{B}$ with identity $\mathbb{1}$, the operator $\mathcal{P}_a$ from $\mathcal{W}_p^r([a,b]; \mathbb{B})$ into itself is analytic (holomorphic) everywhere.

Proof. This follows from Part II, Theorem 5.17. ☐

For definitions and some facts on infinite-dimensional holomorphy see Section 5.3 in part II. We give in Part II, Theorem 5.17 a positive lower bound depending on F for the radius of uniform convergence of the power series representing $\mathcal{P}_a$ in a $\| \cdot \|_{[p]}$-ball with center F. The radius is infinite by inequalities of Lyons (1994). For example, the Taylor series $\sum_n z_n^n$ defines an entire holomorphic function on the whole Hilbert space ℓ^2 of square-summable sequences $\{z_n\}$, but the radius of uniform convergence is 1.

Freedman (1983, Theorem 4.1) showed that the indefinite product integral with respect to a continuous function of bounded p-variation with $1 \leq p < 2$ solves a linear Riemann-Stieltjes integral equation. His proof is based on the Banach fixed point theorem. The existence of a unique solution of a non-linear Riemann-Stieltjes integral equation with respect to a continuous function of bounded p-variation with $1 \leq p < 2$ was proved by Lyons (1994) via Picard iterations. Our proof of the above Theorem 5.7 rests on an extension of a Duhamel formula to functions of bounded p-variation with $0 < p < 2$ (Part II, Theorem 5.5). This formula shows also that the indefinite product integral with respect to a possibly discontinuous function solves a linear extended Riemann-Stieltjes integral equation. For example, if $F \in W_p([a,b]; \mathbb{B})$ and $0 < p < 2$ then $G := \mathcal{P}_a F$ is a solution of the integral equation $G(x) = \mathbb{I} + (LY) \int_a^x dF\, G$ (Part II, Theorem 5.21). This result applies to sample paths of an α-stable Lévy process with $\alpha \in (0,2)$, as well as to sample paths of a fractional Brownian motion B_H with $H \in (1/2, 1)$ (cf. Corollaries 5.23 and 5.24 in Part II, respectively). Since fractional Brownian motion B_H with $H \neq 1/2$ is not a semimartingale, the problem of giving a meaning to a linear integral equation driven by B_H has attracted considerable attention in the literature (cf. Ciesielski, Kerkyacharyan and Roynette, 1993; Cutland, Kopp and Willinger, 1995; Lin, 1995; Dai and Heyde, 1996; Zähle, 1998).

Product integrals can also be defined for interval functions, as e.g. Gill and Johansen (1990) do. Let K be an interval in $\mathbb{R}$, possibly unbounded, and $\mathbb{B}$ a Banach algebra with identity $\mathbb{I}$. Let $\mathcal{I}(K)$ be the set of all subintervals of K, open or closed at either end. A function μ from $\mathcal{I}(K)$ into $\mathbb{B}$ will be called an *interval function*. Two disjoint intervals will be called *adjoining* if their union is an interval. An interval function μ will be called *additive* if $\mu(I \cup J) = \mu(I) + \mu(J)$ whenever I and J are adjoining. Thus $\mu(\emptyset) = 0$.

For two disjoint intervals A and B, $A \prec B$ will mean that $x < y$ for all $x \in A$ and $y \in B$. An interval function $\pi : \mathcal{I}(K) \mapsto \mathbb{B}$ will be called *multiplicative* if $\pi(\emptyset) = \mathbb{I}$ and for any adjoining intervals $A, B \in \mathcal{I}(K)$ such that $A \prec B$, we have $\pi(A \cup B) = \pi(A)\pi(B)$. The product integral can be viewed, and may be best viewed, as a mapping from suitable additive interval functions to multiplicative ones.

For any closed interval $K := [a,b]$ and regulated function $f \in \mathcal{R}[a,b]$ an additive interval function $\mu := \mu_f$ is defined on $\mathcal{I}[a,b]$ by, for $a \leq c \leq d \leq b$, $\mu([c,d]) := f_+^{(b)}(d) - f_-^{(a)}(c)$, $\mu([c,d)) := f_-^{(a)}(d) - f_-^{(a)}(c)$, $\mu((c,d]) := f_+^{(b)}(d) - f_+^{(b)}(c)$, and for $c < d$, $\mu((c,d)) := f_-^{(a)}(d) - f_+^{(b)}(c)$. The interval function μ_f doesn't depend on values $f(x)$ different from both $f_-^{(a)}(x)$ and $f_+^{(b)}(x)$, and in this way interval functions simplify the situation. We plan to return to interval functions in more detail in future work.

6. Probability and p-variation.

Let P be a probability measure on $\mathbb{R}$ with distribution function F, $F(x) \equiv P((-\infty, x])$. Let $X_1, X_2, \cdots$ be independent random variables with distribution P. Let $\delta_x(B) := 1_B(x) := 1$ for $x \in B$, 0 for $x \notin B$, for any set B. Let $P_n := \frac{1}{n} \sum_{i=1}^n \delta_{X_i}$, called the *empirical measure* for P. Let F_n be its distribution function, called the *empirical distribution function* for F, $F_n(x) := P_n((-\infty, x])$ for all x. Then $\alpha_n := \alpha_{n,F} := n^{1/2}(F_n - F)$ is called the *empirical process*. A Brownian bridge process b is given by $b(t) = x(t) - tx(1)$, $0 \leq t \leq 1$, where $x = \{x(t): t \geq 0\}$ is a standard Brownian motion process, e.g. Dudley

(1993, Section 12.3). We say that the central limit theorem holds for the empirical process in the p-variation norm if α_n converges in law in the Hoffmann-Jørgensen sense to $b \circ F$ with respect to $\| \cdot \|_{[p]}$, i.e. $E^*Y(\alpha_n) \to EY(b \circ F)$ for every bounded, continuous functional Y on $\mathcal{W}_p$, where $E^*W := \inf\{EV: V \text{ measurable}, V \geq W\}$. For sequences Y_n of random variables and numbers $a_n > 0$, $Y_n = O_p(a_n)$ means that for every $\varepsilon > 0$ there is an $M < \infty$ such that $\Pr(|Y_n| > Ma_n) < \varepsilon$ for all n (or, for n large enough, which is equivalent except possibly for the choice of M). Let U be the $U[0,1]$ distribution function, $U(x) := \max(0, \min(x, 1))$ for all $x \in \mathbb{R}$. Let U_n be an empirical distribution function for U. It is well known and easily checked that for any distribution function F, $F_n := U_n \circ F$ is indeed an empirical distribution function for F. Then we have:

6.1 Theorem. For $2 < p < \infty$ and any distribution function F on $\mathbb{R}$, the central limit theorem holds for the empirical process α_n in the p-variation norm. In particular, we have $\|\alpha_n\|_{[p]} = O_p(1)$, in other words $\|F_n - F\|_{[p]} = O_p(n^{-1/2})$ as $n \to \infty$.

Proof. Since $F \equiv U \circ F$ we can write $\alpha_{n,F} \equiv \alpha_{n,U} \circ F$. For any real-valued function g on $[0,1]$, $\|g \circ F\|_{[p]} \leq \|g\|_{[p]}$ since F is non-decreasing. Thus we can assume $F \equiv U$.

Let $q := p/(p-1)$, so $1 < q < 2$, and let

$$\|h\|'_{[q]} := \sup\left\{|h(f)|: f \in \mathcal{CW}_q^*, \ \|f\|_{[q]} \leq 1\right\}$$

be the dual norm of a linear functional h on the set $\mathcal{CW}_q^*$ of continuous functions f such that $V_q^*(f) = 0$ (cf. definition (B.1) in Appendix B). The unit ball of $\mathcal{CW}_q^*$, being a subset of that of $\mathcal{W}_q$, is a dominated uniform Donsker class by Theorem 2.2 of Dudley (1992). So, $\nu_n := \sqrt{n}(P_n - P)$ converges in law in the Hoffmann-Jørgensen sense with respect to $\| \cdot \|'_{[q]}$. By the almost sure representation Theorem 4.1 of Dudley (1985), we can construct a sequence $\tilde{\nu}_n$ convergent almost surely in $\| \cdot \|'_{[q]}$ and such that each $\tilde{\nu}_n$ has the same distribution as ν_n, in the sense defined in Dudley (1985). Let $\tilde{\alpha}_n(t) := \tilde{\nu}_n([0,t])$ for $0 \leq t \leq 1$. Then by Proposition 3.7 of Dudley (1992), $\|\tilde{\alpha}_m - \tilde{\alpha}_n\|_{[p]} \leq 2^{3+1/p}\|\tilde{\nu}_m - \tilde{\nu}_n\|'_{[q]} \to 0$ a.s. as $m, n \to \infty$. Thus for some process b, $\|\tilde{\alpha}_n - b\|_{[\infty]} \leq \|\tilde{\alpha}_n - b\|_{[p]} \to 0$ a.s. as $n \to \infty$. Since α_n converges in law with respect to the supremum norm, b has the finite-dimensional distributions of the Brownian bridge. Moreover, by way of almost surely convergent realizations, one can conclude that $\tilde{\alpha}_n$ are asymptotically equicontinuous with respect to the usual metric. It follows then that b has continuous paths almost surely and is the usual Brownian bridge. Since almost all sample paths of Brownian motion are of bounded p'-variation on $[0,1]$ for each $p' > 2$, being Hölder of any order $< 1/2$ (e.g. Itô and McKean, 1974, pp. 36-38), it follows by Lemma 2.14 from Part II that the Brownian bridge b has almost all its sample paths in $\mathcal{CW}_p^*$, which is separable by Lemma 2 of Kisliakov (1984) and Lemma B.1 from Appendix B. An application of the almost sure representation Theorem 4.1 of Dudley (1985) once again yields that α_n converges in law in the Hoffmann-Jørgensen sense to the Brownian bridge b with respect to the p-variation norm.   □

Theorem 6.1 is both strengthened and simplified by the following. Y.-C. Huang (1994, 1997) proved that there exists a sequence b_n of Brownian bridges such that

$$E\|\alpha_n - b_n \circ F\|_{[p]} = O(n^{\frac{1}{p}-\frac{1}{2}}) \text{ as } n \to \infty,$$

where also if F is continuous, so that $X_1, \cdots, X_n, \cdots$ are all distinct with probability 1, $\|\alpha_n - g\|_{[p]} \geq n^{\frac{1}{p} - \frac{1}{2}}$ for any continuous function g, e.g. $g = b_n \circ F$, so as Huang noted, his upper bound is sharp.

The conclusion of Theorem 6.1 no longer holds for $p \leq 2$ if F has a non-constant continuous component. But $p < 2$ is needed for the product integral (Section 5 above) and $p < 2$ or $q < 2$ for the integration operator (Section 4 above). Let $Ln := \max(1, \log n)$. Then we have:

6.2 Theorem. For $1 \leq p \leq 2$ and any distribution function F on $\mathbb{R}$,

$$\|F_n - F\|_{[p]} = O_p(n^{(1-p)/p}(LLn)^{1/2}),$$

in other words $\|\alpha_n\|_{[p]} = O_p(n^{(2-p)/(2p)}(LLn)^{1/2})$. Conversely if F is continuous, then almost surely for all n, $\|F_n - F\|_{[p]} \geq n^{(1-p)/p}$.

Proof. This is Theorem 2 of Dudley (1997). $\qquad\qquad\qquad\qquad\qquad$ □

Remark. Qian (1998) has proved that the upper bound with $(LLn)^{1/2}$ is sharp for $p = 2$, while for $1 \leq p < 2$ the $(LLn)^{1/2}$ factor can be completely omitted and the lower bound is sharp.

If T is a nonlinear functional or operator defined on distribution functions, Fréchet differentiable at F with respect to some $\| \cdot \|_{[p]}$, $p > 1$, with derivative $DT(F)(\cdot)$, we can then write $F_n = F + (F_n - F)$ and so

$$T(F_n) = T(F) + DT(F)(F_n - F) + o(\|F_n - F\|_{[p]}). \qquad (6.1)$$

Now, let's see how the bounds of Theorems 6.1 and 6.2 combine with the differentiability facts of Sections 2-5 to give good bounds when the operators are applied to empirical distribution functions F_n (and G_m).

Here there are some general observations to be made. The quantity $n^{1/2}\|F_n - F\|_{[p]}$ is bounded in probability for $p > 2$ but not for $p \leq 2$. Yet, the product integral operator has been shown to be differentiable with respect to $\| \cdot \|_{[p]}$ only for $p < 2$ and if it is defined and not 0 for a real-valued function f then $f \in \mathcal{W}_2$. In the integration and convolution operators, $\frac{1}{p} + \frac{1}{q} > 1$ requires that at least one of p and q must be < 2. Thus, it seems, p-variation differentiability might fall short of being effective.

On closer examination, however, the apparent problem for $p < 2$ disappears. Suppose the operator T in (6.1) has a Fréchet remainder bound of order $\gamma > 1$ at F (cf. (1.4) above). The p-variation is not needed for the first derivative term(s) or even for higher order derivatives when they exist. The p-variation is only needed to bound the remainders, which are smaller than the derivative terms by a power $\gamma > 1$ for each operator as mentioned above in Table 2. Since $\|F_n - F\|_{[p]}$ is (in probability) of order $n^{(1-p)/p}$ for $p < 2$, the remainder is $O_p(n^{-\beta})$, or

$$\sqrt{n}[T(F_n) - T(F)] = DT(F)(\sqrt{n}(F_n - F)) + O_p(n^{-\beta+1/2}),$$

where $\beta := \beta(\gamma, p) := \gamma(p-1)/p \rightarrow \gamma/2 > 1/2$ as $p \uparrow 2$. If $DT(F)(\sqrt{n}(F_n - F))$ is asymptotically normal then $T \in FRB_\gamma(F)$ with $\gamma > 1$ satisfies the central limit theorem with remainder bound of order $O_p(n^{-\beta+1/2})$.

Now, let us see how the above general observations apply to our six operators. For the integration operator, evaluated at empirical distribution functions F_n, G_m for F, G, respectively, we have

$$\int F_n \, dG_m \;=\; \int F \, dG + \int (F_n - F) \, dG + \int F \, d(G_m - G) + \int (F_n - F) \, d(G_m - G).$$

Here $n^{1/2} \int (F_n - F) \, dG$ and $m^{1/2} \int F \, d(G_m - G)$ have asymptotic normal distributions as $m \to \infty$ and $n \to \infty$ by the classical 1-dimensional central limit theorem. Then, for any $\varepsilon > 0$, take $p < 2$ such that $(1 - p)/p < \varepsilon - 1/2$. Taking $q > 2$ with $\frac{1}{p} + \frac{1}{q} > 1$, we have

$$\left| \int (F_n - F) \, d(G_m - G) \right| \;\le\; C_{p,q} \| F_n - F \|_{[p]} \| G_m - G \|_{[q]} \;=\; O_p(n^{\varepsilon - 1/2} m^{-1/2}).$$

If $n > m$, we can get a better bound by interchanging n with m and p with q. If $m = n$ we get $O_p(n^{\varepsilon - 1})$ where compact differentiability would only give $o_p(n^{-1/2})$. Thus, if $n = m$ then $\sqrt{n}[\int F_n \, dG_m - \int F \, dG]$ converges in distribution to a sum of two normal random variables and the remainder $\sqrt{n}[\int (F_n - F) \, d(G_m - G)]$ has bound $O_p(n^{\varepsilon - 1/2})$ for any $\varepsilon > 0$. See Section 13.3 in Dudley (1997) for further results.

For convolution, evaluated at empirical distribution functions F_n, G_m for F, G respectively, the first derivative terms $(F_n - F)*G$ and $F*(G_m - G)$ are function-valued (stochastic processes) rather than numerical random variables. But since G and F are of bounded 1-variation, one can approximate $n^{1/2}(F_n - F)*G$ by $(B_n \circ F)*G$ for Brownian bridges B_n via the results of Huang (1994, 1997) and likewise for $F*(G_m - G)$.

The product integral operator $\mathcal{P}_a$ is holomorphic. If $\| H_n - H \|_{[p]} \to 0$ as $n \to \infty$ for some $p < 2$ and some right-continuous functions H_n, H (not necessarily distribution functions) we will have for any $k = 1, 2, \ldots$,

$$\mathcal{P}_a(H_n) \;=\; \mathcal{P}_a(H) + \left[\sum_{j=1}^{k} Q_j(H_n - H) \right] + R_k(H_n - H)$$

where each Q_j is an explicit homogeneous functional polynomial of order j (Part II, proof of Theorem 5.17). The remainder R_k satisfies $\| R_k(h) \|_{[p]} = O(\| h \|_{[p]}^{k+1})$ as $\| h \|_{[p]} \to 0$, $h = H_n - H$. So for k large, $\| R_k(H_n - H) \|_{[p]}$ is of arbitrarily small order relative to $\| H_n - H \|_{[p]}$.

For the quantile operator, as shown in Dudley (1994), we get via p-variation, and letting $p \downarrow 2$, for F smooth enough, that the remainder in differentiating $f \mapsto (F + f)^{\leftarrow}$ at $f = F_n - F$ is (in L^2 norm) of order $O_p(n^{\varepsilon - 3/4})$ for any $\varepsilon > 0$, where $-3/4$ is the Bahadur-Kiefer correct exponent. It can be shown by other methods that the ε is unnecessary and that the L^2 norm can be strengthened to the sup norm (Kiefer, 1970).

For the composition operator, the interest has been not so much in $F_n \circ G_m$ but rather in the so-called procentile-procentile or P-P plot $F_n \circ G_m^{\leftarrow} : (0,1) \mapsto [0,1]$, where F_n, G_m are empirical distribution functions for F, G, respectively, see e.g. Beirlant and Deheuvels (1990). Let U_m and V_n be empirical distribution functions for the $U[0,1]$ distribution function U. Then as noted just before Theorem 6.1, we can write $F_m \equiv U_m \circ F$ and $G_n \equiv V_n \circ G$. The following is known:

6.3 Theorem. If $F = G$ is continuous, then the P-P process $F_m \circ G_n^{\leftarrow}$ for $0 < y < 1$ has the same distribution as when $F = G = U$ for $0 \le x \le 1$.

Proof. By Propositions 3.5 and 3.6,

$$F_m \circ G_n^{\leftarrow} \equiv U_m \circ F \circ (G^{\leftarrow} \circ V_n^{\leftarrow}) \equiv U_m \circ V_n^{\leftarrow},$$

and the conclusion follows. $\square$

So, for $F = G$ continuous, only the case $F = G = U$ needs to be studied: Reeds (1976), Fernholz (1983).

Suppose H has values in $[0,1]$ and h is in a set A of functions such that $H + h$ also has values in $[0,1]$, as occurs if $H = G^{-1}$ and $h = V_n^{\leftarrow} - G^{-1}$ on $(0,1)$. Then one of the usual two remainder terms in differentiating the composition operator, $R_2 := F \circ (G + g) - F \circ G - (F' \circ G)g$, reduces to 0 since F is linear on $[0,1]$. So we have for $0 < y < 1$

$$(F_n \circ G_m^{\leftarrow})(y) = y + [F_n(y) - y] + [G_m^{\leftarrow}(y) - y] + R_1(y),$$

where $R_1(y) := (F_n - F)(G_m^{\leftarrow}(y)) - (F_n - F)(y)$. For $1 \le p < \infty$ and $s = \infty$, Theorem 2.5 and the fact that $\|G_m^{\leftarrow} - G^{-1}\|_\infty \le \|G_m - G\|_\infty$ for $G(x) \equiv x$ (e.g. Dudley, 1997, Proposition 11) give the bound

$$\|R_1\|_p \le C_p \|F_n - F\|_{[p]} \|G_m - G\|_\infty^{1/p}.$$

Applying Theorem 6.2 for $\|F_n - F\|_{[p]}$ with $p = 2$ and a classical theorem of Kolmogorov (1933, 1941) implying $\|G_m - G\|_\infty = O_p(m^{-1/2})$ gives

$$\|R_1\|_2 = O_p((LLn)^{1/2} n^{-1/2} m^{-1/4}), \tag{6.2}$$

which is $O_p((LLn)^{-1/2} n^{-3/4})$ if m and n are of the same order of magnitude. For $0 \le s, t \le 1$ we have $E\{[(F_n - F)(t) - (F_n - F)(s)]^2\} = [|s - t| - (s - t)^2]/n$ and $E|G_m^{\leftarrow}(y) - y| \ge C_1 m^{-1/2}$, $E(G_m^{\leftarrow}(y) - y)^2 \le C_2 m^{-1}$ for some $0 < C_1 < C_2 < \infty$ if $1/4 \le y \le 3/4$. So if the empirical distribution functions F_m and G_n are independent, then $ER_1(y)^2 \ge C/(nm^{1/2})$ for some $C > 0$ if $1/4 \le y \le 3/4$. Thus in (6.2) the exponents $-1/2$ and $-1/4$ are sharp.

By a theorem of Komlós, Major and Tusnády (1975), see also Bretagnolle and Massart (1989, Theorem 1), on the rate of approximation of the empirical processes by Brownian bridges, F_n can be defined on some probability space on which there exist Brownian bridges B_n with $F_n - F = n^{-1/2} B_n + V_n$ where $\|V_n\|_\infty = O_p(\log n/n)$. For $0 < \alpha < 1/2$, the sample functions of a Brownian bridge are a.s. in $H_\alpha([0,1], \mathbb{R})$ (e.g. Itô and McKean, 1974, p. 36). Thus we have by Proposition 1.2

$$\|R_1\|_\infty = O_p(\|G_m - G\|_\infty^\alpha n^{-1/2} + \log n/n) = O_p(m^{-\alpha/2} n^{-1/2} + \log n/n).$$

If m and n are of the same order of magnitude this gives for any $\varepsilon > 0$ that $\|R_1\|_\infty = O_p(n^{\varepsilon - 3/4})$ where $-3/4$ cannot be improved, since it cannot for $\|R_1\|_p$ with $p < \infty$.

So, for all the operators in our list, p-variation differentiability implies bounds, when applied to empirical distribution functions F_n approaching F, which improve on the $o(n^{-1/2})$ bounds from compact differentiability and come within n^ε for any $\varepsilon > 0$ of being precise.

7. General and C-differentiability. Recall that three of our six operators are bilinear, so they and the product integral are holomorphic. The two others, composition and the quantile operator, are less smooth. In this section we will have in mind the latter two operators.

We recall C-differentiability on a set A as defined in Section 1: let X be a normed space with norm $\|\cdot\|$ and let T be a mapping from $A \subset X$ into another normed space Y with norm also written $\|\cdot\|$. Given a collection C of bounded subsets of A, T is called C-differentiable at $F \in A$ if there is a bounded linear operator L from X into Y such that, for each $C \in C$,

$$\|T(F + tf) - T(F) - tL(f)\| = o(|t|) \tag{7.1}$$

as $t \to 0$ with $F + tf \in A$, uniformly for $f \in C$. Sebastião e Silva (1956) first defined C-differentiability. Averbukh and Smolyanov (1967, 1968) treated it in two surveys.

Both Fréchet differentiability, e.g. for p-variation norms, and compact differentiability for the supremum norm, are special cases of C-differentiability, and neither implies the other. For subsets of $W_p[0, 1]$, specifically, neither compactness for $\|\cdot\|_\infty$, nor boundedness for $\|\cdot\|_{[p]}$, implies the other. But, consider for example the set of indicator functions $1_{[c,1]}$, $0 \le c \le 1$. The whole set is bounded in any $W_p[0, 1]$, but a compact set for $\|\cdot\|_\infty$ can contain only finitely many such indicators. More generally, a norm-compact set is separable, hence in one sense very thin in the non-separable space $D[0, 1]$ with supremum norm.

On the other hand, there are individual functions f in $D[0, 1]$ but not in $W_p[0, 1]$ for any $p < \infty$. Failure of Fréchet differentiability for $\|\cdot\|_\infty$ means that the $o(|t|)$ in (7.1), uniformly over each $\|\cdot\|_\infty$-compact set, may be arbitrarily slow: Dudley (1994), Proposition 2.1. In this sense, an $\|\cdot\|_\infty$-compact set K can, as a subset of some separable subspace of $D[0, 1]$, be much larger than a bounded set in p-variation with $p < \infty$. In other words K can have "semiaxes" of lengths going to 0 so slowly that they prevent improving the $o(|t|)$ in (7.1) to $O(|t|^r)$ for any $r > 1$.

So, when differentiability holds compactly for $\|\cdot\|_\infty$ and in the Fréchet sense for some $\|\cdot\|_{[p]}$, each is contributing information beyond the other. Although individually the compact sets for the sup norm are thin, collectively they give the uniformity of differentiability over a very large class of separable subsets. How can all this information about an operator be synthesized and extended further?

For a given operator T at some F, it's then natural to ask: in what directions f is T differentiable, i.e. is $t \mapsto T(F + tf)$ differentiable at $t = 0 \in \mathbb{R}$. This question makes sense whenever $T(F + tf)$ is defined for $|t|$ small enough, whether or not f belongs to a predetermined class of functions such as $D[0, 1]$ or the space of regulated functions.

A second question is over what classes of f is the $o(|t|)$ uniform. These questions are addressed in Part III for the composition and quantile operators, briefly as follows.

Let X be a vector space and T an operator, differentiable at some F along all f in X, let M be the class of all sets $C \subset X$ such that (7.1) holds uniformly for $f \in C$. Then T is M-differentiable at F and M is the largest class C of sets for which T is C-differentiable at F. The following will give some indications of how to find quite large classes C for which C-differentiability holds for some operators, although we do not settle whether they equal M.

A real-valued function f on a possibly unbounded interval $J \subset \mathbb{R}$ will be called a *Riemann function* if it is continuous almost everywhere for Lebesgue measure and bounded on each bounded subinterval of J. For $0 < \tau < \infty$, let $R_\tau(\mathbb{R})$ be the set of all Riemann functions f on $\mathbb{R}$ which are universally measurable and satisfy $\|f\|_{\{\tau\}} := \sup_x |f(x)|/(1 + |x|^\tau) < \infty$.

Recall that on $[0,1]$, for $1 \le p < \infty$, the quantile operator $f \mapsto (U + f)^\leftarrow$ is differentiable into L^p along g and along h with $h(x) \equiv g(1-x)$ if and only if g is a Riemann function (Theorem 3.3 and Proposition 3.4 above). Recall also that the composition operator $(f, g) \mapsto (F + f) \circ (G + g)$ is jointly Fréchet differentiable at $f = g = 0$ from $V \times L^s$ into L^p for $1 \le p < s$, for suitable F and G, where V is the one-dimensional space spanned by a function h, if and only if $h \in R_{s/p}(\mathbb{R})$ (Theorem 2.5 above). Every regulated function, and thus every function in any $\mathcal{W}_p$, is Borel measurable, thus universally measurable, and a Riemann function.

Let λ be Lebesgue measure. A class $\mathcal{F}$ of Riemann functions on a bounded interval $[a, b]$ will be called *uniformly Riemann* if, for every $\varepsilon > 0$, there is a $\delta > 0$ such that for each $f \in \mathcal{F}$ there is a set $B \subset [a, b]$ such that $\lambda(B) < \varepsilon$ and if $x \notin B$ and $|y - x| < \delta$, then $|f(x) - f(y)| < \varepsilon$, and

$$\sup\{|f(x) - f(y)| : \ a \le x \le y \le b, \ f \in \mathcal{F}\} \ < \ \infty.$$

Let $Ri[a, b]$ be the set of all Riemann functions from $[a, b]$ into $\mathbb{R}$.

7.1 Theorem. Let $1 \le p < \infty$ and let F be an increasing diffeomorphism from an interval $[a, b]$ onto an interval $[c, d]$. Then $f \mapsto (F + f)^\leftarrow$ is C-differentiable at $f = 0$ from $Ri[a, b]$ into $L^p([c, d], \lambda)$, where C is the class of all uniformly Riemann sets of functions on $[a, b]$.

Proof. This follows from Theorem 4.5 of Part III.   □

If $0 < \tau < \infty$, a set $\mathcal{F} \subset R_\tau(\mathbb{R})$ will be called *uniformly τ-Riemann* if for any bounded interval J, the class of restrictions of functions in $\mathcal{F}$ to J is uniformly Riemann, and $\sup\{\|f\|_{\{\tau\}} : f \in \mathcal{F}\} < \infty$. For $1 \le p < s$, the composition operator $(f, g) \mapsto (F + f) \circ (G + g)$ is C-differentiable at $f = g = 0$ from $R_{s/p}(\mathbb{R}) \times L^s$ into L^p for suitable F, G, where C is the class of all sets $\mathcal{F} \times B_1$ where $\mathcal{F}$ is uniformly (s/p)-Riemann and B_1 is the unit ball of L^s. For further details of the formulation and consequences see Part III, Theorem 5.1.

Thus, for differentiability just to hold, with remainders in (7.1) being $o(|t|)$ at an arbitrarily slow rate, as for compact differentiability when Fréchet differentiability fails, one can have C-differentiability for classes C of sets much larger than the compact sets for the supremum norm.

To give some feeling about how large uniformly Riemann sets are, we shall compare some of them with compact sets in the space $(\mathcal{R}[0,1], \|\cdot\|_\infty)$ of all regulated functions on $[0,1]$. A subset $\mathcal{F} \subset \mathcal{R}[0,1]$ is called *uniformly regulated* if $\mathcal{F}$ is uniformly bounded and, for every $\epsilon > 0$, $\sup_{f \in \mathcal{F}} \sup\{n : |f(x_{2i}) - f(x_{2i-1})| > \epsilon, \ i = 1, \cdots, n, \ 0 \le x_0 \le \cdots \le x_{2n} \le 1\} < \infty$. By Proposition 3.3 of Part III, uniformly regulated sets are uniformly Riemann. Hildebrandt (1966) proved that a set $\mathcal{G} \subset \mathcal{R}[0,1]$ is compact for the supremum norm if and only if (1) $\mathcal{G}$ is uniformly bounded and (2) for each $x \in [0,1)$ or $(0,1]$, the relations (4.1), (4.2) respectively hold uniformly in $f \in \mathcal{G}$. *Claim:* if $\mathcal{G}$ is compact for $\|\cdot\|_\infty$ then, for each $\epsilon > 0$, there is a finite set $A \subset [0,1]$ such that functions in $\mathcal{G}$ can have jumps bigger than ϵ only at points in A. Suppose not. Then, for some $\epsilon > 0$, the set of points where some function in $\mathcal{G}$ has a jump bigger than ϵ is infinite. Since $[0, 1]$ is compact, either (4.1) or (4.2) fails to hold uniformly at some point, proving the Claim. On the other hand, the set of jumps larger than 1 for a uniformly regulated class of functions could be all of $[0, 1]$. For any $\epsilon > 0$, and $n = n(\varepsilon)$ as in the definition of "uniformly regulated," each function in $\mathcal{F}$ has no more than n jumps of magnitude larger than ϵ.

There are cases in probability theory where stochastic processes converge in distribution to a limit process having discontinuous trajectories, e.g. Skorohod (1957), Bingham (1975, Section 9). In such cases, most often, convergence does not hold in the supremum norm, and then it will also not hold in any p-variation norm. Convergence often does hold in a Skorohod topology as treated e.g. by Billingsley (1968, Chapter 14). Compact sets in the Skorohod (J_1) topology satisfy uniformity properties for differentiability of the composition and quantile operators (Part III, Theorem 2.6, Proposition 3.3, Theorems 4.5 and 5.1). Thus, the $\mathcal{C}$-differentiability of Part III can apply when p-variation and (compact) differentiability in the sup norm do not. Whitt (1980) surveys continuity properties of a number of operators, including several of our six, in Skorohod topologies. Notably, addition $(f, g) \mapsto f + g$ and multiplication $(f, g) \mapsto f \cdot g$ are not jointly continuous for Skorohod topologies, but can be at pairs (f, g) where f and g do not have common discontinuities.

8. The chain rule and concluding remarks. Differentiability can be extended from given operators to others by the chain rule. For example, composing the composition operator $(f, h) \mapsto (F + f) \circ (H + h)$ with the quantile operator $g \mapsto H + h := (G + g)^{\leftarrow}$ we get the operator $(f, g) \mapsto (F + f) \circ (G + g)^{\leftarrow}$, as in P-P plots, Section 6 above. Differentiating this operator was a main aim of Reeds (1976). So, let's consider chain rules for different types of differentiability.

In one sense, compact differentiability is the weakest form of differentiability for which a chain rule holds, but in other senses, chain rules do hold for weaker forms such as compact differentiability restricted to a subset or tangential to a subspace. It is known that there is no real contradiction, but we will try to clarify matters further in this section.

Let X, Y and Z be normed spaces, let $A \subset X$ and $B \subset Y$, and let $T : A \mapsto B$ and $S : B \mapsto Z$, so that $S \circ T : A \mapsto Z$. Let $x \in A$ and $y := T(x) \in B$. Suppose T has a derivative DT_x at x in some sense (e.g., Gateaux) and likewise S has a derivative DS_y at y. If also $S \circ T$ has a derivative $D(S \circ T)_x$ at x, then the *chain rule formula* is

$$D(S \circ T)_x = DS_y \circ DT_x, \tag{8.1}$$

which in general may or may not hold. A chain rule is a theorem saying that if DT_x and DS_y exist in specified senses, then $D(S \circ T)_x$ also exists in a specified sense and (8.1) holds.

It is well known that the composition of Gateaux differentiable mappings need not be Gateaux differentiable in general. For $\alpha, \beta = 0$ or 1 let $S_\alpha(x, y) := x^\alpha 1_{\{0 < y = x^2\}}$, where $0^0 := 1$, and $T_\beta(u, v) := (u, u^2 + \beta v)$. Then each T_β is Fréchet differentiable and analytic everywhere: $\mathbb{R}^2 \mapsto \mathbb{R}^2$. At $(0, 0)$, we have: S_0 is not continuous, but S_0 and S_1 are Gateaux differentiable. $S_\alpha \circ T_1$ is not Gateaux differentiable, by discontinuity for $\alpha = 0$ and nonlinearity for $\alpha = 1$. $S_1 \circ T_0$ is Gateaux differentiable but (8.1) fails. Such examples were given by Fréchet (1937), Averbukh and Smolyanov

(1968, p. 240) and Reeds (1976). The work on infinite-dimensional analysis by J. Hadamard, R. Gateaux, M. Fréchet and P. Lévy led to the notion of differentiability to be defined next. A function f from an open interval $J \subset \mathbb{R}$ into a normed vector space $(X, \| \cdot \|)$ will be said to be *simply differentiable* at $t \in J$ if for some $u \in X$, $\|f(t+s) - f(t) - su\|/|s| \to 0$ as $s \to 0$, and then one sets $f'(t) := u$. At $t \in J$, clearly all forms of $\mathcal{C}$-differentiability such as Gateaux, compact and Fréchet, coincide with simple differentiability. Fréchet (1937) suggested the following definition (see also §2 in Averbukh and Smolyanov, 1968, for historical comments).

Definition. A map T between normed spaces X and Y is called *Hadamard differentiable* at $x \in X$ if it is Gateaux differentiable at x and for every function $f : (-1, 1) \mapsto X$ with $f(0) = x$, simply differentiable at 0, the composition $T \circ f$ is simply differentiable at 0 with the derivative $(T \circ f)'(0) = DT_x(f'(0))$.

Hadamard differentiability is equivalent to compact differentiability by Theorem 1.1.8 of Sova (1966; see also Theorem 3.2 in Averbukh and Smolyanov, 1968). It is clear from the definition that a composition $S \circ T$ of two Hadamard differentiable maps S, T is Hadamard differentiable, and since every element x of X is $f'(0)$ for some differentiable f, e.g. $f(t) := tx$, (8.1) holds. Hadamard differentiability is the weakest form of differentiability in the class defined by Averbukh and Smolyanov (1968) as follows. An $\mathfrak{R}$-*form of differentiation* is given if a set $\mathcal{R} = \mathcal{R}(X, Y)$ of maps, called infinitesimal maps, is given between any two normed spaces X and Y. Let $\mathcal{L}(X, Y)$ be the set of all bounded linear operators from X into Y. Then, the map T from $U \subset X$ into Y is $[\mathcal{R}]$-*differentiable* at $x \in U$ if there are maps $DT_x \in \mathcal{L}(X, Y)$ and $r \in \mathcal{R}(X, Y)$ such that for $h \in X$ and $x + h \in U$, we have $T(x + h) - T(x) = DT_x h + r(h)$. Let two $\mathfrak{R}$-forms of differentiation be given by two sets $\mathcal{R}_1$ and $\mathcal{R}_2$ of infinitesimal maps. The first $\mathfrak{R}$-form is *weaker* than the second one, or $[\mathcal{R}_1]$-differentiability is weaker than $[\mathcal{R}_2]$-differentiability, if $\mathcal{R}_2(X, Y) \subset \mathcal{R}_1(X, Y)$ for any normed spaces X, Y. An $\mathfrak{R}$-form of differentiation is called *quasiregular* if the following all hold:

1. For any normed space Y, if $r \in \mathcal{R}(\mathbb{R}, Y)$, then $r(x)/x \to 0$ as $x \to 0$;
2. For any normed spaces X, Y, Z, if $r \in \mathcal{R}(X, Y)$, $q \in \mathcal{R}(Y, Z)$, $A \in \mathcal{L}(X, Y)$, and $B \in \mathcal{R}(Y, Z)$, then
   a) $B \circ r \in \mathcal{R}(X, Z)$,
   b) $q \circ (r + A) \in \mathcal{R}(X, Z)$.
3. For any normed spaces X, Y, $\mathcal{R}(X, Y)$ is a vector space and $\mathcal{R}(X, Y) \cap \mathcal{L}(X, Y) = \{0\}$.

A quasiregular $\mathfrak{R}$-form of differentiability is called *regular* if also

5. For any normed space Y such that $\mathcal{R}(\mathbb{R}, Y) \neq \emptyset$, if $r : \mathbb{R} \to Y$ is such that $r(x)/x \to 0$ as $x \to 0$, then $r \in \mathcal{R}(\mathbb{R}, Y)$.

Averbukh and Smolyanov (1968) give the preceding definitions and show in Section 3 that: the chain rule holds for every quasiregular $\mathfrak{R}$-form of differentiability; for every quasiregular $\mathfrak{R}$-form there is a weaker regular $\mathfrak{R}$-form; and that Hadamard, equivalently compact, differentiability is the weakest quasiregular $\mathfrak{R}$-form of differentiability (has the largest $\mathcal{R}(X, Y)$ for all X, Y), and *a fortiori* the weakest quasiregular $\mathfrak{R}$-form for which the chain rule holds.

Definition. Let two $\mathfrak{R}$-forms of differentiation be given by two sets $\mathcal{R}_1$ and $\mathcal{R}_2$ of infinitesimal maps. We say that $[\mathcal{R}_2]$-differentiability *preserves* $[\mathcal{R}_1]$-differentiability if whenever $T: X \mapsto Y$ is $[\mathcal{R}_1]$-differentiable at $x \in X$ and $S: Y \mapsto Z$ is $[\mathcal{R}_2]$-differentiable at $y = T(x)$ then $S \circ T$ is $[\mathcal{R}_1]$-differentiable at x and the chain rule formula (8.1) holds.

We will next see in Theorem 8.1 that there are chain rules for $S \circ T$ where S is indeed Fréchet or compactly differentiable but T, and $S \circ T$, can have any form of differentiability, weaker than that of S, in a certain general class. Then, in Theorem 8.2, both S and T can have differentiability weaker than compact, but under more restrictions.

Let X be a normed space, let U be an open subset of X and let $\mathcal{S}$ be a set of pairs $(t, h) := (\{t_n : n \geq 1\}, \{h_n : n \geq 1\})$ where real numbers $0 \neq t_n \to 0$ and h is a bounded sequence of elements of X. We say that an $\mathfrak{S}$-form of differentiation is given if a set $\mathcal{S}$ is given for each normed space X. If $x \in U$ and T is a mapping of U into another normed space $(Y, \| \cdot \|)$, then T will be called $(\mathcal{S})$-*differentiable* at x if there is $DT_x \in \mathcal{L}(X, Y)$ such that for all $(t, h) \in \mathcal{S}$, $x + t_n h_n \in U$,

$$\|T(x + t_n h_n) - T(x) - t_n (DT_x)(h_n)\| = o(|t_n|) \tag{8.2}$$

as $n \to \infty$. Let $\mathcal{R}_\mathcal{S}(X, Y)$ be the set of all maps $r : X \mapsto Y$ such that $r(0) = 0$ and $r(t_n h_n)/t_n \to 0$ as $n \to \infty$ for each $(t, h) \in \mathcal{S}$. Then the $\mathfrak{R}$-form of differentiation given by the set $\mathcal{R}_\mathcal{S}$ of infinitesimal maps, namely $[\mathcal{R}_\mathcal{S}]$-differentiability, is equivalent to $(\mathcal{S})$-differentiability.

In the strongest form of $(\mathcal{S})$-differentiability, $\mathcal{S} = \mathcal{S}_F$, the set of all pairs (t, h) for any $t_n \to 0$ and any bounded sequence h. Then $(\mathcal{S}_F)$-differentiability is easily seen to be equivalent to Fréchet differentiability.

It is also easy to verify that for any class $\mathcal{C}$ of bounded subsets of X, containing all finite sets, $\mathcal{C}$-differentiability is equivalent to $(\mathcal{S}_C)$-differentiability, where $\mathcal{S}_C := \{(t, h) : t_n \to 0 \text{ and } \{h_n : n \geq 1\} \subset B \text{ for some } B \in \mathcal{C}\}$. Moreover, compact differentiability is equivalent to $(\mathcal{S}_c)$-differentiability where $\mathcal{S}_c := \{(t, h) : t_n \to 0 \text{ and } h_n \to h_0 \text{ for some } h_0 \in X\}$, as is known and easily to proved. We then have:

8.1 Theorem. Let X, Y and Z be normed spaces, let $U \subset X$ and $V \subset Y$ be open sets and let $T : U \mapsto V$ be $(\mathcal{S})$-differentiable at $x \in U$ with $y := T(x)$. Consider the following two cases:

(a) $\mathcal{S} \subset \mathcal{S}_c$ and S is compactly differentiable at y from V into Z;
(b) $\mathcal{S} \subset \mathcal{S}_F$ and S is Fréchet differentiable at y from V into Z.
In either case, $S \circ T$ is $(\mathcal{S})$-differentiable at x from U into Z and the chain rule holds.

Proof. Let $x \in U$ and $(t, h) \in \mathcal{S}$. We want to show, as $n \to \infty$,

$$\Delta_n := \|(S \circ T)(x + t_n h_n) - (S \circ T)(x) - (DS_y \circ DT_x)(t_n h_n)\| = o(|t_n|). \tag{8.3}$$

Since T is $\mathcal{S}$-differentiable, (8.2) holds as $n \to \infty$. Thus $u_n := t_n^{-1}[T(x + t_n h_n) - T(x)] = DT_x(h_n) + o(1)$ as $n \to \infty$. In case (b), since $\{h_n : n \geq 1\}$ is bounded, so is $\{u_n : n \geq 1\}$. Since S is Fréchet differentiable, it follows that

$$\Delta_n \leq \|S(y + t_n u_n) - S(y) - t_n DS_y(u_n)\| + |t_n| \|(DS_y)(u_n - DT_x(h_n))\|$$
$$= o(|t_n|) \tag{8.4}$$

as $n \to \infty$. Thus (8.3) holds. In case (a), since $h_n \to h_0$, $u_n \to DT_x(h_0)$. Since S is compactly differentiable, (8.4) follows and again (8.3) holds. The theorem is proved. $\square$

Compact differentiability restricted to a subset A is $(\mathcal{S}_{c,\restriction A})$-differentiability where $\mathcal{S}_{c,\restriction A} := \{(t,h) \in \mathcal{S}_c : x + t_n h_n \in A \text{ for all } n\}$. Here is a chain rule for compact differentiability restricted to subsets, if one subset A is duly mapped into another, B.

8.2 Theorem. Let X, Y and Z be normed spaces, $A \subset X$ and $B \subset Y$. Let T be $(\mathcal{S}_{c,\restriction A})$-differentiable: $A \mapsto B$ at $x \in A$ and let S be $(\mathcal{S}_{c,\restriction B})$-differentiable: $B \mapsto Z$ at $y := T(x) \in B$. Then $S \circ T$ is $(\mathcal{S}_{c,\restriction A})$-differentiable: $A \mapsto Z$ and (8.1) holds.

Proof. Let $x \in A$ and $(t,h) \in \mathcal{S}_{c,\restriction A}$. That is $t_n \to 0$, $h_n \to h_0 \in X$ and $x + t_n h_n \in A$ for each $n \geq 1$. Since T is $(\mathcal{S}_{c,\restriction A})$-differentiable as $n \to \infty$, (8.2) holds. Then $u_n := t_n^{-1}[T(x + t_n h_n) - T(x)] \to DT_x(h_0)$ in Y, $y + t_n u_n \in B$ and $\|S(y + t_n u_n) - S(y) - t_n DS_y(u_n)\| = o(|t_n|)$. Since DS_y is a bounded linear operator, $|t_n| \|(DS_y)(u_n - DT_x(h_n))\| = o(|t_n|)$. Therefore (8.4) and (8.3) hold as $n \to \infty$. The theorem is proved. $\square$

If $h_n \to h_0$ and DT_x is a bounded linear operator, then in equation (8.2), the definition of (S)-differentiability, in the argument of DT_x, h_n can be replaced equivalently by h_0, giving

$$\|T(x + t_n h_n) - T(x) - t_n(DT_x)(h_0)\| = o(|t_n|). \tag{8.5}$$

On the other hand, given a linear subspace H of X and $x \in A \subset X$, T will be said to be *compactly differentiable restricted to a subset A at x tangentially to H*, or $(\mathcal{S}_{c,\restriction A})$-differentiable at x tangentially to H, if (8.5) is only assumed to hold when $x + t_n h_n \in A$ for all n and $h_n \to h_0 \in H$. Thus, the linear transformation DT_x need only be defined on H. One is likely to be interested in differentiability tangentially to H when T is not actually differentiable in (some) directions outside of H (see Appendix E).

Andersen, Borgan, Gill and Keiding (1993, Definition II.8.2) define compact differentiability tangential to a linear subspace H in the same way except that they assume DT_x extends to a bounded linear operator on X. The extension is not always possible. A bounded linear operator from a Banach space X onto a closed linear subspace Y, equal to the identity on Y, is called a *projection* of X onto Y. It is known that if X is a Banach space, then it has a projection onto every closed subspace Y if and only if X is isomorphic to a Hilbert space (Lindenstrauss and Tzafriri, 1973). In particular, $X := D[0,1]$ with the supremum norm has no projection onto $Y := C[0,1]$ (Corson, 1961 p. 12, Example 2, (ii)). Thus the identity from Y to itself has no bounded extension from X to Y. If a bounded linear extension of DT_x to L on X exists, it will not be unique (unless $H = X$), by the Hahn-Banach theorem. When the extensions are not unique, the chain rule formula (8.1) will not hold for them unless the extension is chosen in a compatible way. Thus, we prefer to define the derivative operator for tangential differentiability only on the subspace H. Here is a chain rule for tangential differentiability (as in van der Vaart and Wellner, 1995, Lemma 3.9.3):

8.3 Theorem. Let X, Y and Z be normed spaces, $A \subset X$ and $B \subset Y$, and let $H \subset X$ and $G \subset Y$ be linear subspaces. If T is $(\mathcal{S}_{c,\restriction A})$-differentiable: $A \mapsto B$ at $x \in A$ tangential to H, S is $(\mathcal{S}_{c,\restriction B})$-differentiable: $B \mapsto Z$ at $y := T(x) \in B$ tangential to G, and DT_x takes H into P, then SoT is $(\mathcal{S}_{c,\restriction A})$-differentiable: $A \mapsto Z$ at x tangential to H and (8.1) holds on H.

Proof. The proof is much like the two previous proofs. Let $x \in X$, $t_n \to 0$, $h_n \in X$, $h_n \to h_0 \in H$ and $x + t_n h_n \in A$ for each n. Then (8.5) holds as $n \to \infty$. So $u_n := t_n^{-1}[T(x + t_n h_n) - T(x)] \to DT_x(h_0) \in H$. Thus $\|S(y + t_n u_n) - S(y) - t_n DS_y(DT_x(h_0))\| = o(|t_n|)$, which implies the conclusion. □

Recall that for several of the six operators, with respect to the sup norm, compact differentiability fails, but forms of differentiation as in Theorem 8.3 hold (see the "$\mathcal{W}_{1,M}$" entries in Table 3 and the "CTF" entries in Table 4, Section 1 above). Thus, apparently, chain rules for such forms would be needed.

It has often been said that compact differentiability is the weakest form for which a chain rule holds. But, Theorems 8.2 and 8.3 are chain rules for SoT where T in Theorem 8.2, and both S, T in Theorem 8.3, may have forms of differentiability weaker than compact. We can then state some facts as follows:

• compact differentiability is the weakest form which preserves simple differentability (due to equivalence to Hadamard differentiability);

• compact differentiability also preserves every form of differentiability weaker than or equal to itself (Theorem 8.1(a));

• Fréchet differentiability preserves all forms of differentiability (of interest, all of which are weaker than or equal to itself; Theorem 8.1(b));

• Under further "matching" hypotheses, a form of differentiability weaker than compact can preserve another such form (Theorems 8.2 and 8.3).

If S and T are each Fréchet differentiable with remainder bounds of orders γ and β respectively, then SoT is Fréchet differentiable with remainder bound of order $\min(\gamma, \beta)$, but, even if γ and β are best for T and S respectively, $\min(\gamma, \beta)$ may not be best for SoT: cf. Dudley (1997, end of Section 13.5).

A weaker mode of differentiability can be sufficient for one's purposes and relatively easy to prove. Our proof of Fréchet differentiability of the product integral operator $\mathcal{P}_a$ in $\mathcal{W}_p$ for $0 < p < 2$ in Part II is, indeed, long. Even those segments of Part II needed for the proof of the first Fréchet derivative are much longer than Gill and Johansen's (1990) proof of compact differentiability on a set A, and at a function F, of bounded 1-variation. We do go further: we show that $\mathcal{P}_a : \mathcal{W}_p^r \mapsto \mathcal{W}_p^r$ is holomorphic everywhere on $\mathcal{W}_p^r$, where $\mathcal{W}_p^r$ is the class of all right-continuous functions in $\mathcal{W}_p$, with no further "on A" or "at F" restriction needed. Also, in Part II we give a detailed development of Young integrals $\int_a^b F\,dG\,H$ where F, G and H may have non-commuting values. In Gill and Johansen (1990) and Gill (1994), the definition of such integrals is only sketched, and only for F and H of bounded 1-variation. Young integrals, specifically (LY) integrals as defined in Section 5 above, appear in the derivative of $\mathcal{P}_a$.

In the use of modified forms of compact differentiability, one needs to take account of the process of choosing the modifications, and fitting them together for different operators, as well as proving them once chosen.

One proof of compact differentiability is apparently a bit longer than had been noted (Appendix D), although this could be avoided by choosing a suitable set A.

In at least one case, Proposition 1.2 above, compact differentiability can be drastically weakened, to uniformity over one compact (convex, symmetric) set K; it is still applicable (Section 6) in a more modest but sufficient way; it becomes extremely easy to prove; and it is Fréchet differentiability with a sharp remainder bound, for the norm with closed unit ball K, specifically a Hölder norm.

Summing up then, we suggest that our six operators are better viewed as operating on actual functions rather than on elements of abstract Banach spaces, and that norms are technical tools for studying the operators. Different norms, or sets of functions not necessarily defined by norms, can and should be chosen depending on their usefulness in given situations.

Appendix A. Convolution.

Let g be a real-valued regulated function on $[a, b]$ and let h be a real-valued function on $[c, d] \times [a, b]$ such that $h(x, \cdot)$ is a regulated function on $[a, b]$ for each $x \in [c, d]$. Define a function f^C on $[c, d]$ by

$$f^C(x) := (CY) \int_a^b h(x, y) \, dg(y) \tag{A.1}$$

if the (CY) integral exists for each $x \in [c, d]$. An inequality for the p-variation of f^C will be proved.

Given positive real numbers r and p, let $\mathcal{W}_{r,p} = \mathcal{W}_{r,p}([c, d] \times [a, b])$ be the class of functions $h \colon [c, d] \times [a, b] \mapsto \mathbb{R}$ such that, for some finite constants A and B,

$$\sum_{i=1}^n |h(x_i, y) - h(x_{i-1}, y)|^r \le A^r \quad \text{and} \quad \sum_{j=1}^m |h(x, y_j) - h(x, y_{j-1})|^p \le B^p \tag{A.2}$$

for all $y \in [a, b]$, all partitions $c = x_0 < \cdots < x_n = d$, all $x \in [c, d]$, and all partitions $a = y_0 < \cdots < y_m = b$. Let $A_{r,\infty} = A_{r,\infty}(h; [c, d], [a, b])$ and $B_{\infty,p} = B_{\infty,p}(h; [c, d], [a, b])$ be the minimal numbers A and B such that (A.2) holds. The following is the main statement of this appendix.

A.1 Theorem. Let $p, q, r \in [1, \infty)$ and $1/p + 1/q > 1$. Let $g \in \mathcal{W}_q([a, b])$ and $h \in \mathcal{W}_{r,p}([c, d] \times [a, b])$. There exists a finite constant $C = C(p, q, r)$ such that, for the function $\bar{f}^C := f^C - h(\cdot, a)[g(b) - g(a)]$ with f^C defined by (A.1), the inequality

$$V_\nu(\bar{f}^C; [c, d]) \le C A_{r,\infty}(h; [c, d], [a, b])^{\frac{r}{\nu}} B_{\infty,p}(h; [c, d], [a, b])^{1 - \frac{r}{\nu}} V_q(g; [a, b]) \tag{A.3}$$

holds with ν given by $1/\nu = (p/r)(1/p + 1/q - 1)$.

Remark. To bound the ν-variation of the function f^C given by (A.1), notice that $\nu \ge r$ in Theorem A.1. Thus, by (A.3), it follows that

$$\begin{aligned} V_\nu(f^C; [c, d]) &\le C A_{r,\infty}(h; [c, d], [a, b])^{r/\nu} B_{\infty,p}(h; [c, d], [a, b])^{1 - \frac{r}{\nu}} V_q(g; [a, b]) \\ &\quad + A_{r,\infty}(h; [c, d], [a, b])|g(b) - g(a)|. \end{aligned}$$

Let F and G be regulated functions on $[a - b, b - a]$ and $[a, b]$, respectively. The main application we have in mind is to convolution $F * G$ defined on $[a, b]$ by $(F * G)(x) := (CY) \int_a^b F(x - y) \, dG(y)$ if the (CY) integral exists for each $x \in [a, b]$. The following statement is a special case of Theorem A.1 with $h(x, y) = F(x - y)$ for $x, y \in [a, b]$ and $r = p$.

A.2 Corollary. Let $p, q \in [1, \infty)$ and $1/p + 1/q > 1$. There exists a finite constant $C = C(p, q)$ such that the inequality

$$V_\nu(F * G; [a, b]) \le C V_p(F; [a - b, b - a]) V_q(G; [a, b])$$

holds with ν given by $1/\nu = 1/p + 1/q - 1$.

Before proving Theorem A.1 we will prove Theorem 1.1.

Proof of Theorem 1.1. For $a < b$ and $t \in \mathbb{R}$, let $(F*G)_{a,b}(t) := (CY)\int_a^b F(t - x)\,dG(x)$. The integral is defined and finite, and bounded uniformly in $a < b$ by (1.2), since for all t, a and b, $\|F_{[a,b]}(t - \cdot)\|_{[p]} \leq \|F\|_{[p]}$, where $F_{[a,b]}(t - \cdot)$ is $F(t - \cdot)$ restricted to $[a, b]$. Also, the integrals converge as $a \to -\infty$ and $b \to +\infty$ since the q-variation of G restricted to $(-\infty, a]$ or $[b, \infty)$ goes to 0. So $(F*G)(t)$ is well-defined for all t and $\|F*G\|_\infty \leq C_{p,q}\|F\|_{[p]}\|G\|_{(q)}$. Let $(F*G)^{a,b}$ be $(F*G)_{a,b}$ restricted to $[a, b]$. Then Corollary A.2 implies $\|(F*G)^{a,b}\|_{(\nu)} \leq C(p, q)\|F\|_{(p)}\|G\|_{(q)}$. Letting $a \to -\infty$ and $b \to +\infty$, and noting that $\|\cdot\|_{(\nu)}$ is a supremum over finite sets, Theorem 1.1 follows with $K(p, q) = \max(C_{p,q}, C(p, q))$. $\square$

We will prove Theorem A.1 in three steps. First, inequality (A.3) with the (CY) integral replaced by the Riemann-Stieltjes integral will be established in Theorem A.3 assuming g and each $h(x, \cdot)$, $x \in [c, d]$, to be periodic, and g to be continuous. The continuity assumption on g will be removed in Theorem A.8 below by using the (RYS) integral defined in Section 4 above, see also Section 3.3 of Part II. At the final step we prove inequality (A.3) without restrictions as stated in Theorem A.1.

Suppose that g and $h(x, \cdot)$, for each $x \in [c, d]$, are real-valued functions of a real variable, with period $\Lambda > 0$. Let

$$f(x) := \frac{1}{\Lambda}\int_0^\Lambda h(x, y)\,dg(y) \tag{A.4}$$

if the integral exists in the Riemann-Stieltjes sense for each $x \in [c, d]$. The following result is due to Gehring (1954, Theorem 3.2). We include its proof here, with minor modifications, for completeness.

A.3 Theorem. Let the numbers p, q and r be as in Theorem A.1. Let real-valued functions g and $h(x, \cdot)$, $x \in [c, d]$, defined for all real numbers, have period $\Lambda > 0$, let g be continuous and let h, when restricted to $[c, d] \times [0, \Lambda]$, be in $\mathcal{W}_{r,p}([c, d] \times [0, \Lambda])$. Then, for the function f defined by (A.4), the inequality

$$V_\nu(f; [c, d]) \leq C A_{r,\infty}(h; [c, d], [0, \Lambda])^{r/\nu} B_{\infty,p}(h; [c, d], [0, \Lambda])^{1 - \frac{r}{\nu}} V_q(g; [0, \Lambda]) \tag{A.5}$$

holds with ν given by $1/\nu = (p/r)(1/p + 1/q - 1)$ and $C = \Lambda^{\frac{1}{q}-1}[2^{\frac{1}{p}} + 1/(2^{\frac{r}{p\nu}} - 1)]$.

We first show how the Riemann-Stieltjes integrals for periodic functions can be approximated by way of Lebesgue convolutions. So, such convolutions appear not only as a special case (Corollary A.2 above) but also in treating the general case of Theorem A.3.

Suppose that h, g are functions of a real variable, with period $\Lambda > 0$. Let

$$S := \frac{1}{\Lambda}\int_0^\Lambda h(y)\,dg(y) \quad \text{and} \quad f^L(z) := \frac{1}{\Lambda}\int_{(\Lambda)} h(y - z)g(y)\,dy \tag{A.6}$$

for each real z, if the first integral exists in the Riemann-Stieltjes sense and the second is a Lebesgue integral over the interval $[0, \Lambda]$. Notice that the second integral in (A.6) is a convolution, as for "densities," in a different sense than we have been using the term up to now; also, usually, such a convolution is defined with $z - y$ rather than $y - z$.

We will argue as did Young (1938a) except that the function $h(y + z)$ in his Lebesgue convolution is replaced by the function $h(y - z)$ as in (A.6). The derivative at 0 of f^L will then be shown to equal to the Riemann-Stieltjes integral S. A somewhat more general statement is proved by Young (1970, Proposition 5.4.(i)).

A.4 Lemma.   Let real-valued functions h and g, defined for all real numbers, have period Λ. Let g be continuous, let h be bounded and measurable, and let the integrals (A.6) exist. Then f^L has a derivative at 0 equal to S.

Proof. Since $(f^L)'(0) = \Lambda \lim_{t \to 0}\{f^L(t/\Lambda) - f^L(0)\}/t$, by the change of variables $y \mapsto y/\Lambda$ in (A.6), it is enough to prove the lemma for $\Lambda = 1$. Thus, we will show that

$$\lim_{t \downarrow 0} \{f^L(t) - f^L(0)\}/t = S = \lim_{t \downarrow 0} \{f^L(0) - f^L(-t)\}/t. \qquad (A.7)$$

Denoting by $[t^{-1}]$ the integer part of t^{-1}, for each $y \in [0, 1]$ and $t \in (0, 1)$, let

$$U_t(y) := \sum_{i=1}^{[t^{-1}]} h(y + (i - 1)t)[g(y + it) - g(y + (i - 1)t)].$$

By periodicity, the Lebesgue integrals of $h(\cdot - z)g$ over $[x, x + 1]$ and $[0, 1]$ coincide for each z and x. Thus, for each $t \in (0, 1)$, we have

$$\{f^L(t) - f^L(0)\}/t = (t[t^{-1}])^{-1} \int_{(1)} U_t(y)\,dy.$$

To prove the first equality in (A.7) it is enough to show that, given $\epsilon > 0$,

$$|U_t(y) - S| < \epsilon + 3\epsilon\|h\|_\infty \qquad (A.8)$$

for all $y \in [0, 1]$ and all sufficiently small t.

Since S exists in the Riemann-Stieltjes sense, there is a $\delta \in (0, 1)$ such that

$$|\sum_{j=1}^{m} h(z_j)[g(x_j) - g(x_{j-1})] - S| < \epsilon$$

for any partition $\{0 = x_0 < \cdots < x_m = 1\}$ of $[0, 1]$ such that $\max_j(x_j - x_{j-1}) < \delta$ and for all $z_j \in [x_{j-1}, x_j]$, $j = 1, \ldots, m$. By uniform continuity of g over $[0, 1]$, we can also suppose that $|g(y) - g(y + t)| < \epsilon$ for all $y \in [0, 1]$ and all $t \in [-\delta, \delta]$. To prove (A.8) we split $U_t(y)$ into two parts by separating terms with $y + it \leq 1$ and $y + it > 1$, and use periodicity. Let $i_y := [(1 - y)/t] + 1$ and $m = m(t) := [t^{-1}]$. So we have $y + (i_y - 1)t \leq 1$ and $y + i_y t > 1$. If $i_y > 1$ then let $z_1 := y + i_y t - 1$, $z_j := x_{j-1} := y + (j + i_y - 1)t - 1$ for $j = 2, \ldots, m - i_y$ and $z_j := x_{j-1} := y + (j - m + i_y - 1)t$ for $j = m - i_y + 1, \ldots, m$. Then we have

$$U_t(y) = \Big(\sum_{i=i_y+1}^{m} + \sum_{i=1}^{i_y} \Big) h(y + (i - 1)t)[g(y + it) - g(y + (i - 1)t)]$$

$$= \sum_{j=1}^{m-i_y} h(y+(j+i_y-1)t-1)[g(y+(j+i_y)t-1)-g(y+(j+i_y-1)t-1)]$$

$$+ \sum_{j=m-i_y+1}^{m} h(y+(j-m+i_y-1)t)[g(y+(j-m+i_y)t)-g(y+(j-m+i_y-1)t)]$$

$$= \sum_{j=1}^{m} h(z_j)[g(x_j)-g(x_{j-1})]+h(y+i_yt)[g(0)-g(y+i_yt)]$$

$$+h(y+(i_y-1)t)[g(y+i_yt)-g(1)]+h(y+(m-1)t)[g(y+mt)-g(y)].$$

Thus, (A.8) holds for all $t < \delta$ whenever $i_y > 1$. If $i_y = 1$ then let $z_j := x_{j-1} := y+jt-1$ for $j = 2,\ldots,m-1$, $z_1 := y+t-1$ and $z_m = x_{m-1} := y$. Then, similarly to the case $i_y > 1$, we have

$$U_t(y) = \sum_{i=2}^{m} h(y+(i-1)t)[g(y+it)-g(y+(i-1)t)]+h(y)[g(y+t)-g(y)]$$

$$= \sum_{j=1}^{m-1} h(y+jt-1)[g(y+(j+1)t-1)-g(y+jt-1)]+h(y)[g(y+t)-g(y)]$$

$$= \sum_{j=1}^{m} h(z_j)[g(x_j)-g(x_{j-1})]+h(y+t)[g(0)-g(y+t)]$$

$$+h(y)[g(y+t)-g(1)]+h(y+(m-1)t)[g(y+mt)-g(y)].$$

Hence (A.8) holds for all $t < \delta$ and all $y \in [0,1]$.

To prove the second equality in (A.7), by periodicity, for all $t \in (0,1)$, we have

$$\{f^L(0)-f^L(-t)\}/t = (t[t^{-1}])^{-1}\int_{(1)} W_t(y)\,dy,$$

where, for each $y \in [0,1]$, $W_t(y) := \sum_{i=1}^{[t^{-1}]} h(y+it)[g(y+it)-g(y+(i-1)t)]$. Choosing i_y as in the proof of the first equality in (A.7), one can find that $W_t(y)$ differs from the corresponding Riemann-Stieltjes sum by

$$h(y+i_yt)[g(0)-g(y+(i_y-1)t)]+h(y+(i_y-1)t)[g(y+(i_y-1)t)-g(1)]$$

$$+h(y+mt)[g(y+mt)-g(y)].$$

Thus, the second equality in (A.7) follows similarly as the first did. The proof is complete. □

Suppose again that h, g are functions of a real variable, with period $\Lambda > 0$. Recalling (A.6), for each $k = 0,1,\ldots$, let

$$f_k := \frac{2^k}{\Lambda}\{f^L(2^{-k}\Lambda)-f^L(0)\} = \frac{2^k}{\Lambda^2}\int_{(\Lambda)} [h(y-2^{-k}\Lambda)-h(y)]g(y)\,dy. \qquad (A.9)$$

By periodicity, we have $f_0 \equiv 0$.

A.5 Lemma. For each $k \geq 1$, the relation

$$f_k - f_{k-1} = \frac{2^{k-1}}{\Lambda^2} \int_{(\Lambda)} \left[h(y - 2^{-k}\Lambda) - h(y) \right] \left[g(y) - g(y + 2^{-k}\Lambda) \right] dy \qquad (A.10)$$

holds.

Proof. By periodicity, we have

$$\int_{(\Lambda)} \left[h(y - 2^{-k}\Lambda) - h(y) \right] g(y) \, dy = \int_{(\Lambda)} \left[h(y) - h(y + 2^{-k}\Lambda) \right] g(y + 2^{-k}\Lambda) \, dy$$

and

$$\int_{(\Lambda)} \left[h(y - 2^{-k+1}\Lambda) - h(y) \right] g(y) \, dy = \int_{(\Lambda)} \left[h(y - 2^{-k}\Lambda) - h(y + 2^{-k}\Lambda) \right] g(y + 2^{-k}\Lambda) \, dy.$$

Then (A.10) follows from the representation $f_k - f_{k-1} = f_k/2 + f_k/2 - f_{k-1}$ by using the first equality for $f_k/2$ and the second one for f_{k-1}. The proof is complete. $\square$

The following is essentially Theorem 1.2.1 of Gehring (1954, cf. the proof of his Theorem 3.2).

A.6 Proposition. Let $1 \leq q < \infty$ and let g be a continuous function with period Λ and such that the q-variation of g over any interval of length Λ is less than 1. There exists a continuous increasing function ϕ of a real variable such that $\phi(0) = 0$, $\phi(t + \Lambda) = \phi(t) + \Lambda$ and $|g(\phi(t)) - g(\phi(s))|^q \leq t - s$ for each $s < t$.

Proof. First suppose $\Lambda = 1$. Let $x_0 \in (0, 1]$ be a point where g attains its maximum. Let $\bar{g}(y) := g(y + x_0)$ for each real number y. Then $\bar{g}$ is a continuous function with period 1 and attains its maximum at 0. For each $x \geq 0$, let $\gamma(x) := v_q(\bar{g}; [0, x]) = v_q(g; [x_0 - 1, x_0 - 1 + x])$. By Lemma 3 of Marcinkiewicz (1934, p. 39) or by Lemma 4.3 of Young (1937, p. 454), γ is a continuous function on $[0, \infty)$. To prove $\gamma(x + 1) = \gamma(x) + \gamma(1)$ for $x \geq 0$, it is enough to show that

$$\gamma(x + 1) \leq \gamma(x) + \gamma(1) \qquad (A.11)$$

because the reverse inequality always holds. Given $x > 0$ and $\epsilon > 0$, choose a partition $0 = x_0 < \cdots < x_n = x + 1$ of $[0, x + 1]$ in such a way that

$$\gamma(x + 1) < \sum_{i=1}^{n} |\bar{g}(x_i) - \bar{g}(x_{i-1})|^q + \epsilon.$$

If the index j is such that $x_{j-1} < 1 \leq x_j$ then $|\bar{g}(x_j) - \bar{g}(x_{j-1})|^q \leq |\bar{g}(1) - \bar{g}(x_j)|^q \vee |\bar{g}(1) - \bar{g}(x_{j-1})|^q \leq |\bar{g}(1) - \bar{g}(x_j)|^q + |\bar{g}(1) - \bar{g}(x_{j-1})|^q$ because $\bar{g}$ has its maximum at 1. Hence $\gamma(x + 1) < \gamma(x) + \gamma(1) + \epsilon$. Since $\epsilon > 0$ is arbitrary, (A.11) follows.

Let $\theta(x) := \gamma(x + 1 - x_0) - \gamma(1 - x_0) + x[1 - \gamma(1)]$ for each $x \geq 0$. Then θ is a continuous and increasing function on $[0, \infty)$. Let ϕ be the inverse function of θ.

Then $\phi(0) = 0$ and $\phi(t + 1) = \phi(t) + 1$ for each $t \geq 0$ because $\theta(\phi(t) + 1) = t + 1$ by (A.11). We extend the definition of $\phi(t)$ to negative t by using the relation $\phi(t) = \phi(t + 1) - 1$ recursively. Then, for each $0 \leq s < t \leq 1$, we have

$$
\begin{aligned}
|g(\phi(t)) - g(\phi(s))|^q &\leq v_q(g; [x_0 - 1, \phi(t)]) - v_q(g; [x_0 - 1, \phi(s)]) \\
&= \gamma(\phi(t) + 1 - x_0) - \gamma(\phi(s) + 1 - x_0) \\
&= t - s - [\phi(t) - \phi(s)][1 - \gamma(1)] \leq t - s.
\end{aligned}
$$

The inequality for arbitrary $s < t$ then follows by periodicity.

For $\Lambda \neq 1$, let $\bar{g}(y) := g(\Lambda y)$ for each real number y. Then $\bar{g}$ has period 1 and $v_q(\bar{g}; [x, x + 1]) = v_q(g; [\Lambda x, \Lambda x + \Lambda]) \leq 1$. Let $\bar{\phi}$ be the function constructed in the previous paragraph. Then the function ϕ defined by $\phi(t) := \Lambda \bar{\phi}(t/\Lambda)$ for each t has the desired properties. The proof is complete. □

We need the following statement. More general results are due to Marcinkiewicz (1934, Theorem 4) and Gehring (1954, Theorem 1.3.3).

A.7 Lemma. Let $1 \leq p < \infty$ and let h be a measurable real-valued function of a real variable. If the p-variation of h over any interval of length Λ never exceeds 1, then

$$
\frac{1}{\Lambda} \int_{(\Lambda)} |h(t + \Lambda/n) - h(t)|^p \, dt \leq 1/n \tag{A.12}
$$

for every positive integer n.

Proof. Let $\bar{h}(t) := h(\Lambda t)$ for each real number t. Then $\bar{h}$ is a measurable real-valued function with $v_p(\bar{h}; [t, t + 1]) = v_p(h; [\Lambda t, \Lambda t + \Lambda]) \leq 1$ for each real number t. Thus, using a change of variables in (A.12) we can suppose that $\Lambda = 1$. Let $n \geq 1$ be an integer. Then, by a change of variables again, we have

$$
\int_{(1)} |h(t + 1/n) - h(t)|^p \, dt = \sum_{i=1}^{n} \int_{A_{in}} |h(t + 1/n) - h(t)|^p \, dt
$$

$$
= \int_{(1/n)} \sum_{i=1}^{n} |h(t + i/n) - h(t + (i-1)/n)|^p \, dt \leq \int_{(1/n)} v_p(h; [t, t + 1]) \, dt \leq 1/n,
$$

where $A_{in} := [(i-1)/n, i/n]$ for $i = 1, \ldots, n$. The proof of Lemma A.7 is complete. □

Now we are ready to prove Theorem A.3.

Proof of Theorem A.3. If either $V_q(g; [0, \Lambda]) = 0$ or $= +\infty$ then (A.5) holds. Moreover, (A.5) also holds if $B_{\infty,p}(h; [c, d], [0, \Lambda]) = 0$ because $g(\Lambda) = g(0)$. Let x be a real number and let m be an integer such that $x \leq m\Lambda \leq x + \Lambda$. Then, by periodicity, we have

$$
\begin{aligned}
V_q(g; [x, x + \Lambda]) &\leq 2^{1 - 1/q} [v_q(g; [(m - 1)\Lambda, m\Lambda]) + v_q(g; [m\Lambda, (m + 1)\Lambda])]^{1/q} \\
&= 2V_q(g; [0, \Lambda]).
\end{aligned}
$$

PART I. DIFFERENTIABILITY AND p-VARIATION

49

Similarly we have $B_{\infty,p}(h;[c,d],[x,x+\Lambda]) \le 2B_{\infty,p}(h;[c,d],[0,\Lambda])$ for any x. Thus, by homogeneity of the inequality (A.5), we can and do assume that, for each x,

$$B_{\infty,p}(h;[c,d],[x,x+\Lambda]) = 1 \quad \text{and} \quad V_q(g;[x,x+\Lambda]) \le 1. \tag{A.13}$$

Let ϕ be the continuous increasing function constructed in Proposition A.6. For each real number t, let $\bar{g}(t) := g(\phi(t))$ and, for each $x \in [c,d]$, let $\bar{h}_x(t) = \bar{h}(x,t) := h(x,\phi(t))$. Then $\bar{g}$ and $\bar{h}_x$ have period Λ, their q- and p-variations respectively over any interval of length Λ do not exceed 1, $f(x) = \Lambda^{-1} \int_{(\Lambda)} \bar{h}(x,t)\,d\bar{g}(t)$ for each $x \in [c,d]$ and

$$|\bar{g}(t) - \bar{g}(s)|^q \le t - s \tag{A.14}$$

for each $s < t$.

Since g is continuous and $1/p + 1/q > 1$, the function f in (A.4) is well defined by a theorem on Stieltjes integrability (cf. p. 164 in Young, 1936 or Theorem 3.27 in Part II). For each $k = 0,1,\ldots$ and each $x \in [c,d]$, let

$$f_k(x) := \frac{2^k}{\Lambda^2} \int_{(\Lambda)} [\bar{h}(x,y-2^{-k}\Lambda) - \bar{h}(x,y)]\bar{g}(y)\,dy$$
$$= \frac{2^k}{\Lambda^2} \int_{(\Lambda)} \bar{h}(x,y)[\bar{g}(y+2^{-k}\Lambda) - \bar{g}(y)]\,dy \tag{A.15}$$

(cf. (A.9) above). By Lemma A.5, applied to $\bar{h}_x$ and $\bar{g}$ in place of h and g, (A.14), and Lemma A.7, for each $k \ge 1$ and each $x \in [c,d]$, we have

$$|f_k(x) - f_{k-1}(x)| \le \frac{2^{k-1}}{\Lambda^2} \int_{(\Lambda)} |\bar{h}_x(t-2^{-k}\Lambda) - \bar{h}_x(t)||\bar{g}(t) - \bar{g}(t+2^{-k}\Lambda)|\,dt$$
$$\le \Lambda^{\frac{1}{q}-1} 2^{k(1-\frac{1}{q})-1} \left\{ \frac{1}{\Lambda} \int_{(\Lambda)} |\bar{h}_x(t-2^{-k}\Lambda) - \bar{h}_x(t)|^p\,dt \right\}^{1/p}$$
$$\le \Lambda^{\frac{1}{q}-1} 2^{-\frac{kr}{p\nu}-1},$$

because $1/q + 1/p - 1 = r/(p\nu)$. By Lemma A.4, also for $\bar{h}_x$ and $\bar{g}$, $f(x) = \lim_{k\to\infty} f_k(x)$ for each $x \in [c,d]$. Thus, for each $k = 0,1,\ldots$ and each $x \in [c,d]$, we get

$$|f(x) - f_k(x)| \le \Lambda^{\frac{1}{q}-1} \sum_{l=k+1}^{\infty} 2^{-\frac{lr}{p\nu}-1} = C_1 2^{-\frac{kr}{p\nu}}, \tag{A.16}$$

where $C_1 = 2^{-1}\Lambda^{\frac{1}{q}-1}/(2^{r/p\nu}-1)$. Let $k \ge 1$ and $c \le x' < x'' \le d$. By (A.15), (A.14) and Jensen's inequality, we have

$$|f_k(x'') - f_k(x')| \le \frac{2^k}{\Lambda^2} \int_{(\Lambda)} |\bar{h}(x'',y) - \bar{h}(x',y)||\bar{g}(y+2^{-k}\Lambda) - \bar{g}(y)|\,dy$$
$$\le \Lambda^{\frac{1}{q}-1} 2^{k(1-\frac{1}{q})} \Delta_r(x',x''), \tag{A.17}$$

where

$$\Delta = \Delta_r(x',x'') := \left(\frac{1}{\Lambda} \int_{(\Lambda)} |\bar{h}(x'',y) - \bar{h}(x',y)|^r\,dy \right)^{1/r}.$$

By considering three cases, we will show next that

$$|f(x'') - f(x')| \le C_2 \Delta^{r/\nu}, \tag{A.18}$$

where $C_2 = \Lambda^{\frac{1}{q}-1}[2^{1/p} + 1/(2^{r/(p\nu)} - 1)]$.

Suppose first that $1 < \Delta < \infty$. Then (A.18) follows from (A.16) with $k = 0$ because $2C_1 < C_2$. Second suppose that $0 < \Delta \le 1$. Choose $k \ge 1$ so that $2^{-k/p} < \Delta \le 2^{-(k-1)/p}$. By (A.16), it follows that

$$|f(x) - f_k(x)| \le C_1 \Delta^{r/\nu}$$

for $x \in [c, d]$. Next, from (A.17) we get

$$|f_k(x'') - f_k(x')| \le \Lambda^{\frac{1}{q}-1} 2^{1/p} \Delta^{r/\nu}.$$

Then (A.18) is a consequence of the last two inequalities because $C_2 = 2C_1 + \Lambda^{\frac{1}{q}-1} 2^{\frac{1}{p}}$. Third suppose that $\Delta = 0$. Since $f_k \to f$ pointwise as shown just before (A.16),

$$|f(x'') - f(x')| = \lim_{k\to\infty} |f_k(x'') - f_k(x')| = 0.$$

Thus (A.18) holds in this case also.

To complete the proof, let $c = x_0 < \cdots < x_n = d$ be any partition of $[c, d]$. Then, by (A.18), we have

$$\left(\sum_{i=1}^{n} |f(x_i) - f(x_{i-1})|^{\nu}\right)^{1/\nu} \le C_2 \left(\frac{1}{\Lambda} \int_{(\Lambda)} \sum_{i=1}^{n} |\bar{h}(x_i, y) - \bar{h}(x_{i-1}, y)|^r \, dy\right)^{1/\nu}$$

$$\le C_2 A_{r,\infty}(\bar{h}; [c, d], [0, \Lambda])^{r/\nu} = C_2 A_{r,\infty}(h; [c, d], [0, \Lambda])^{r/\nu}.$$

The last equality holds by periodicity of h_x for each $x \in [c, d]$. By (A.13), the proof of Theorem A.3 is complete. □

Turning to the second step of proving Theorem A.1, suppose that g is a regulated real-valued function on $[a, b]$ and h is a real-valued function on $[c, d] \times [a, b]$. Let

$$f^R(x) := (RYS) \int_a^b h(x, y) \, dg(y) \tag{A.19}$$

if the (RYS) integral (defined following Theorem 4.3 above and in subsection 3.3 of Part II) exists for each $x \in [c, d]$. The proof of the following theorem is based on arguments of Young (1937, p. 459).

A.8 Theorem. Let the numbers p, q, r and functions h, g be as in Theorem A.1. In addition, let $g(a) = g(b)$ and $h(x, a) = h(x, b)$ for each $x \in [c, d]$. Then, for the function f^R defined by (A.19), the inequality

$$V_\nu(f^R; [c, d]) \le C A_{r,\infty}(h; [c, d], [a, b])^{\frac{r}{\nu}} B_{\infty,p}(h; [c, d], [a, b])^{1-\frac{r}{\nu}} V_q(g; [a, b]) \tag{A.20}$$

holds with ν given by $1/\nu = (p/r)(1/p + 1/q - 1)$ and C being the same as in (A.5).

For the proof we need the following two auxiliary statements.

A.9 Lemma. Let g be a regulated real-valued function on $[a, b]$ and let g_1 be a step function on $[a, b]$ which agrees with g at the endpoints, and is equal to either $g(x)$ or $g(x\pm)$ for some x in each interval of constancy of g_1. Then $v_q(g_1; [a, b]) \le v_q(g; [a, b])$ for any $q > 0$.

Proof. We can and do assume that $v_q(g) < \infty$. Let $\kappa = \{a = x_0 < \cdots < x_n = b\}$ be a partition of $[a, b]$. Then, since g is regulated, the sum $v_q(g_1; \kappa) := \sum_{i=1}^{n} |g_1(x_i) - g_1(x_{i-1})|^q$ is a limit of similarly defined sums $v_q(g; \kappa')$ with partitions κ' such that their points either agree with or approach points of κ. Thus $v_q(g_1; \kappa) \le v_q(g; [a, b])$. Since κ is an arbitrary partition this completes the proof. $\square$

A.10 Lemma. Let g be a step function on $[a, b]$ having values $c_1, \ldots, c_K$ on its non-degenerate intervals of constancy, so that for some $a = t_0 < t_1 < \cdots t_K = b$, $g(t) = c_i$ for $t_{i-1} < t < t_i$ for each i. Let $c_0 := g(a)$ and $c_{K+1} := g(b)$. Let $s_0 := a$, $s_i := (t_{i-1} + t_i)/2$ for $i = 1, \cdots, K$ and $s_{K+1} := b$. Let g_1 be a piecewise linear continuous function on $[a, b]$ having the pairs (s_i, c_i) for $i = 0, 1, \cdots, K+1$ as endpoints of line segments in its graph for $i = 0, 1, \cdots, K+1$. Then $v_q(g_1; [a, b]) \le v_q(g; [a, b])$ whenever $1 \le q < \infty$.

Proof. We have

$$v_q(g; [a, b]) \ge \max \Big\{ \sum_{l=1}^{m} |c_{j_l} - c_{j_{l-1}}|^q : 0 = j_0 < \cdots < j_m = K+1 \Big\}. \qquad (A.21)$$

Here there is a possibility of inequality because the contribution to $v_q(g)$ coming from the values of g at single points is not included in the right side of (A.21). Let $\kappa := \{a = x_0 < \cdots < x_n = b\}$ be a partition of $[a, b]$ and let $v_q(g_1; \kappa) := \sum_{i=1}^{n} |g_1(x_i) - g_1(x_{i-1})|^q$.

In evaluating $\sup_{\kappa} v_q(g_1; \kappa)$, we can assume that consecutive increments of g_1 have opposite signs, since $u^q + v^q \le (u + v)^q$ for any $u, v \ge 0$. (This holds for $u = 1$ by differentiating with respect to v, then for all $u \ge 0$ by homogeneity.) Then if $g_1(x_{i-1}) < g_1(x_i) > g_1(x_{i+1})$, we can assume that g_1 has a local maximum at x_i, since otherwise x_i could be replaced by such a point, increasing $v_q(g_1; \kappa)$. So $g_1(x_i) = c_j$ for some j. Likewise if $g_1(x_{i-1}) > g_1(x_i) < g_1(x_{i+1})$ then $g_1(x_i) = c_j$ for some j since we can take a relative minimum of g_1. Then by (A.21), it follows that $v_q(g_1; \kappa) \le v_q(g; [a, b])$. The proof of the lemma is complete. $\square$

Proof of Theorem A.8. Let $\phi(y) := a + y(b - a)$ for $y \in [0, 1]$. For each $x \in [c, d]$ and $y \in [0, 1]$, let $\bar{g}(y) := g(\phi(y))$ and $\bar{h}_x(y) = \bar{h}(x, y) := h(x, \phi(y))$. Then

$$A_{r,\infty}(\bar{h}; [c, d], [0, 1]) = A_{r,\infty}(h; [c, d], [a, b]), \qquad V_q(\bar{g}; [0, 1]) = V_q(g; [a, b]),$$

$B_{\infty,p}(\bar{h}; [c, d], [0, 1]) = B_{\infty,p}(h; [c, d], [a, b])$ and $f^R(x) = (RYS) \int_0^1 \bar{h}_x \, d\bar{g}$ for each $x \in [c, d]$. By extending the functions so defined we can and do assume that g and h_x are defined for all real numbers and have period 1. Thus we can use Theorem A.3 with $\Lambda = 1$ whenever the function g is continuous.

Suppose g is not continuous. Then we will approximate the function f^R in two steps by similar functions which have continuous integrators as follows. Since $1/p + 1/q > 1$, the function f^R is well defined by Theorem 5.1 of Young (1938b). Given a partition $\kappa = \{0 = y_0 < \cdots < y_m = 1\}$ of $[0, 1]$ and points $\eta_j \in (y_{j-1}, y_j)$ for $j = 1, \ldots, m$, let

$$f^\kappa(x) := \sum_{j=1}^{m} h(x, \eta_j)[g_-(y_j) - g_+(y_{j-1})] + \sum_{j=0}^{m} h(x, y_j)[g_+^{(1)}(y_j) - g_-^{(0)}(y_j)]$$

for each $x \in [c, d]$; recall that for a regulated function g on an interval $[a, b]$,

$$g_-(y) := g(y-) := \lim_{x \uparrow y} g(x), \quad g_+(y) := g(y+) := \lim_{x \downarrow y} g(x),$$

$$g_-^{(a)}(y) := g_-(y) \quad \text{for } a < y \le b, \quad g_-^{(a)}(a) := g(a),$$

$$g_+^{(b)}(y) := g_+(y) \quad \text{for } a \le y < b, \quad g_+^{(b)}(b) := g(b).$$

Let $c = x_0 < \cdots < x_n = d$ be any partition of $[c, d]$. Then, by the definition of (RYS) integral, we have

$$\sum_{i=1}^{n} |f^R(x_i) - f^R(x_{i-1})|^\nu = \lim_{(\kappa)} \sum_{i=1}^{n} |f^\kappa(x_i) - f^\kappa(x_{i-1})|^\nu, \qquad (A.22)$$

where the limit is taken under refinements of partitions κ. Let κ be a partition of $[0, 1]$ as above. For each $j = 1, \cdots, m$, let $\eta_j := (y_{j-1} + y_j)/2$, $y_{j-1} < u_j < \eta_j < v_j < y_j$, and let $F^\kappa(x)$ be the sum

$$\sum_{j=1}^{m} h(x, y_{j-1})[g(u_j) - g(y_{j-1})] + h(x, \eta_j)[g(v_j) - g(u_j)] + h(x, y_j)[g(y_j) - g(v_j)]$$

$$= h(x, 0)[g(u_1) - g(0)] + h(x, 1)[g(1) - g(v_m)] + \sum_{j=1}^{m} h(x, \eta_j)[g(v_j) - g(u_j)]$$

$$+ \sum_{j=1}^{m-1} h(x, y_j)[g(u_{j+1}) - g(v_j)] = (RS) \int_0^1 h_1(x, y) dg_1(y),$$

where $g_1(y) = g(y)$ for $y = y_i$, u_j or v_j for any $i = 0, 1, \cdots, m$ or $j = 1, \cdots, m$, and g_1 is linear in between, so g_1 is piecewise linear and continuous, while

$$h_1(x, y) = \begin{cases} h(x, y_{j-1}), & \text{for } y_{j-1} \le y < u_j, \\ h(x, \eta_j), & \text{for } u_j \le y < v_j \\ h(x, y_j), & \text{for } v_j \le y < y_j, \end{cases}$$

$j = 1, \cdots, m$, and $h_1(x, 1) := h(x, 1)$. For fixed x, $F^\kappa(x) \to f^\kappa(x)$ as $u_j \downarrow y_{j-1}$ and $v_j \uparrow y_j$ for each $j = 1, \cdots, m$. Thus for such choices of u_j and v_j, (A.22) also holds with F^κ in place of f^κ.

Now $V_\nu(F^\kappa; [c, d])$ can be bounded above by Theorem A.3 with $g = g_1$ and $h = h_1$. The right side of (A.5) can be bounded as follows: $V_q(g_1; [0, 1]) \leq V_q(g; [0, 1])$ by Lemmas A.9 and A.10, $A_{r,\infty}(h_1; [c, d]; [0, 1]) \leq A_{r,\infty}(h; [c, d], [0, 1])$ and $B_{\infty,p}(h_1; [c, d], [0, 1]) \leq B_{\infty,p}(h; [c, d], [0, 1])$ by Lemma A.9. The proof of Theorem A.8 is complete. $\qquad\square$

Now we are ready for the final step.

Proof of Theorem A.1. Let $h_x(y) := h(x, y)$ for each $x \in [c, d]$. By the (Y_1) definition of the (CY) integral, for each $x \in [c, d]$, we have

$$
\begin{aligned}
\bar{f}^C(x) &= f^C(x) - h_x(a)[g(b) - g(a)] \\
&= (MPS) \int_a^b \left[(h_x)_+^{(b)} - h_x(a)\right] dg_-^{(a)} + G(x) - H(x),
\end{aligned}
\tag{A.23}
$$

where $G(x) := [h_x(b) - h_x(a)]\Delta^- g(b)$ and

$$
H(x) := [\Delta^+ h_x \Delta^+ g](a) + \sum_{(a,b)} \Delta^+ h_x [g_+ - g_-].
\tag{A.24}
$$

Here $\sum_{(a,b)} F$ is the sum of all $F(y)$ for $y \in (a, b)$ provided that the sum is absolutely convergent, so that all but countably many terms are 0, and $\Delta^+ F(y) := F(y+) - F(y)$. Let $\bar{g}(y) := g(y-)$ for $y \in (a, b]$ with $\bar{g}(a) := g(b-)$ and, for each $x \in [c, d]$, let $\bar{h}_x(y) := h_x(y+)$ for $y \in [a, b)$ with $\bar{h}_x(b) := h_x(a+)$. Then $\bar{g}(a) = \bar{g}(b)$ and $\bar{h}_x(a) = \bar{h}_x(b)$ for each $x \in [c, d]$. Since $(h_x)_+^{(b)}$ is right-continuous at a and $g_-^{(a)}$ is left-continuous at b, for each $x \in [c, d]$, it follows that

$$
\begin{aligned}
(MPS) \int_a^b \bar{h}_x \, d\bar{g} &= (MPS) \int_a^b \bar{h}_x \, dg_-^{(a)} - h_x(a)[g_-(b) - g_-^{(a)}(a)] \\
&= (MPS) \int_a^b \left[(h_x)_+^{(b)} - h_x(a)\right] dg_-^{(a)},
\end{aligned}
\tag{A.25}
$$

where all three (MPS) integrals exist whenever one of them does. By Part II, Theorems 3.6 and 3.7 and Propositions 3.15 and 3.17, if an integrand is right-continuous and an integrator is left-continuous then the (RYS) and (MPS) integrals exist whenever one of them does and the two are equal. Writing $\bar{f}(x) := (MPS) \int_a^b \bar{h}_x \, d\bar{g}$, by Theorem A.8, it then follows that

$$
V_\nu(\bar{f}; [c, d]) \leq 4C A_{r,\infty}(h; [c, d], [a, b])^{\frac{r}{\nu}} B_{\infty,p}(h; [c, d], [a, b])^{1-\frac{r}{\nu}} V_q(g; [a, b])
\tag{A.26}
$$

because

$$
V_q(\bar{g}) \leq 2V_q(g), \quad A_{r,\infty}(\bar{h}; [c, d], [a, b]) \leq A_{r,\infty}(h; [c, d], [a, b]),
$$

and $B_{\infty,p}(\bar{h}; [c, d], [a, b]) \leq 2B_{\infty,p}(h; [c, d], [a, b])$.

To bound the ν-variation of G, let $c = x_0 < \cdots < x_n = d$ be a partition of $[c, d]$ and let $\Delta_i^2 h := h(x_i, b) - h(x_i, a) - h(x_{i-1}, b) + h(x_{i-1}, a)$ for $i = 1, \ldots, n$. Then we have

$$\left(\sum_{i=1}^n |\Delta_i^2 h|^\nu \right)^{\frac{1}{\nu}} \le \left(\sum_{i=1}^n |\Delta_i^2 h|^r \right)^{\frac{1}{\nu}} \max_{1 \le i \le n} |\Delta_i^2 h|^{1 - \frac{r}{\nu}}$$

$$\le 2 A_{r,\infty}(h; [c, d], [a, b])^{r/\nu} B_{\infty,p}(h; [c, d], [a, b])^{1 - \frac{r}{\nu}}.$$

Therefore it follows that

$$V_\nu(G; [c, d]) \le 2 A_{r,\infty}(h; [c, d], [a, b])^{\frac{r}{\nu}} B_{\infty,p}(h; [c, d], [a, b])^{1 - \frac{r}{\nu}} \mathfrak{S}_q(g; [a, b]). \quad \text{(A.27)}$$

Let H be the function defined by (A.24) above. We show next that

$$V_\nu(H; [c, d]) \le 2 A_{r,\infty}(h; [c, d], [a, b])^{\frac{r}{\nu}} B_{\infty,p}(h; [c, d], [a, b])^{1 - \frac{r}{\nu}} \mathfrak{S}_q(g; [a, b]). \quad \text{(A.28)}$$

Let $\kappa := \{x_i \colon i = 0, 1, \ldots, n\}$ be a partition of $[c, d]$, let $\Delta_i^{(+)} h := \Delta^+ h(x_i, \cdot) - \Delta^+ h(x_{i-1}, \cdot)$ for $i = 1, \ldots, n$, where $\Delta^+ h(x, y) := h(x, y+) - h(x, y)$, $a \le y < b$, and let $\Delta g(x) := \Delta^+ g_-^{(a)}(x)$ for $a \le x < b$, and $:= \Delta^- g(b)$ for $x = b$. Suppose first that $q \in (1, \infty)$. Let α, β, γ be positive numbers satisfying $1/\alpha + 1/\beta + 1/\gamma = 1$. For each $i = 1, \ldots, n$, writing

$$|\Delta_i^{(+)} h| |\Delta g| = \left(|\Delta_i^{(+)} h|^{\frac{r}{\alpha}} |\Delta g|^{\frac{q}{\alpha}} \right) |\Delta_i^{(+)} h|^{1 - \frac{r}{\alpha}} |\Delta g|^{1 - \frac{q}{\alpha}}$$

and applying Hölder's inequality with exponents α, β, γ, cf. Theorem 12 on p. 24 of Hardy, Littlewood and Polya (1959), or the usual Hölder inequality for α, ζ with $\alpha^{-1} + \zeta^{-1} = 1$, then for $\beta/\zeta, \gamma/\zeta$ with $\beta^{-1} + \gamma^{-1} = \zeta^{-1}$, we get

$$|H(x_i) - H(x_{i-1})| \le \sum_{[a,b)} |\Delta_i^{(+)} h| |\Delta g|$$

$$\le \left(\sum_{[a,b)} |\Delta_i^{(+)} h|^r |\Delta g|^q \right)^{1/\alpha} \left(\sum_{[a,b)} |\Delta_i^{(+)} h|^{\beta(1 - \frac{r}{\alpha})} \right)^{1/\beta} \left(\sum_{[a,b)} |\Delta g|^{\gamma(1 - \frac{q}{\alpha})} \right)^{1/\gamma}. \quad \text{(A.29)}$$

If $q > 1$ we use this bound with $\alpha = \nu$, $1/\beta = 1 - 1/q$ and $1/\gamma = 1/q - 1/\nu$. By (A.2), it follows that for each $i = 1, \ldots, n$, $\sum_{[a,b)} |\Delta_i^{(+)} h|^p \le 2^p B_{\infty,p}^p$ and, for each $y \in [a, b)$,

$$\sum_{i=1}^n |\Delta_i^{(+)} h(y)|^r \le 2^r A_{r,\infty}^r. \quad \text{(A.30)}$$

Since $\nu = \alpha$, $\beta(1 - r/\alpha) = p$ and $\gamma(1 - q/\alpha) = q$, by (A.29), using a change of the order of summation over $i \in \{1, \ldots, n\}$ and $[a, b)$, it follows that

$$V_\nu(H; \kappa) := \left(\sum_{i=1}^n |H(x_i) - H(x_{i-1})|^\nu \right)^{1/\nu}$$

$$\le \left(\sum_{[a,b)} |\Delta g|^q \left(\sum_{i=1}^n |\Delta_i^{(+)} h|^r \right) \right)^{1/\nu} 2^{p/\beta} B_{\infty,p}^{p/\beta} \mathfrak{S}_q(g)^{q/\gamma}$$

$$\le 2^{r/\nu} A_{r,\infty}^{r/\nu} \mathfrak{S}_q(g)^{q/\nu} 2^{p/\beta} B_{\infty,p}^{p/\beta} \mathfrak{S}_q(g)^{q/\gamma} = 2 A_{r,\infty}^{r/\nu} B_{\infty,p}^{1 - \frac{r}{\nu}} \mathfrak{S}_q(g)$$

whenever $q > 1$. If $q = 1$ then applying Hölder's inequality for r, s with $r^{-1} + s^{-1} = 1$, we have

$$|H(x_i) - H(x_{i-1})| \leq \sum_{[a,b)} (|\Delta_i^{(+)}h||\Delta g|^{\frac{1}{r}})|\Delta g|^{\frac{1}{s}} \leq \Big(\sum_{[a,b)} |\Delta_i^{(+)}h|^r |\Delta g|\Big)^{\frac{1}{r}} \mathfrak{S}_1(g)^{\frac{1}{s}}$$

for $i = 1, \ldots, n$. Since $r = \nu$, by (A.30), using a change of the order of summation as above, we get

$$V_\nu(H; \kappa) \leq \Big(\sum_{[a,b)} |\Delta g|\big(\sum_{i=1}^n |\Delta_i^{(+)}h|^r\big)\Big)^{\frac{1}{r}} \mathfrak{S}_1(g)^{\frac{1}{s}} \leq 2A_{r,\infty}\mathfrak{S}_1(g)$$

whenever $q = 1$. Thus (A.28) is proved because κ is an arbitrary partition of $[c, d]$. Collecting the bounds (A.26), (A.27) and (A.28), by (A.23) and (A.25), we get the desired inequality (A.3). The proof of Theorem A.1 is complete.   □

Appendix B. Local p-variation.

Here we treat relations between two local variants of p-variation. Let f be a real-valued function on a closed interval $[a, b]$. The class of all partitions $\kappa = \{a = x_0 < x_1 < \cdots < x_n = b\}$ of $[a, b]$ will be denoted by $Q([a, b])$. For $\lambda, \kappa \in Q([a, b])$, we write $\lambda \supset \kappa$ whenever all points of the partition κ are in the partition λ and let $|\kappa| := \max_{1 \le i \le n}(x_i - x_{i-1})$. Given $0 < p < \infty$ and $\kappa \in Q([a, b])$, let $v_p(f; \kappa) := \sum_{i=1}^{n} |f(x_i) - f(x_{i-1})|^p$. The p-variation of f over $[a, b]$ is defined by $V_p(f) = V_p(f; [a, b]) := \sup\{v_p(f; \kappa)^{1/p}: \kappa \in Q([a, b])\}$. To define two other related quantities, for $\kappa \in Q([a, b])$, let $v_p^*(f; \kappa) := \sup\{v_p(f; \lambda): \lambda \in Q([a, b]), \lambda \supset \kappa\}$ and, for $\epsilon > 0$, let $v_p(f; \epsilon) := \sup\{v_p(f; \kappa): |\kappa| \le \epsilon\}$. Let

$$V_p^*(f) := \inf_{\kappa \in Q([a,b])} v_p^*(f; \kappa)^{1/p} \quad \text{and} \quad \overline{V}_p(f) := \lim_{\epsilon \downarrow 0} v_p(f; \epsilon)^{1/p}. \qquad (B.1)$$

Since $|\lambda| \le |\kappa|$ for $\lambda \supset \kappa$, we have $v_p^*(f; \kappa) \le v_p(f; \epsilon)$ for each $\kappa \in Q([a, b])$ such that $|\kappa| \le \epsilon$. This implies the first inequality in

$$V_p^*(f) \le \overline{V}_p(f) \le V_p(f) \qquad \forall f, \qquad (B.2)$$

while the second one is clear. Strict inequality between $V_p^*(f)$ and $\overline{V}_p(f)$ may hold if a function f has jumps on both sides of the same point and $p > 1$. For example, if $f(x) := 0$ for $x \in [0, 1/2)$, $f(1/2) := 1/2$ and $f(x) := 1$ for $x \in (1/2, 1]$ then $V_p^*(f) = 2^{1-p} < 1 = \overline{V}_p(f)$ whenever $p > 1$. However, by the following statement, jumps on both sides of a point are the only reason the two quantities in (B.1) can differ.

B.1 Lemma. If $0 < p \le 1$ or if, for each $x \in (a, b)$, either $\Delta^- f(x) = 0$ or $\Delta^+ f(x) = 0$, then $V_p^*(f) = \overline{V}_p(f)$.

Proof. If $0 < p \le 1$, we can apply Propositions 2.12 and 2.13 from Part II. For $1 < p < \infty$, by (B.2), it is enough to prove that $V_p^*(f) \ge \overline{V}_p(f)$. Suppose $V_p^*(f) < \overline{V}_p(f) =: C$. Then there exist $\kappa = \{x_i: i = 0, \ldots, n\} \in Q([a, b])$ and a sequence $\{\kappa_m: m \ge 1\} \subset Q([a, b])$ such that $|\kappa_m| \downarrow 0$ and

$$v_p^*(f; \kappa) < C^p \le v_p(f; \kappa_m) \qquad \text{for } m = 1, 2, \cdots. \qquad (B.3)$$

Since f has only one-sided jumps, $f(y_i^m) \to f(x_i)$ for each $i = 1, \ldots, n-1$ and some $y_i^m \in \kappa_m$ chosen in such a way that no $y \in \kappa_m$ is between y_i^m and x_i. Let κ_m' be the same as κ_m except that y_i^m is replaced by x_i. Then $|v_p(f; \kappa_m) - v_p(f; \kappa_m')| \to 0$ as $m \to \infty$. Since $\kappa_m' \supset \kappa$, this contradicts (B.3) and, hence, completes the proof. $\square$

The quantity $\overline{V}_p(f)$ was introduced by Wiener (1924) and later used by many authors. The quantity $V_p^*(f)$ first appeared in Love and Young (1938, p. 29) and is extensively used in Part II. For $1 \le p < \infty$ there exist continuous $f \in W_p([a, b])$ with $V_p^*(f) > 0$, e.g. Bruneau (1974).

Appendix C. Necessary conditions on F
for differentiability of the quantile operator at F.

Let $F\colon [a,b] \mapsto \mathbb{R}$, and for $a \le x \le b$ set $\overline{F}(x) := \max(F(x), \limsup_{u \to x} F(u))$. Then $\overline{F}$ is upper semicontinuous, i.e. for $a \le x \le b$, $\overline{F}(x) \ge \limsup_{u \to x} \overline{F}(u)$. Let $F^{\leftarrow} := F^{\leftarrow}_{[a,b]}$ and $\overline{F}^{\leftarrow} := (\overline{F})^{\leftarrow}_{[a,b]}$ as defined in Section 3.

C.1 Proposition. (a) For all $y \in \mathbb{R}$, $F^{\leftarrow}(y-) \le \overline{F}^{\leftarrow}(y) \le F^{\leftarrow}(y)$.
(b) $F^{\leftarrow}(y) = \overline{F}^{\leftarrow}(y)$ except for at most countably many y where $F^{\leftarrow}$ and $\overline{F}^{\leftarrow}$ have jumps.

Proof. (a) is easy to check, and implies (b). □

If f is continuous then for any F, $\overline{F+f} \equiv \overline{F} + f$. So, in considering differentiability of $f \mapsto (F+f)^{\leftarrow}$ into L^p spaces along continuous f we can assume F is upper semicontinuous.

C.2 Proposition. If F is upper semicontinuous and $f \mapsto (F+f)^{\leftarrow}$ is differentiable at $f = 0$ into L^p for some $p > 1$ along constant functions $f \equiv t \to 0$ then F is strictly increasing.

Proof. On any interval $[a,d]$ with $a \le d \le b$, by upper semicontinuity, F attains its maximum at some point c. Suppose there is such a $c < d$. Let $\eta := F(c) \ge F(x)$ for $a \le x \le d$. Then $(F+t)^{\leftarrow}(y) \ge d$ if $y > \eta+t$, while $(F+t)^{\leftarrow}(y) \le c$ if $y \le \eta+t$. Then $\|(F+t)^{\leftarrow} - F^{\leftarrow}\|_p \ge [t(d-c)^p]^{1/p} = t^{1/p}(d-c)$. Thus for $p > 1$, $t \mapsto (F+t)^{\leftarrow}$ is not Lipschitz into L^p, so not differentiable at $t = 0$, a contradiction. So $c = d$ and the conclusion follows. □

A strictly increasing function is upper semicontinuous if and only if it is right-continuous.

Appendix D. A special case of compact differentiability.

This Appendix will prove the following fact, stated but apparently not fully proved by Reeds (1976) and Fernholz (1983). The seemingly very special function F here can be applied rather generally by way of Theorem 6.3.

D.1 Proposition. Let $F(x) := \max(0, \min(x, 1))$ for all $x \in \mathbb{R}$ and $H(x) := x$ for $0 \le x \le 1$. Then for $1 \le p < \infty$, $h \mapsto F \circ (H + h)$ is compactly differentiable from $L^p[0, 1]$ into itself.

Proof. For $h \in L^p[0, 1]$ and $t \in \mathbb{R}$ we have

$$F \circ (H + th) = F \circ H + (F' \circ H) \cdot th + R_2 \equiv H + th + R_2(t, h)$$

where the remainder

$$R_2(t, h)(x) = \begin{cases} 0, & \text{if } x + th(x) \in [0, 1] \\ -x - th(x), & \text{if } x + th(x) < 0 \\ 1 - x - th(x), & \text{if } x + th(x) > 1 \end{cases}$$

except that $R_2(t, h)(x)$ is undefined for $x = 0$ or 1. Let K be compact in $L^p[0, 1]$. Claim 1: the functions $|f|^p$ for $f \in K$ are uniformly integrable. If not, then for some $\varepsilon > 0$ and sequence $\{f_n\} \subset K$, there are sets $A(n) \subset [0, 1]$ with $\lambda(A_n) < 1/n$ and $\int_{A(n)} |f_n|^p dx > \varepsilon$. Then there is a subsequence $f_{n(k)}$ converging to some $f \in L^p$, so as $k \to \infty$, $\int_{A(n(k))} |f_{n(k)} - f|^p dx \to 0$, and $\int_{A(n(k))} |f|^p dx \to 0$ by dominated convergence, which gives a contradiction (taking pth roots), proving Claim 1.

Let $j^- := -\min(j, 0)$ for any real-valued function j. Claim 2: $\|(H + th)^-\|_p = o(|t|)$ as $t \to 0$, uniformly for $h \in K$. To prove this is equivalent to showing that $\|(\frac{H}{t} + h)^-\|_p \to 0$ as $t \downarrow 0$, uniformly for $h \in K$, since $K \cup -K$ is compact. Then $H/t > 0$ and increases to $+\infty$ as $t \downarrow 0$. We have $(\frac{H}{t} + h)^- \downarrow 0$ and $(\frac{H}{t} + h)^- \le h^-$ for all t, so the functions $((\frac{H}{t} + h)^-)^p$ are uniformly integrable by Claim 1. If Claim 2 fails there are $t(n) \downarrow 0$ and $h_n \in K$ such that $\|(\frac{H}{t(n)} + h_n)^-\|_p$ doesn't go to 0 as $n \to \infty$. Let $s(n) := t(n)^{1/2}$. We then have

$$\int_0^{s(n)} \left(\left(\frac{x}{t(n)} + h_n(x) \right)^- \right)^p dx \le \int_0^{s(n)} |h_n(x)|^p dx \to 0$$

by uniform integrability, while

$$\int_{s(n)}^1 \left(\left(\frac{x}{t(n)} + h_n(x) \right)^- \right)^p dx \le \int_{|h_n| \ge 1/s(n)} |h_n(x)|^p dx \to 0$$

by another form of uniform integrability. This contradiction proves Claim 2.

Dealing with the case $x + th(x) > 1$ symmetrically, interchanging x and $1 - x$, the Proposition follows. $\square$

Proposition D.1 extends, with a similar proof, to the case where on $[0, 1]$, H is an increasing diffeomorphism from $[0, 1]$ onto itself. The proof of Proposition D.1 yields $\mathcal{C}$-differentiability of the composition operator where $\mathcal{C}$ is the class of sets C of measurable functions f such that $\{|f|^p : f \in C\}$ is uniformly integrable. Norvaiša (1997a) used this observation to prove a central limit theorem for L-statistics.

Appendix E. Statistical applications
of compact and tangential differentiability

Recall the notion of compact differentiability tangential to a subspace defined in Section 8 before Proposition 8.3. As noted before Proposition 1.2, there are cases where tangential compact differentiability holds and neither "compact" nor "tangential" can be deleted (for given norms).

Van der Vaart (1991) shows that under suitable conditions, tangentially compactly differentiable operators preserve, in the following sense, efficiency of estimators. Let κ be a mapping from a set of probability measures $\mathcal{P}$ into a normed space X and let T be a mapping from X into another normed space Y. Given an efficient sequence $\{\theta_n: n \geq 1\}$ of estimators for $\kappa(P)$, $P \in \mathcal{P}$, one needs to estimate the value $T(\kappa(P))$. The notion of efficiency presupposes the existence of a certain subspace H of X. Suppose T is compactly differentiable at $\kappa(P)$ tangentially to H. Then under the conditions of Theorem 3.1 of van der Vaart (1991), $\{T \circ \theta_n: n \geq 1\}$ is an efficient sequence of estimators for $T(\kappa(P))$.

Van der Vaart and Wellner (1996) show that validity of the bootstrap is preserved by tangential compact differentiability. Given a distribution function F, let F_n and $\check{F}_n$ be the empirical distribution function for F and the bootstrap empirical distribution function for F, respectively. Suppose that F_n and $\hat{F}_n$ are weakly consistent estimators for F on a normed space X in the following sense: the sequence $\{\sqrt{n}(F_n - F): n \geq 1\}$ converges in law and the sequence $\{\sqrt{n}(\hat{F}_n - F_n): n \geq 1\}$ converges conditionally in law to a limit G with almost all sample functions in a subspace $H \subset X$. By Theorem 3.9.11 of Van der Vaart and Wellner (1996), if T is a mapping from X into another normed space Y, compactly differentiable at F tangentially to H, then $T(F_n)$ and $T(\hat{F}_n)$ are also weakly consistent estimators for $T(F)$ on Y.

Gill, van der Laan and Wellner (1995) prove a compact differentiability property for a product integral operator in two-dimensional time.

For some stochastic processes on $\mathbb{R}$ other than empirical processes of i.i.d. variables, central limit theorems have been proved in the supremum norm but not, at this writing, in p-variation norms. For such cases, compact (possibly tangential) differentiability of some operators in the supremum norm is the available, applicable mode of differentiability, e.g. Andersen et al. (1993).

Appendix F. The Ward–Perron–Stieltjes and gauge integrals

Ward (1936) defined a Perron–Stieltjes integral which includes both the Lebesgue-Stieltjes and Moore–Pollard–Stieltjes integrals. Given two real-valued functions F, G on $[a,b]$, say M is a *major function* of F with respect to G if $M(a) = 0$, M has finite values on $[a,b]$, and for each $x \in [a,b]$ there exists $\delta(x) > 0$ such that

$$M(z) \geq M(x) + F(x)[G(z) - G(x)] \quad \text{if} \quad x \leq z \leq \min\{b, x + \delta(x)\},$$

$$M(z) \leq M(x) + F(x)[G(z) - G(x)] \quad \text{if} \quad \max\{a, x - \delta(x)\} \leq z \leq x.$$

Let $\mathcal{U}(F,G)$ be the class of all major functions of F with respect to G, and let

$$(UPS) \int_a^b F \, dG := \begin{cases} \inf\{M(b): M \in \mathcal{U}(F,G)\} & \text{if } \mathcal{U}(F,G) \neq \emptyset, \\ +\infty & \text{if } \mathcal{U}(F,G) = \emptyset. \end{cases}$$

A function m is a *minor function* of F with respect to G if $-m \in \mathcal{U}(-F,G)$. Let $\mathcal{L}(F,G)$ be the class of all minor functions of F with respect to G, and let

$$(LPS) \int_a^b F \, dG := \begin{cases} \sup\{m(b): m \in \mathcal{L}(F,G)\} & \text{if } \mathcal{L}(F,G) \neq \emptyset, \\ -\infty & \text{if } \mathcal{L}(F,G) = \emptyset. \end{cases}$$

If $(UPS)\int_a^b F \, dG = (LPS)\int_a^b F \, dG$ is finite then denote the common value by $(WPS)\int_a^b F \, dG$ and call it the Ward–Perron–Stieltjes integral, or the (WPS)-integral. By Theorem 5 of Ward (1936), if $(MPS)\int_a^b F \, dG$ exists then so does $(WPS)\int_a^b F \, dG$, with the same value. Ward (1936) stated and Saks (1937, Theorem VI.8.1) gave a proof of the fact that $(WPS)\int_a^b F \, dG$ is defined provided the corresponding Lebesgue-Stieltjes integral is defined. Then $(WPS)\int_a^b F \, dG = (LS)\int_a^b F \, dG$ as defined after Theorem 4.1.

Kurzweil (1957, Section 1.2) suggested an equivalent definition of the (WPS) integral based on an extension of the limit as the mesh of partitions tends to zero in the definition of the (RS) integral. For a partition $a = x_0 < x_1 < \cdots < x_n = b$, points $y_i \in [x_{i-1}, x_i]$, $i = 1, \ldots, n$, are called *tags* and the set $\{(y_i, [x_{i-1}, x_i]): i = 1, \ldots, n\}$ is called a *tagged partition* of $[a,b]$. A *gauge function* is any function with strictly positive values. Given a gauge function $\delta(\cdot)$ on $[a,b]$, a tagged partition $\{(y_i, [x_{i-1}, x_i]): i = 1, \ldots, n\}$ is δ-*fine* if $y_i - \delta(y_i) \leq x_{i-1} \leq y_i \leq x_i \leq y_i + \delta(y_i)$ for $i = 1, \ldots, n$. The *gauge* or *Henstock-Kurzweil* integral $(HK)\int_a^b F \, dG$ is defined as the number I, whenever it exists, such that for each $\epsilon > 0$ there is a gauge function $\delta(\cdot)$ on $[a,b]$ such that $|\sum_{i=1}^n F(y_i)[G(x_i) - G(x_{i-1})] - I| < \epsilon$ for each δ-fine tagged partition $\{(y_i, [x_{i-1}, x_i]): i = 1, \ldots, n\}$ of $[a,b]$. Kurzweil (1957, Theorem 1.2.1) proved the following statement:

F.1 Theorem. The integral $(WPS)\int_a^b F \, dG$ exists if and only if $(HK)\int_a^b F \, dG$ exists, and then their values are equal.

The integral defined as the (HK) integral, but without the inequalities $x_{i-1} \leq y_i \leq x_i$, is called the McShane integral (see Pfeffer, 1993). Kurzweil (1957, 1958, 1959), Henstock (1961, 1963, 1993) and McShane (1969) consider extended integrals for functions $U(\cdot,\cdot)$ of two variables, the second and third authors also on general spaces X, where $(HK)\int_a^b F\,dG$ is the special case $U(x,y) \equiv F(y)G(x)$ and $X = [a,b]$.

For notational convenience let $\mathcal{W}_\infty([a,b]) := \mathcal{W}_\infty^*([a,b]) := \mathcal{R}([a,b])$ be the class of all regulated functions on $[a,b]$. Schwabik (1973) proves the following statement for $G \in \mathcal{W}_1([a,b])$ and F bounded.

F.2 Theorem. Let $1/p + 1/q \geq 1$ with $1 \leq p < \infty$, and let either $F \in \mathcal{W}_p^*([a,b])$ and $G \in \mathcal{W}_q([a,b])$, or $F \in \mathcal{W}_p([a,b])$ and $G \in \mathcal{W}_q^*([a,b])$. If $(RYS)\int_a^b F\,dG$ exists then so does $(WPS)\int_a^b F\,dG$, and the two integrals are equal.

Proof. Since $\mathcal{W}_p$ increases with p, we can assume $1/p + 1/q = 1$. We prove the statement when $F \in \mathcal{W}_p^*$ for $1 < p < \infty$. The proof for the other cases is similar and is omitted. Let $\epsilon > 0$. By definition of the (RYS) integral, Proposition 2.23 of Part II, and since F, G are regulated, there exist $\lambda = \{z_j: j = 0,\dots,m\} \in Q([a,b])$ and numbers v_{j-1}, u_j, $j = 1,\dots,m$, such that $a = z_0 < v_0 < u_1 < z_1 < \cdots < z_{m-1} < v_{m-1} < u_m < z_m = b$ and the following properties (A), (B), (C) hold:

(A) for any refinement $\kappa = \{\tilde{x}_i: i = 0,\dots,n\}$ of λ and for any $\tilde{y}_i \in (\tilde{x}_{i-1},\tilde{x}_i)$, $i = 1,\dots,n$,

$$\left| \sum_{i=1}^n \Phi(\tilde{y}_i;[\tilde{x}_{i-1},\tilde{x}_i]) - (RYS)\int_a^b F\,dG \right| < \epsilon.$$

Here and below Φ denotes the quantity (4.6) in the sum defining the (RYS) integral.

(B) $\sum_{j=1}^m v_p(F;[z_{j-1}+,z_j-]) < (\epsilon/V_q(G))^p$;

(C) for any $\{r_{j-1},\eta_{j-1},p_j,\xi_j: j = 1,\dots,m\}$ such that $z_0 < r_0 < \eta_0 \leq v_0 < \cdots < u_j \leq \xi_j < p_j < z_j < r_j < \eta_j \leq v_j < \cdots < u_m \leq \xi_m < p_m < z_m$,

$$\left|\Phi(r_0;[z_0,\eta_0]) - F(z_0)[G(\eta_0) - G(z_0)]\right| + \left|\Phi(p_m;[\xi_m,z_m]) - F(z_m)[G(z_m) - G(\xi_m)]\right|$$

$$+ \sum_{j=1}^{m-1} \left|\Phi(p_j;[\xi_j,z_j]) + \Phi(r_j;[z_j,\eta_j]) - F(z_j)[G(\eta_j) - G(\xi_j)]\right| < \epsilon.$$

Define the gauge function $\delta(\cdot)$ on $[a,b]$ as follows: $\delta(a) := v_0 - a$, $\delta(z_j) := \min\{v_j - z_j, z_j - u_j\}$, $j = 1,\dots,m-1$, $\delta(b) := b - u_m$ and $\delta(y) := \min\{z_j - y, y - z_{j-1}\}/2$ if $y \in (z_{j-1},z_j)$ for some $j = 1,\dots,m$. Let $\{(y_i,[x_{i-1},x_i]): i = 1,\dots,n\}$ be a δ-fine tagged partition of $[a,b]$. By definition of $\delta(\cdot)$, all elements of λ must be among the tags $\{y_i: i = 1,\dots,n\}$; that is, for each $j = 0,\dots,m$, there exists $i(j) \in \{1,\dots,n\}$ such that $y_{i(j)} = z_j$. In particular, $i(0) = 1$, $y_1 = x_0 = z_0 = a$ and $i(m) = n$, $y_n = x_n = z_m = b$. By joining adjacent intervals if necessary, we can and do assume that $y_{i(j)} \in (x_{i(j)-1},x_{i(j)})$ for $j = 1,\dots,m-1$. Let $\kappa := \{x_i: i = 0,\dots,n\} \cup \{z_j: j = 1,\dots,m-1\}$. Then κ is a refinement of λ. For each $i \in I := \{2,\dots,n-1\}\setminus\{i(j): j = 1,\dots,m-1\}$, let $y_i' := y_i$ if $y_i \in (x_{i-1},x_i)$ and $y_i' := (x_{i-1} + x_i)/2$ otherwise. Also, for each $j = 1,\dots,m$, let $r_{j-1} := (z_{j-1} + \eta_{j-1})/2$, $p_j := (\xi_j + z_j)/2$, where $\eta_{j-1} := x_{i(j-1)}$ and $\xi_j := x_{i(j)-1}$ for $j = 1,\dots,m$. By the definition of the gauge function $\delta(\cdot)$, $\eta_{j-1} \leq v_{j-1}$ and $\xi_j \geq u_j$ for $j = 1,\dots,m$. Therefore by property (C), we have

$$\left|\Phi(r_0;[z_0,\eta_0]) + \Phi(p_m;[\xi_m,z_m]) + \sum_{j=1}^{m-1}\left[\Phi(p_j;[\xi_j,z_j]) + \Phi(r_j;[z_j,\eta_j])\right]\right.$$

$$+ \sum_{i \in I} \Phi(y_i'; [x_{i-1}, x_i]) - \sum_{i=1}^{n} F(y_i)[G(x_i) - G(x_{i-1})] \Big| < \epsilon + \sum_{j=1}^{m} \sum_{i \in I(j)} |D_i|,$$

where $D_i := \Phi(y_i'; [x_{i-1}, x_i]) - F(y_i)[G(x_i) - G(x_{i-1})]$ for $i \in I(j) := \{i(j-1) + 1, \dots, i(j) - 1\}$ and $j = 1, \dots, m$. If $y_i \in (x_{i-1}, x_i)$ then

$$D_i = [F(x_{i-1}) - F(y_i)]\Delta^+ G(x_{i-1}) + [F(x_i) - F(y_i)]\Delta^- G(x_i).$$

If $y_i = x_{i-1}$ then

$$D_i = [F(y_i') - F(x_{i-1})][G(x_i-) - G(x_{i-1}+)] + [F(x_i) - F(x_{i-1})]\Delta^- G(x_i).$$

If $y_i = x_i$ then

$$D_i = [F(y_i') - F(x_i)][G(x_i-) - G(x_{i-1}+)] + [F(x_{i-1}) - F(x_i)]\Delta^+ G(x_{i-1}).$$

By property (B) and Hölder's inequality, it then follows that

$$\sum_{j=1}^{m} \sum_{i \in I(j)} |D_i| \le 2 \sum_{j=1}^{m} V_p(F; [z_{j-1}+, z_j-]) V_q(G; [z_{j-1}, z_j])$$

$$\le 2 \Big(\sum_{j=1}^{m} v_p(F; [z_{j-1}+, z_j-]) \Big)^{1/p} V_q(G) < 2\epsilon.$$

By property (A), we then have

$$\Big| \sum_{i=1}^{n} F(y_i)[G(x_i) - G(x_{i-1})] - (RYS) \int_a^b F \, dG \Big| < 4\epsilon$$

for any δ-fine tagged partition $\{(y_i, [x_{i-1}, x_i]): i = 1, \dots, n\}$ of $[a, b]$. Since $\epsilon > 0$ is arbitrary, Theorem F.2 follows from Theorem F.1. $\quad\square$

Next, we say that the Young sums for F and G are unbounded on any subinterval of $[a, b]$ if, for all $a \le c < d \le b$, $\sup\{|\sum_{i=1}^{n} \Phi(y_i; [x_{i-1}, x_i])|\} = +\infty$, where the supremum is taken over all partitions $\{x_i: i = 0, \dots, n\}$ of $[c, d]$ and all $y_i \in (x_{i-1}, x_i)$, $i = 1, \dots, n$. Then the sums based on partitions which are refinements of any given partition are also unbounded.

F.3 Proposition. Given F left- or right-continuous on (a, b) and $G \in \mathcal{R}([a, b])$, suppose that the Young sums for F and G are unbounded on any subinterval $[c, d]$ of $[a, b]$. Then $(WPS) \int_c^d F \, dG$ doesn't exist for any such c, d.

Proof. It will suffice to show that $(HK) \int_a^b F \, dG$ doesn't exist. Suppose it does. Let $\delta(\cdot)$ be a gauge function and I a number such that for any δ-fine tagged partition $\{(y_i, [x_{i-1}, x_i]): i = 1, \dots, n\}$ of $[a, b]$,

$$\Big| \sum_{i=1}^{n} F(y_i)[G(x_i) - G(x_{i-1})] - I \Big| < 1. \tag{F.1}$$

By the category theorem (e.g. Theorem 2.5.2 in Dudley, 1993), there is an interval $[c, d]$ and a positive integer m such that $\delta(x) > 1/m$ for all x in a dense set S in $[c, d]$. We can assume that $0 < d - c < \epsilon$ and that $c, d \in S$. Thus every tagged partition $\{(\eta_j, [\xi_{j-1}, \xi_j]): j = 1, \ldots, k\}$ of $[c, d]$ with $\eta_j \in S$ for $j = 1, \ldots, k$ is δ-fine. Take a δ-fine tagged partition as in (F.1), where we can take some $y_i = c < y_l = d$, and also take $x_i = c$ and $x_l = d$, as follows. If initially $y_i = c < x_i$, then let $x'_r := x_r$ and $y'_r := y_r$ for $r = 0, 1, \ldots, i - 1$. Let $x'_i := y'_i := c$ and for $r = i, \ldots, n$, let $y'_{r+1} := y_r$ and $x'_{r+1} := x_r$. Then $\{(y'_i, [x'_{i-1}, x'_i]): i = 1, \ldots, n + 1\}$ is a δ-fine tagged partition of $[a, b]$ giving the same sum as in (F.1). We can take d as some x_l likewise. Let κ be the resulting δ-fine tagged partition of $[a, b]$. We show next that the part of κ corresponding to $[c, d]$ can be replaced by a tagged partition of $[c, d]$ such that the corresponding Riemann-Stieltjes sum is arbitrarily large and the resulting tagged partition of $[a, b]$ is δ-fine.

For any $M < \infty$, by assumption, there is a tagged partition $\{(v_j, [u_{j-1}, u_j]): j = 1, \ldots, k\}$ of $[c, d]$ such that $|\sum_{j=1}^k \Phi(v_j; [u_{j-1}, u_j])| > M$. Since G is regulated, there exist $t_j < u_j$, $j = 1, \ldots, k$, and $w_j > u_j$, $j = 0, \ldots, k - 1$, such that $u_{j-1} < w_{j-1} < v_j < t_j < u_j$ for $j = 1, \ldots, k$ and

$$|F(c)[G(w_0) - G(c)] + \sum_{j=1}^{k-1} F(u_j)[G(w_j) - G(t_j)] + \sum_{j=1}^k F(v_j)[G(t_j) - G(w_{j-1})]$$

$$+ F(d)[G(d) - G(t_k)]| > M.$$

Now since F is right- or left-continuous, and S is dense, we can replace v_j by some $v'_j \in S$ as close to v_j as desired, making a small change in the sum. Likewise we can replace u_j by some $u'_j \in S$, where $w_{j-1} < v'_j < t_j < u'_j < w_j$ for each $j = 1, \ldots, k - 1$. Recall that $c, d \in S$; the endpoints u_0, u_k are not replaced. We thus obtain a Riemann-Stieltjes sum Σ for a δ-fine tagged partition ζ of $[c, d]$ with $|\Sigma| > M$. Then joining ζ with the δ-fine tagged partitions of $[a, c]$ (if $a < c$) and $[d, b]$ (if $d < b$) given by the fixed tagged partition κ, we get unbounded Riemann-Stieltjes sums for δ-fine tagged partitions of $[a, b]$, contradicting (F.1). $\qquad \square$

The following pair of lacunary Fourier series provides an example of functions satisfying the conditions of the preceding proposition. For $0 \leq x \leq 1$, an integer $m \geq 2$ and $1 < p, q < \infty$, let

$$S_p(x) := \sum_{k=1}^{\infty} m^{-k/p} \sin(m^{k+1} \pi x) \quad \text{and} \quad C_q(x) := \sum_{k=1}^{\infty} m^{-k/q} \cos(m^{k+1} \pi x).$$

Then $|S_p(x+h) - S_p(x)| \leq 8\pi h^{1/p}$ and $|C_q(x+h) - C_q(x)| \leq 8\pi h^{1/q}$ for $0 \leq h < 1$ and $0 \leq x \leq 1 - h$ as shown by Hardy (1916, pp. 310, 311), see Zygmund (1959, Theorem II.4.9). On the other hand, the Riemann-Stieltjes sums $\sum_{i=1}^{m^n} S_p(im^{-n})[C_q(im^{-n}) - C_q((i - 1)m^{-n})]$ increase to infinity with $n \to \infty$ provided $1/p + 1/q = 1$ (see pp. 99–102 in Leśniewicz and Orlicz, 1973). For $N = 1, 2, \ldots$, let S_p^N and C_q^N be the partial sums of the first N terms of the series S_p and C_q, respectively, and let $R_p^N := S_p - S_p^N$ and $R_q^N := C_q - C_q^N$. Since S_p^N and C_q^N are smooth (of bounded variation), while R_p^N and R_q^N are of period m^{-N}, the Riemann-Stieltjes sums for

S_p and C_q are unbounded on any subinterval of $[0, 1]$ and $1/p + 1/q = 1$. Due to continuity of S_p and C_q so are the Young sums. It follows from Proposition F.3 that $(WPS) \int_0^1 S_p \, dC_q$ also does not exist.

In Schwabik (1973), examples of pairs (F_i, G_i), $i = 1, 2$, are given, where each G_i is of bounded variation and each F_i is unbounded, such that $(WPS) \int_a^b F_1 \, dG_1$ exists while $(RYS) \int_a^b F_1 \, dG_1$ does not, and $(RYS) \int_a^b F_2 \, dG_2$ exists while $(WPS) \int_a^b F_2 \, dG_2$ does not. Thus neither the (WPS) nor the (Y) integral includes the other. Further examples will be given to show that either the (WPS) or the (CY) integral can exist when the other does not, even when both F and G are regulated, and hence bounded functions.

For a set E of real numbers, let $c_0(E)$ be the set of all functions $G \colon E \mapsto \mathbb{R}$ for which exists a set $\{\xi_i \colon i \geq 1\} \subset E$ of distinct numbers such that $G(\xi_i) \to 0$ as $i \to \infty$ and $G(x) = 0$ if $x \in E \setminus \{\xi_i \colon i \geq 1\}$. Then

$$c_0(E) = \{\sum_i c_i 1_{\{\xi_i\}} \colon \xi_i \neq \xi_j \text{ for } i \neq j, \, \xi_i \in E, \, c_i \to 0 \text{ as } i \to \infty\}.$$

If $G \in c_0(J)$ for an open interval J then $G \in \mathcal{R}(J)$ with $G_+ \equiv G_- \equiv 0$. If $G \in c_0((a, b))$ then $G_+(a) = G_-(b) = 0$. Thus defining $G(a) := G(b) := 0$, we have $G_+^{(b)} \equiv G_-^{(a)} \equiv 0$ on $[a, b]$.

F.4 Proposition. For any $G \in c_0((a, b))$ and $F \in \mathcal{R}([a, b])$, $(RYS) \int_a^b F \, dG = (CY) \int_a^b F \, dG = 0$.

Proof. This follows by definitions of the integrals because $G_-^{(a)} \equiv G_+^{(b)} \equiv 0$ and $\Delta^{\pm} G \equiv 0 = (\Delta^+ G)(a) = (\Delta^- G)(b)$. $\qquad\square$

We will show in Proposition F.6 that the analogous statement for the (WPS) integral doesn't hold.

F.5 Lemma. For each $G \in c_0((a, b))$ there exists a right-continuous $F \in \mathcal{R}([a, b])$ such that $\Delta^{\pm} F \equiv G$ on (a, b).

Proof. Let $G = \sum_i c_i 1_{\{\xi_i\}}$. For each $i \geq 1$, let $f_i(x) := 0$ if $x \in [a, \xi_i] \cup [\xi_i + \delta_i, b]$ for some $\delta_i > 0$, let $f_i(\xi_i) := c_i$ and let f_i be linear on $[\xi_i, \xi_i + \delta_i]$. If $\delta_i \to 0$ fast enough as $i \to \infty$, $F := \sum_i f_i$ converges uniformly on $[a, b]$ and F has the stated property. $\qquad\square$

F.6 Proposition. There exist $G \in c_0((0, 1))$ and $F \in \mathcal{R}([0, 1])$ such that $(WPS) \int_0^1 F \, dG$ is undefined.

Proof. For k odd, $k = 1, 3, \ldots, 2^n - 1$, $n = 1, 2, \ldots$, let $G(k/2^n) := 1/2^{n/3}$ and $G = 0$ elsewhere. Then $G \in c_0((0, 1))$. By Lemma F.5, take $F \in \mathcal{R}([0, 1])$ such that $\Delta^{\pm} F(k/2^n) = 1/2^{n/3}$ for $k = 1, 3, \ldots, 2^n - 1$ and $n = 1, 2, \ldots$.

Let $\delta(\cdot)$ be any strictly positive function on $[0, 1]$. As in the proof of Proposition F.3, there is an interval J of length $\epsilon > 0$ such that $\delta(y) > \epsilon$ for a dense set S of y in J, and the endpoints of J are not dyadic rationals. So, all tagged partitions of J with tags $y_i \in S$ are δ-fine. Take any fixed δ-fine tagged partition $\kappa = \{(v_i, [u_{i-1}, u_i]) \colon i = 1, \ldots, n\}$ of $[0, 1]$ with the endpoints of J among the u_i. Consider other tagged partitions of $[0, 1]$ which differ only inside of J. For any finite set M of binary rationals $k/2^n$ in J, let $\zeta = \{(y_i, [x_{i-1}, x_i]): i = 1, \ldots, n\}$ be a tagged partition of J such that $M \subset \{x_i: i = 1, \ldots, n-1\}$; if $x_i \in M$, $x_l \in M$, $i \neq l$, then $|i - l| \geq 2$; if $x_r \notin M$, then $G(x_r) = 0$ (x_r is not a dyadic rational). For each $x_i \in M$, choose $y_i \in S \cap (x_{i-1}, x_i)$ and $y_{i+1} \in S \cap (x_i, x_{i+1})$ close enough to x_i so that $|[F(y_{i+1}) - F(y_i) - \Delta^{\pm} F(x_i)]G(x_i)| < 1/|M|$ where $|M| = \operatorname{card}(M)$. Then

$$\left| -\sum_{i=1}^{n} F(y_i)[G(x_i) - G(x_{i-1})] - \sum_{x_i \in M} \Delta^{\pm} F(x_i) G(x_i) \right| < 1.$$

Since the latter sums are unbounded as M increases, so are the former sums, and $(WPS) \int_0^1 F \, dG$ doesn't exist by Theorem F.1. $\square$

We will show next that $(WPS) \int_0^1 F \, dF$ may exist while $(CY) \int_0^1 F \, dF$ doesn't.

F.7 Theorem. If $F: [0,1] \mapsto \mathbb{R}$ is continuous and everywhere differentiable on $(0,1)$ then $(WPS) \int_0^1 F'(x) \, dx = F(1) - F(0)$.

Proof. For $G(x) \equiv x$ the (WPS) integral $\int_0^1 F' \, dG$ reduces to the Perron integral for which the theorem is known (e.g. pp. 201, 202 in Saks, 1937). $\square$

F.8 Proposition. For F as in the last theorem, $(WPS) \int_0^1 F \, dF$ exists and equals $[F^2(1) - F^2(0)]/2$.

Proof. We can assume that $|F(x)| \leq 1$ for all $0 \leq x \leq 1$. The function $F^2/2$ is everywhere differentiable with the derivative FF'. Let $\epsilon > 0$. By Theorem F.7 and Theorem F.1, there is a gauge function $\delta_1(\cdot): [0,1] \mapsto (0,1)$ such that for any δ_1-fine tagged partition $\{(y_i, [x_{i-1}, x_i]): i = 1, \ldots, n\}$,

$$\left| \sum_{i=1}^{n} (FF')(y_i)(x_i - x_{i-1}) - [F^2(1) - F^2(0)]/2 \right| < \epsilon/2.$$

For each $y \in [0,1]$ there is a $\delta_2(y) > 0$ such that $|F(y+h) - F(y) - hF'(y)| \leq \epsilon|h|/4$ for $h \in \mathbb{R}$ such that $\max\{0, y - \delta_2(y)\} \leq y + h \leq \min\{1, y + \delta_2(y)\}$. Let $\delta := \min\{\delta_1, \delta_2\}$. Then for any δ-fine tagged partition $\{(y_i, [x_{i-1}, x_i]): i = 1, \ldots, n\}$ we have

$$S_n := \sum_{i=1}^{n} F(y_i)[F(x_i) - (x_{i-1})] = \sum_{i=1}^{n} F(y_i)[F(x_i) - F(y_i) + F(y_i) - F(x_{i-1})]$$

$$= \sum_{i=1}^{n} F(y_i) F'(y_i)[x_i - y_i + y_i - x_{i-1}] + F(y_i)[\theta_i(x_i - y_i) + \rho_i(y_i - x_i)],$$

where $|\theta_i| \leq \epsilon/4$ and $|\rho_i| \leq \epsilon/4$. Thus $|S_n - [F^2(1) - F^2(0)]/2| < \epsilon$ and the conclusion holds. $\square$

F.9 Proposition. If F is a continuous real-valued function on $[0,1]$ then $(A) \int_0^1 F \, dF$ exists for $(A) = (RS), (MPS), (RYS)$ or (CY) if and only if $F \in \mathcal{W}_2^*([0,1])$.

Proof. By Theorem II.10.9 in Hildebrandt (1963) and Proposition 3.17 in Part II, for F continuous, existence of any one of the four integrals is equivalent to that of the others. Let $F \in \mathcal{W}_2^*$ and $\epsilon > 0$. By Proposition 2.23 in Part II, there is a partition λ of $[0,1]$ such that $\sum_{i=1}^{n}[F(x_i) - F(x_{i-1})]^2 < (\epsilon/V_2(F))^2$ for any refinement $\kappa = \{x_i \colon i = 0, \ldots, n\}$ of λ. Let κ be any such partition and let $y_i', y_i'' \in [x_{i-1}, x_i]$ for $i = 1, \ldots, n$. Then we have

$$\left| \sum_{i=1}^{n}[F(y_i') - F(y_i'')][F(x_i) - F(x_{i-1})] \right| \le \left(\sum_{i=1}^{n}[F(y_i') - F(y_i'')]^2 \right)^{1/2} V_2(F) < \epsilon/2.$$

Thus taking $y_i' = x_{i-1}$, the integral $(MPS) \int_0^1 F \, dF$ exists and equals $[F(1)^2 - F(0)^2]/2$. The converse implication follows using the Cauchy condition for the sums $\sum_i F(x_i)[F(x_i) - F(x_{i-1})]$ and $\sum_i F(x_{i-1})[F(x_i) - F(x_{i-1})]$ and Proposition 2.23 in Part II. $\square$

Example. Let $F(x) := x^2 \sin(\pi/(2x^4))$ for $0 < x \le 1$ and $F(0) := 0$. Then $(WPS) \int_0^1 F \, dF$ exists by Proposition F.8, but the four integrals of Proposition F.9 do not.

REFERENCES

Andersen, P. K., Borgan, Ø., Gill, R. D. and Keiding, N. (1993). *Statistical Models based on Counting Processes.* Springer-Verlag, New York.

Appell, J. (1983). Upper estimates for superposition operators and some applications. *Ann. Acad. Sci. Fenn. Ser. A I. Math.* **8**, 149-159.

Appell, J. (1988). The superposition operator in function spaces - A survey. *Expositiones Math.* **6**, 209-270.

Appell, J. and Zabrejko, P. P. (1990). *Nonlinear superposition operators.* Cambridge University Press.

Averbukh, V. I. and Smolyanov, O. G. (1967). The theory of differentiation in linear topological spaces. (Russian) *Math. Surveys* **22** no. 6, 201-258 = *Uspekhi Mat. Nauk* **22** no. 6, 201-260 (1967).

Averbukh, V. I. and Smolyanov, O. G. (1968). The various definitions of the derivative in linear topological spaces. (Russian) *Math. Surveys* **23** no. 4, 67-113 = *Uspekhi Mat. Nauk* **23** no. 4, 67-116.

Beirlant, J., and Deheuvels, P. (1990). On the approximation of P-P and Q-Q plot processes by Brownian bridges. *Statist. Probab. Letters* **9**, 241-251.

Billingsley, P. (1968). *Convergence of Probability Measures.* Wiley, New York.

Bingham, N. H. (1975). Fluctuation theory in continuous time. *Adv. Appl. Prob.* **7**, 705-766.

Bingham, N. H., Goldie, C. M. and Teugels, J. L. (1987). *Regular variation.* Encyclopedia of Mathematics and its Applications, **27**. Cambridge University Press, Cambridge-New York.

Bretagnolle, J. and Massart, P. (1989). Hungarian constructions from the non-asymptotic viewpoint. *Ann. Probab.* **17**, 239-256.

Brokate, M. and Colonius, F. (1990). Linearizing equations with state-dependent delays. *Appl. Math. Optimiz.* **21**, 45-52.

Bruneau, M. (1974). *Variation totale d'une fonction. Lect. Notes in Math.* (Springer-Verlag) **413**.

Ciesielski, Z., Kerkyacharian, G. and Roynette, B. (1993). Quelques espaces fonctionnels associés à des processus gaussiens. *Studia Math.*, **107**, 171-204.

Corson, H. H. (1961). The weak topology of a Banach space. *Trans. Amer. Math. Soc.* **101**, 1-15.

Csörgő, M. (1983). *Quantile Processes with Statistical Applications.* SIAM, Philadelphia.

Cutland, N. J., Kopp, P. E. and Willinger, W. (1995). Stock price returns and the Joseph effect: a fractional version of the Black-Scholes model. *Seminar on Stochastic Analysis, Random Fields and Applications* (Ascona, 1993), *Progr. Probab.*, **36**, 327-351. Birkhäuser, Basel.

Dai, W. and Heyde, C. C. (1996). Itô's formula with respect to fractional Brownian motion and its application. *J. Appl. Math. Stochastic Anal.* **9**, 439-448.

Dieudonné, J. (1960). *Foundations of Modern Analysis.* Academic Press, New York; *Fondements de l'analyse moderne* **1**. Gauthier-Villars, Paris, 1963.

Dudley, R. M. (1985). An extended Wichura theorem, definitions of Donsker class, and weighted empirical distributions. In *Probability in Banach Spaces V* (A. Beck, R. Dudley, M. Hahn, J. Kuelbs and M. Marcus, eds.). *Lect. Notes in Math.* (Springer-Verlag) **1153**, 141-178.

Dudley, R. M. (1992). Fréchet differentiability, p-variation and uniform Donsker classes. *Ann. Probab.* **20**, 1968-1982.

Dudley, R. M. (1993). *Real Analysis and Probability.* Second printing, corrected. Chapman and Hall, New York.

Dudley, R. M. (1994). The order of the remainder in derivatives of composition and inverse operators for p-variation norms. *Ann. Statist.* **22**, 1-20.

Dudley, R. M. (1997). Empirical processes and p-variation. In *Festschrift for Lucien Le Cam* (D. Pollard, E. Torgersen and G. L. Yang, eds.) 219-233. Springer-Verlag, New York.

Ebin, D. G. and Marsden, J. (1970). Groups of diffeomorphisms and the motion of an incompressible fluid. *Ann. Math.* **92**, 102-163.

Esty, W., Gillette, R., Hamilton, M. and Taylor, D. (1985). Asymptotic distribution theory of statistical functionals: the compact derivative approach for robust estimators. *Ann. Inst. Statist. Math.* **37**, 109-129.

Fernholz, L. T. (1983). *von Mises calculus for statistical functionals. Lect. Notes in Statist.* (Springer-Verlag) **19**.

de Finetti, B. and Jacob, M. (1935)(*). Sull' integrale di Stieltjes-Riemann. *Giorn. Istituto Ital. Attuari* **6**, 303-319.

Fréchet, M. (1937). Sur la notion de différentielle. *J. Math. Pures Appl.* **16**, 233-250.

Freedman, M. A. (1983). Operators of p-variation and the evolution representation problem. *Trans. Amer. Math. Soc.* **279**, 95-112.

Garay, B. M. (1996). Hyperbolic structures in ODE's and their discretization. In *Nonlinear Analysis and Boundary Value Problems for ODEs* (Proc. CISM, Udine, 1995; F. Zanolin, ed.). *CISM Courses and Lectures* **371**, 149-174. Springer-Verlag, Vienna.

Gateaux, R. (1913). Sur les fonctionelles continues et les fonctionelles analytiques. *C. R. Acad. Sci. Paris* **157**, 325-327.

Gateaux, R. (1919). Fonctions d'une infinité de variables indépendantes. *Bull. Soc. Math. France* **47**, 70-96.

Gateaux, R. (1922). Sur diverses questions de calcul fonctionnel. *Bull. Soc. Math. France* **50**, 1-37.

Gehring, F. W. (1954). A study of α-variation. I. *Trans. Amer. Math. Soc.* **76**, 420-443.

Gill, R. D. (1989). Non- and semi-parametric maximum likelihood estimators and the von Mises method (Part 1). *Scand. J. Statist.* **16**, 97-128.

Gill, R. D. (1994). Lectures on Survival Analysis. In *Ecole d'été de Probabilités de Saint-Flour* (P. Bernard, ed.). *Lect. Notes in Math.* (Springer-Verlag) **1581**, 115-241.

Gill, R. D., and Johansen, S. (1990). A survey of product-integration with a view toward application in survival analysis. *Ann. Statist.* **18**, 1501-1555.

Gill, R. D., van der Laan, M. J., and Wellner, J. A. (1995). Inefficient estimators of the bivariate survival function for three models. *Ann. Inst. Henri Poincaré* **31**, 545-597.

Gray, A. (1975). Differentiation of composites with respect to a parameter. *J. Austral. Math. Soc.* (Ser. A) **19**, 121-128.

Hardy, G. H. (1916). Weierstrass's non-differentiable function. *Trans. Amer. Math. Soc.* **17**, 301-325. Repr. in *Collected papers of G. H. Hardy*, Vol. IV, Oxford University Press, London.

Hardy, G. H., Littlewood, J. E. and Pólya, G. (1959). *Inequalities.* Cambridge University Press, Cambridge.

Hartung, F. and Turi, J. (1997). On differentiability of solutions with respect to parameters in state-dependent delay equations. *J. Differential Equations* **135**, 192-237.

Henstock, R. (1961). Definitions of Riemann type of the variational integrals. *Proc. London Math. Soc.* (3) **11**, 402-418.

Henstock, R. (1963). *Theory of integration.* Butterworths, London.

Henstock, R. (1991). *The general theory of integration.* Oxford University Press, New York.

Hewitt, E. and Stromberg, K. (1969). *Real and Abstract Analysis.* Springer-Verlag, New York.

Hewitt, E. (1960). Integration by parts for Stieltjes integrals. *Amer. Math. Monthly* **67**, 419-423.

Hildebrandt, T. H. (1938). Definitions of Stieltjes integrals of the Riemann type. *Amer. Math. Monthly* **45**, 265-278.

Hildebrandt, T. H. (1963). *Introduction to the theory of integration.* Academic Press, New York.

Hildebrandt, T. H. (1966). Compactness in the space of quasi-continuous functions. *Amer. Math. Monthly* **73**, Part 2, 144-145.

Huang, Yen-Chin (1994). *Empirical distribution function statistics, speed of convergence, and p-variation.* Ph. D. thesis, Massachusetts Institute of Technology.

Huang, Yen-Chin (1997). Speed of convergence of classical empirical processes in p-variation norm. Preprint.

Itô, K. and McKean, H. P. (1974). *Diffusion Processes and their Sample Paths.* Springer-Verlag, Berlin.

Kiefer, J. (1970). Deviations between the sample quantile process and the sample df. In *Nonparametric Techniques in Statistical Inference* (M. L. Puri, ed.) 299-319. Cambridge University Press.

Kisliakov, S. V. (1984). A remark on the space of functions of bounded p-variation. *Math. Nachr.* **119**, 137-140.

Kolmogorov, A. N. (1933). Sulla determinazione empirica di una legge di distribuzione. *Giorn. Istituto Ital. Attuari* 4, 83-91.

Kolmogorov, A. N. (1941). Confidence limits for an unknown distribution function. *Ann. Math. Statist.* **12**, 461-463.

Komlós, J., Major, P., and Tusnády, G. (1975). An approximation of partial sums of independent rv's and the sample df. *Z. Wahrsch. verw. Gebiete* **32**, 111-231.

Krabbe, G. L. (1961a). Integration with respect to operator-valued functions. *Bull. Amer. Math. Soc.* **67**, 214-218.

Krabbe, G. L. (1961b). Integration with respect to operator-valued functions. *Acta Sci. Math. (Szeged)* **22**, 301-319.

Krasnosel'skiĭ, M. A., Zabreĭko, P. P., Pustyl'nik, E. I. and Sobolevskiĭ, P. E. (1966). *Integral operators in spaces of summable functions.* (Russian) Nauka, Moscow; transl. by T. Ando, Noordhoff, Leyden, 1976.

Kurzweil, J. (1957). Generalized ordinary differential equations and continuous dependence on a parameter. (Russian) *Czechoslovak Math. J.* **7**, 568-583.

Kurzweil, J. (1958). On integration by parts. *Czechoslovak Math. J.* **8**, 356-359.

Kurzweil, J. (1959). Addition to my paper "Generalized ordinary differential equations and continuous dependence on a parameter". *Czechoslovak Math. J.* **9**, 564-573.

Lehmann, E. L. (1986). *Testing Statistical Hypotheses*, 2d ed. Wiley, New York; Repr. 1997.

Leśniewicz, R. and Orlicz, W. (1973). On generalized variations (II). *Studia Math.* **45**, 71-109.

Lévy, P. (1922). *Leçons d'analyse fonctionelle.* Gauthier-Villars, Paris.

Lin, S. J. (1995). Stochastic analysis of fractional Brownian motion. *Stochastics Stochastics Rep.* **55**, 121-140.

Lindenstrauss, J. and Tzafriri, L. (1971). On the complemented subspaces problem. *Israel J. Math.* **9**, 263-269.

Love, E. R. (1993). The refinement-Ross-Riemann-Stieltjes (R^3S) integral. In: *Analysis, geometry and groups: a Riemann legacy volume*, Eds. H. M. Srivastava, Th. M. Rassias, Part I. Hadronic Press, Palm Harbor, FL, pp. 289-312.

Love, E. R. and Young, L. C. (1938). On fractional integration by parts. *Proc. London Math. Soc. (Ser. 2)* **44**, 1-35.

Lyons, T. (1994). Differential equation driven by rough signals (I): An extension of an inequality of L. C. Young. *Math. Res. Lett.* **1**, 451-464.

Marcinkiewicz, J. (1934). On a class of functions and their Fourier series. *C. R. Soc. Sci. Varsovie* **26**, 71-77. Repr. in J. Marcinkiewicz, *Collected Papers*, (A. Zygmund, ed.) 36-41. Państwowe Wydawnictwo Naukowe, Warszawa, 1964.

McShane, E. J. (1969). *A Riemann-type integral that includes Lebesgues-Stieltjes, Bochner and stochastic integrals*. Memoirs AMS, **88**; Providence, Rhode Island.

Milnor, J. (1984). Remarks on infinite-dimensional Lie groups. In *Relativité, groupes et topologie II*, (B. S. DeWitt and R. Stora, eds.) 1007-1057.

Norvaiša, R. (1997a). The central limit theorem for L-statistics. *Studia Sci. Math. Hung.* **33**, 209-238.

Norvaiša, R. (1997b). Chain rule and p-variation. Preprint.

Pfeffer, W. F. (1993). *The Riemann approach to integration. Local geometric theory*. Cambridge University Press, New York.

Phillips, E. R. (1971). *An introduction to analysis and integration theory*. Intext Educational Publ., Scranton.

Qian, Jinghua (1998). The p-variation of partial sum processes and the empirical process. *Ann. Probab.* **26**, 1370-1383.

Reeds, J. A. III (1976): *On the definition of von Mises functionals*. Ph. D. thesis, Statistics, Harvard University.

Rosenthal, P. (1995). Review of Shapiro (1993) and Singh and Manhas (1993). *Bull. Amer. Math. Soc.* **32**, 150-153.

Saks, S. (1937). *Theory of the Integral*. 2d. ed., transl. by L. C. Young. Hafner, New York. Repr. by Dover, New York 1964.

Schwabik, Š. (1973). On the relation between Young's and Kurzweil's concept of Stieltjes integral. *Časopis Pěst. Mat.* **98**, 237-251.

Sebastião e Silva, J. (1956). Le calcul différentiel et intégral dans les espaces locale- ment convexes, réels ou complexes I, II. *Rend. Accad. Lincei Sci. Fis. Mat. Nat.* (Ser. 8) **20**, 743-750, **21**, 40-46 (1956).

Shapiro, J. H. (1993). *Composition Operators and Classical Function Theory*. Springer-Verlag, New York.

Singh, R. K. and Manhas, J. S. (1993). *Composition Operators on Function Spaces*. North-Holland, Amsterdam.

Skorohod, A. V. (1957). Limit theorems for stochastic processes with independent increments. *Theory Prob. Appl.* **2**, 138-171.

Sova, M. (1966). Conditions of differentiability in linear topological spaces. (Rus- sian) *Czech. Math. J.* **16**, 339-362.

Stevenson, J. O. (1974). Holomorphy of composition. *Bull. Amer. Math. Soc.* **80**, 300-303.

Stevenson, J. O. (1977). Holomorphy of composition. In *Infinite Dimensional Holo- morphy and Applications*, (M. C. Matos, ed.) 397-424. North-Holland, Amster- dam.

Tvrdý, M. (1989). Regulated functions and the Perron-Stieltjes integral. *Časopis Pěst. Mat.* **114**, 187-209.

van der Vaart, A. (1991). Efficiency and Hadamard differentiability. *Scand. J. Statist.* **18**, 63-75.

van der Vaart, A. W. and Wellner, J. A. (1996). *Weak Convergence and Empirical Processes, with Applications to Statistics.* Springer-Verlag, New York.

Vainberg, M. M. (1964). *Variational methods for the study of nonlinear operators.* With a chapter on Newton's method by L. V. Kantorovich and G. P. Akilov. Translated and supplemented by A. Feinstein, Holden-Day, San Francisco, 1964.

Wang Sheng-Wang (1963). Differentiability of the Nemyckii operator. (Russian) *Doklady Akad. Nauk SSSR* **150**, 1198-1201; transl. in: *Sov. Math. Doklady* **4**, 834-837.

Ward, A. J. (1936). The Perron-Stieltjes integral. *Math. Z.* **41**, 578-604.

Whitt, W. (1980). Some useful functions for functional limit theorems. *Mathematics of Operations Research* **5**, 67-85.

Wiener, N. (1924). The quadratic variation of a function and its Fourier coefficients. *J. Math. and Phys. (MIT)* **3**, 72-94.

Young, L. C. (1936). An inequality of the Hölder type, connected with Stieltjes integration. *Acta Math.* (Sweden) **67**, 251-282.

Young, L. C. (1937). Inequalities connected with bounded p-th power variation in the Wiener sense and with integrated Lipschitz conditions. *Proc. London Math. Soc.* (Ser. 2) **43**, 449-467.

Young, L. C. (1938a). Inequalities connected with binary derivation, integrated Lipschitz conditions and Fourier series. *Math. Z.* **43**, 255-270.

Young, L. C. (1938b). General inequalities for Stieltjes integrals and the convergence of Fourier series. *Math. Ann.* **115**, 581-612.

Young, L. C. (1970). Some new stochastic integrals and Stieltjes integrals. I. Analogues of Hardy-Littlewood classes. In *Advances in Probability and Related Topics*, Vol. II, (P. Ney, ed.) 161-240. Dekker, New York.

Young, R. C. (1929). On Riemann integration with respect to a continuous increment. *Math. Z.* **29**, 217-233.

Young, W. H. (1914). On integration with respect to a function of bounded variation. *Proc. London Math. Soc.* (Ser. 2) **13**, 109-150.

Zähle, M. (1998). Integration with respect to fractal functions and stochastic calculus. I. *Probab. Theory Related Fields* **111**, 333-374.

Zygmund, A. (1959). *Trigonometric Series*, Vol. I. 2d. ed. Cambridge University Press, Cambridge.

(*) We found these references from secondary sources but have not seen them in the original.

PRODUCT INTEGRALS, YOUNG
INTEGRALS AND p-VARIATION

BY R. M. DUDLEY AND R. NORVAIŠA

ABSTRACT. Let $\mathbb{B}$ be a Banach algebra with identity $\mathbb{1}$. Consider a function f defined on an interval $[a, b]$ and with values in $\mathbb{B}$. Let $\{a = x_0 < \cdots < b = x_n\}$ be a partition of $[a, b]$. Then the product integral with respect to f over $[a, b]$ is defined as the limit of the product from $i = 1$ to n of $\mathbb{1} + f(x_i) - f(x_{i-1})$, if it exists, where the limit is taken under refinements of partitions. It is proved that the product integral with respect to f over $[a, b]$ exists if $f \in \mathcal{W}_p([a, b]; \mathbb{B})$, $0 < p < 2$, i.e., if f has bounded p-variation for some $0 < p < 2$, as shown for f continuous by M. A. Freedman, Trans. Amer. Math. Soc. 279 (1983), 95-112. A necessary and sufficient condition for the existence of the product integral is given when $\mathbb{B} = \mathbb{R}$. An operator $\mathcal{P}_a$ from $\mathcal{W}_p([a, b]; \mathbb{B})$ into itself is induced by an indefinite product integral. The main result says that $\mathcal{P}_a$ is Fréchet differentiable. R. D. Gill and S. Johansen (Ann. Statist. 18, 1990, 1501-1555) had shown compact differentiability in the supremum norm, on sets uniformly bounded in 1-variation norm. The present paper shows that when restricted to right- or left-continuous elements of $\mathcal{W}_p([a, b]; \mathbb{B})$, $\mathcal{P}_a$ is analytic. To prove these results a generalized Stieltjes integral due to L. C. Young is developed, as are variants of it called left Young (LY) and right Young (RY) integrals, and the Duhamel formula is extended to (LY) and (RY) integrals. Also using Young integrals a logarithm operator $\mathcal{L}_a$ is defined so that $\mathcal{L}_a(f)$ exists for each $f \in \mathcal{W}_p([a, b]; \mathbb{B})$, $0 < p < 2$, such that the function $x \mapsto f(x)^{-1}$ is in $\ell_\infty([a, b]; \mathbb{B})$. The operators $\mathcal{P}_a$ and $\mathcal{L}_a$ are shown to be inverses of each other. This allows one to determine a unique solution of a linear integral equation and to solve an evolution representation problem whenever the evolution has bounded p-variation for some $0 < p < 2$.

Contents

1. INTRODUCTION

A product integral, like an integral $\int_a^b F(x)\,dG(x)$, can be defined in several different ways. Following the definition of the Riemann integral one can define a product integral as follows. Let g be a $k \times k$ matrix-valued function defined on an interval $[a, b]$ and let $\kappa = \{x_i\colon i = 0, \dots, n\}$ be a partition of $[a, b]$. Volterra (1887; Opere, I, p. 235) defined the product integral of the function g over $[a, b]$ to be the limit

$$\lim_{\kappa} \prod_{i=1}^{n} \left[\mathbb{I} + g(y_i)(x_i - x_{i-1}) \right] \tag{1.1}$$

if it exists, where $\mathbb{I}$ is the identity matrix, y_i are any points from $[x_{i-1}, x_i]$ for $i = 1, \dots, n$ and the limit is taken as the mesh $|\kappa|$ of the partition κ tends to zero. Birkhoff (1937) extended Volterra's definition of the product integral to functions g with values in some infinite-dimensional families of transformations, taking the limit in (1.1) under refinements of the partition κ. Masani (1947, Sec. V) proved among other things that for g with values in a Banach algebra the limit (1.1) exists if and only if the function g is Riemann integrable.

Schlesinger (1931 and 1932) developed product integration as in (1.1) based on approximation by step functions of a matrix-valued function g whose entries are bounded Lebesgue measurable functions. Schmidt (1971) proved existence of such a product integral when g is a Bochner integrable Banach space operator-valued function. Another approach to proving existence of the Volterra product integral of a Lebesgue-integrable matrix-valued function was indicated by Dollard and Friedman (1979, Section 1.8). Moreover, Dollard and Friedman (1979, Theorem 3.5.1) proved that, for a Banach space $\mathbb{B}$ and the space $L(\mathbb{B}) := (L(\mathbb{B}), |\cdot|)$ of bounded linear operators from $\mathbb{B}$ into itself, such a product integral exists for each function $A\colon [a, b] \ni x \mapsto A(x) \in L(\mathbb{B})$ such that, for each $f \in \mathbb{B}$, $A(\cdot)f$ is the pointwise limit of a sequence of simple functions almost everywhere and $|A|$ has finite upper integral on $[a, b]$. Jarník and Kurzweil (1987) initiated a study of the product integral obtained by extending the limit in (1.1) as follows. Let g be a matrix-valued function on $[a, b]$. An invertible matrix Q is called the product integral of g, if for every $\epsilon > 0$ there exists a function $\delta\colon [a, b] \mapsto (0, \infty)$ such that $\| \prod_{i=1}^{n} [\mathbb{I} + g(y_i)(x_i - x_{i-1})] - Q\| < \epsilon$ for all partitions $a = x_0 \le y_1 \le x_1 \le \cdots \le x_{n-1} \le y_n \le x_n = b$ such that $[x_{i-1}, x_i] \subset (y_i - \delta(y_i), y_i + \delta(y_i))$ for all i. Here $\delta(\cdot)$ is a gauge function as in the definition of the gauge integral (see Appendix F in Part I). Further extensions of this product integral are due to Schwabik (1990, 1994).

Another possible definition of product integral is reminiscent of the Riemann-Stieltjes integral. Let f be a function on an interval $[a, b]$ with values in a Banach algebra $\mathbb{B}$ with identity $\mathbb{I}$ and let $\kappa = \{x_i\colon i = 0, \dots, n\}$ be a partition of $[a, b]$. Define the product integral with respect to the function f over $[a, b]$ to be the limit

$$\prod_a^b (\mathbb{I} + df) := \lim_{(\kappa)} \left[\mathbb{I} + f(x_n) - f(x_{n-1}) \right] \cdots \left[\mathbb{I} + f(x_1) - f(x_0) \right] \tag{1.2}$$

if it exists, where the limit is taken under refinements of partitions. The product integral defined in such a way, except that the limit is taken as the mesh $|\kappa|$ tends to zero, was treated by MacNerney (1955), who discovered reciprocal formulas involving so-called additive and product integrals. Earlier analogous formulas, when f is an interval function, were proved by Dobrushin (1953). In particular, MacNerney (1955) proved that (1.2) exists if f is continuous and of bounded variation. Hildebrandt (1959) removed the continuity requirement. Freedman (1983a) replaced the bounded variation condition for a continuous function f by the finiteness of the p-variation $(= V_p(f)^p$ for $V_p(f)$ defined in the next display) for some $p < 2$. He first proved existence of a unique solution of an appropriate linear integral equation and, second, showed that this solution is the limit (1.2).

Product integrals were introduced and used in solving certain types of differential and integral equations. Surveys of and references to such applications, in addition to the book of Dollard and Friedman (1979) already mentioned, can be found in Volterra and Hostinsky (1938), Gantmacher (1960, Section XIV) and Manturov (1990). Also, product integrals proved to be useful in statistics and probability theory. Examples and surveys of such applications can be found in Gill and Johansen (1990) and Gill (1994). It was mainly Gill and Johansen's (1990) paper that spurred our interest in product integrals and introduced us to some of its related constructions, including the Peano series (Section 4.5 below) and the Duhamel formula (Section 5.2 below).

In this paper we consider product integrals defined by (1.2) with respect to possibly discontinuous functions, with values in a Banach algebra $\mathbb{B} = (\mathbb{B}, \|\cdot\|)$ with an identity $\mathbb{I}$, which may have unbounded variation. Namely, by Theorem 4.23 below, (1.2) exists if the function f has bounded p-variation for some $0 < p < 2$, i.e., if

$$V_p(f) := \sup\left\{ \left(\sum_{i=1}^{n} \|f(x_i) - f(x_{i-1})\|^p \right)^{1/p} : \{x_i\} \text{ is a partition of } [a,b] \right\} < \infty.$$

If, in addition, f is either right-continuous or left-continuous at each point of (a, b) then, by Theorem 4.26 below, the limit (1.2) exists as the mesh of the partition κ tends to zero. The class of all $\mathbb{B}$-valued functions on $[a, b]$ with bounded p-variation will be denoted by $\mathcal{W}_p = \mathcal{W}_p([a, b]; \mathbb{B})$. Subsection 2.3 is devoted to describing some properties of the space $\mathcal{W}_p$, $0 < p < \infty$. By Theorem 4.4 below, for a real-valued function f, the limit (1.2) exists and is non-zero if and only if the limit

$$\lim_{(\kappa)} \sum_{i=1}^{n} [f(x_i) - f(x_{i-1})]^2 < \infty \tag{1.3}$$

exists and f has no jumps equal to -1; then the limit in (1.3) is the sum of squares of the left and right jumps of f. Thus $f \in \mathcal{W}_2$ (Theorem 4.3). On the other hand it is well known that there exist continuous real-valued $f \in \mathcal{W}_2$ for which the bounded sums in (1.3) do not converge ("fine" 2-variation, cf. Bruneau, 1974). A weaker form of the limit (1.2), called an exponential semimartingale, exists for suitable random functions with possibly infinite 2-variation (cf. Emery (1978), also Proposition 4.10 and Corollary 4.11 below). Gill and Johansen (1990, p. 1527) asked whether the product integral in the sense of the limit (1.2), or as in their Definition 1, could be defined with respect to semimartingales in a deterministic way. By our Theorem 4.4 (as well as by Proposition 4.9), the class of semimartingales for which such a product integral exists and is non-zero does not contain Brownian motion. If $\mathbb{B}$ is a commutative Banach algebra with an identity $\mathbb{I}$ and f is a $\mathbb{B}$-valued function such that (1.3) holds with the square replaced by the square of the norm in $\mathbb{B}$ then by Theorem 4.17 below, the product integral with respect to f over $[a, b]$ exists and has the following representation:

$$\prod_a^b (\mathbb{I} + df) = \exp(f(b) - f(a)) \prod_{[a,b]} [(\mathbb{I} + \Delta f)\exp(-\Delta f)],$$

where Δf denotes the jumps of the function f and the exponential function $\exp$ is defined, as usual, by its Taylor series.

Our further analysis of product integrals is based on an extension of an integral defined by L. C. Young (1936). This is the main subject of Section 3. Let f, g and h be functions defined on $[a, b]$ with values in a Banach algebra $\mathbb{B}$. Define the Moore-Pollard-Stieltjes (MPS) integral to be the limit

$$(MPS) \int_a^b g \, dh \, f := \lim_{(\kappa)} \sum_{i=1}^n g(y_i) [h(x_i) - h(x_{i-1})] f(y_i)$$

if it exists for any $y_i \in [x_{i-1}, x_i]$, $i = 1, \ldots, n$. Subsection 3.1 is devoted to describing (MPS) integrals. Let f, g and h be, in addition, regulated functions, that is, functions continuous except for jumps. If f is a regulated function on an interval $[a, b]$ let

$$f_-^{(a)}(x) := f_-(x) := \lim_{y \uparrow x} f(y), \quad a < x \le b, \quad f_-^{(a)}(a) := f(a),$$

and

$$f_+^{(b)}(x) := f_+(x) := \lim_{y \downarrow x} f(y), \quad a \le x < b, \quad f_+^{(b)}(b) := f(b).$$

The central Young integral, see also Dudley (1992), will be defined by

$$(Y_1) \int_a^b g \, dh \, f := (MPS) \int_a^b g_+^{(b)} \, dh_-^{(a)} f_+^{(b)} + [g(h - h_-)f](b)$$
$$+ \sum_{a \le x < b} [g(h_+ - h_-^{(a)})f - g_+(h_+ - h_-^{(a)})f_+](x) \qquad (1.4)$$

if the (MPS) integral exists and the sum converges unconditionally in $\mathbb{B}$. In Theorem 3.7, we prove that another variant of the Young integral from Dudley (1992), called the (Y_2) integral, exists whenever the (Y_1) integral (1.4) exists and then the two are equal and will be called the central Young integral or the (CY) integral. Using an extension of inequality (5.1) of Young (1936, p. 254) we prove in Theorem 3.27 below that the (CY) integral (1.4) exists if $h \in \mathcal{W}_p$ and $f, g \in \mathcal{W}_q$ with $p, q > 0$ and $1/p + 1/q > 1$. Moreover, letting $g^{(a+)} := g$ on $(a, b]$ and $g^{(a+)}(a) := g(a+)$ we define (cf. Definition 3.11) the left Young integral

$$(LY) \int_a^b g \, dh \, f := (CY) \int_a^b g^{(a+)} dh \, f_-^{(a)} + \sum_{a < x < b} [g_+(h_+ - h)f - g(h_+ - h)f_-](x)$$

if the (CY) integral exists and the sum converges unconditionally in $\mathbb{B}$. Likewise letting $f^{(b-)} := f$ on $[a,b)$ and $f^{(b-)}(b) := f(b-)$ we define the right Young integral

$$(RY) \int_a^b g\,dh\,f := (CY) \int_a^b g_+^{(b)}\,dh\,f^{(b-)} - \sum_{a < x < b} [g_+(h-h_-)f - g(h-h_-)f_-](x)$$

if the (CY) integral exists and the sum converges unconditionally in $\mathbb{B}$. The (CY), (LY) and (RY) integrals are different in general (cf. Propositions 3.13 and 3.14). However by Theorem 3.12, the (LY) and (RY) integrals are equal whenever both are defined, as they are if $h \in W_p$ and $f,g \in W_q$ with $p,q > 0$ and $1/p + 1/q > 1$, by Corollary 3.28.

Let $f \in W_p([a,b];\mathbb{B})$ with $0 < p < 2$. Then one can define the indefinite product integrals $\mathcal{P}_a(f)$ and $\mathcal{P}^b(f)$ with respect to f by

$$\mathcal{P}_a(f)(y) := \prod_a^y (\mathbb{1} + df) \quad \text{and} \quad \mathcal{P}^b(f)(y) := \prod_y^b (\mathbb{1} + df), \quad y \in [a,b].$$

We show that the (nonlinear) operators $\mathcal{P}_a$ and $\mathcal{P}^b$ take the space W_p into itself for $0 < p < 2$ (Proposition 5.3). A main result of the paper (Theorem 5.16 below) says that both operators $\mathcal{P}_a$ and $\mathcal{P}^b$ are Fréchet differentiable at each $f \in W_p$ with derivatives $D\mathcal{P}_a(f)$ and $D\mathcal{P}^b(f)$ given by, for each $h \in W_p$ and $y \in [a,b]$,

$$(D\mathcal{P}_a(f)h)(y) = (LY) \int_a^y \mathcal{P}^y(f)\,dh\,\mathcal{P}_a(f)$$

and

$$(D\mathcal{P}^b(f)h)(y) = (RY) \int_y^b \mathcal{P}^b(f)\,dh\,\mathcal{P}_y(f).$$

A second main result (Theorem 5.17 below) says that the operator $\mathcal{P}_a$ restricted to right-continuous elements of W_p and the operator $\mathcal{P}^b$ restricted to left-continuous elements of W_p are analytic. To prove this we extend the Duhamel formula (Theorem 5.5) to product integrals with respect to functions in W_p, $0 < p < 2$, using a convergence theorem for (CY) integrals (Proposition 3.33). Gill and Johansen extended the differentiability of the operator $\mathcal{P}_a$ to discontinuous F, apparently for the first time (1990, Theorem 8), in analyzing the asymptotic behavior of certain statistics. They proved in our notation that

$$\|\mathcal{P}_a(f+th) - \mathcal{P}_a(f) - tD\mathcal{P}_a(f)(h)\|_\infty = o(t) \tag{1.5}$$

as $t \to 0$, uniformly for h in sup-norm compact sets, if $f + th$ and f (but not necessarily h) are bounded in total variation. See also Gill (1994, p. 135). Here (1.5) does not follow from Fréchet differentiability in 1-variation norm, in which th may not tend to 0. One can improve on (1.5) using twice Fréchet differentiability of $\mathcal{P}_a$ on W_p, $1 < p < 2$, if suitable bounds on $V_p(th)$ are available, as they are in some cases, e.g. Dudley (1997).

Volterra, who first defined a product integral operator, of the form (1.1), apparently was also the first to differentiate such an operator, see Volterra and Hostinsky

(1938, pp. 86-88, 223), whose form of the derivative is somewhat obscured by their notation. They also allow the "fixed" endpoints a, b to vary. Dollard and Friedman (1979, p. 35) differentiate product integrals of the Volterra form along suitable 1-dimensional curves. Thus far, one would envisage differentiability of $\prod_a^b(1 + df)$ in the 1-variation seminorm $\|f\|_{(1)}$, as actually proved for f absolutely continuous, when $\|f\|_{(1)} = \|f'\|_1$. Dollard and Friedman (1979, p. 36) briefly state analyticity, also along curves, in a form that suggests it also holds with respect to the seminorm $f \mapsto \|f'\|_1$. Gill and Johansen (1990) proved compact differentiability with respect to the supremum norm, although still on $\| \cdot \|_{(1)}$-bounded sets. Freedman (1983a), by showing existence of the product integral for continuous $f \in \mathcal{W}_p$, opened the way to extending the differentiability. There were still some technical problems, relating to jumps, definition of integrals, and infinite-dimensional holomorphy.

Let f be a regulated function on $[a, b]$ with invertible values in a Banach algebra $\mathbb{B}$ such that the function $f^{-1}\colon [a, b] \mapsto \mathbb{B}$, defined by $f^{-1}(x) := [f(x)]^{-1}$, is bounded. We define the function

$$\mathcal{L}_a(f)(y) := (LY) \int_a^y df\, f^{-1}, \quad y \in [a, b],$$

if the (LY) integral exists for all $y \in (a, b]$. Then $\mathcal{L}_a$ is an operator, we call it the logarithm operator, from the subspace $\mathcal{I}_p = \mathcal{I}_p([a, b]; \mathbb{B})$ of elements of $\mathcal{W}_p$ with bounded inverse into $\mathcal{W}_p$ (cf. Lemma 6.6 below). In Theorem 6.10, we will show that for each $f \in \mathcal{W}_p$ with $\mathcal{P}_a(f) \in \mathcal{I}_p$, we have

$$f(y) - f(a) = \mathcal{L}_a(\mathcal{P}_a(f))(y), \quad \forall y \in [a, b],$$

and, in Corollary 6.11, that for each $f \in \mathcal{I}_p$, $\mathcal{L}_a(f)$ is a function, unique up to an additive constant, such that

$$f(y)f^{-1}(a) = \mathcal{P}_a(\mathcal{L}_a(f))(y), \quad \forall y \in [a, b].$$

Analogous relations are proved for the function

$$\mathcal{L}^b(f)(y) := (RY) \int_y^b f^{-1}\, df, \quad y \in [a, b].$$

Using the Duhamel formula and the above relations we prove (in Theorems 5.21 and 5.22 below) that, for each function $f \in \mathcal{W}_p$, $0 < p < 2$, the indefinite integrals $\mathcal{P}_a(f)$ and $\mathcal{P}^b(f)$ are unique solutions in $\mathcal{W}_p$ respectively of the forward linear integral equation in F:

$$F(y) = \mathbb{1} + (LY) \int_a^y df\, F, \quad y \in [a, b], \tag{1.6}$$

and of the backward linear integral equation in G:

$$G(y) = \mathbb{1} + (RY) \int_y^b G\, dg, \quad y \in [a, b].$$

Uniqueness also holds in any $\mathcal{W}_r$ with $1/p + 1/r > 1$, $r \geq p$. Moreover, the above reciprocal formulas involving product integrals and logarithm operators give solutions of the evolution representation problem (cf. Theorem 6.14 below). This extends the main result of Freedman (1983a, Theorem 8.1) to possibly discontinuous functions. Lyons (1994) made an extension in a different direction, replacing $df\,F$ in (1.6) by $df\,\Psi(F)$ for suitable nonlinear Ψ, where $\mathbb{B}$ is finite-dimensional and f continuous in $\mathcal{W}_p$. If $f \in \mathcal{W}_p$, $0 < p < 2$, is right-continuous then, by Definition 3.11, the integral equation (1.6) becomes either the central Young equation

$$F(y) = \mathbb{1} + (CY)\int_a^y df\,F_-^{(a)}, \quad y \in [a,b],$$

or, by the (Y_2) definition of the (CY) integral given in Definition 3.4, becomes the (MPS) integral equation

$$F(y) = \mathbb{1} + (MPS)\int_a^y df\,F_-^{(a)}, \quad y \in [a,b].$$

This agrees with the form of the forward Lebesgue-Stieltjes linear integral equation treated by other authors (cf. Definition 2 on p. 1504 of Gill and Johansen, 1990, and references therein).

Acknowledgments. We are much indebted to Mark van der Laan and especially to Richard Gill for several helpful conversations.

2. p-VARIATION

2.1. Regulated functions. Let τ be a non-degenerate interval, open or closed at either end, and let $\mathbb{B} = (\mathbb{B}, \|\cdot\|)$ be a Banach space. A function f from τ into $\mathbb{B}$ is called *regulated*, say $f \in \mathcal{R} = \mathcal{R}(\tau; \mathbb{B})$, if it has left and right limits everywhere in the interior of τ, a right limit at the left endpoint of τ (even if it is $-\infty$ and/or τ is left open), and a left limit at the right endpoint of τ likewise. Write

$$f(x+) := \lim_{y \downarrow x} f(y) \quad \text{and} \quad f(x-) := \lim_{y \uparrow x} f(y)$$

when these are defined. So, each regulated function f can be redefined on the set $\tau \times \{-, 0, +\}$ by identifying $x-$ with $(x, -)$, x with $(x, 0)$ and $x+$ with $(x, +)$ except for the endpoints: if the endpoints are $-\infty \le a < b \le +\infty$ then f is not defined at $a-$ or $b+$, and it is not defined at a, b if the interval is open at the respective endpoints. Letting $z+ < x- < x < x+ < y-$ for all $z < x < y$, $z, y \in \tau$, gives a linear ordering of points x, $y-$, $z+$.

For each $f \in \mathcal{R}(\tau; \mathbb{B})$, define

$$f_-(x) := f(x-) \quad \text{and} \quad f_+(x) := f(x+), \qquad \forall x \in \tau.$$

The next statement shows that the functions f_- and f_+ are left- and right-continuous modifications of f, respectively.

Lemma 2.1. *Given $f \in \mathcal{R}(\tau; \mathbb{B})$, the functions f_- and f_+ are regulated. Moreover,*

$$f_-(x+) = f(x+) = f_+(x+) \tag{2.1}$$

for any $x \in \tau$ except the right endpoint of τ if it is right closed and

$$f_+(x-) = f(x-) = f_-(x-) \tag{2.2}$$

for any $x \in \tau$ except the left endpoint of τ if it is left closed.

Proof. We prove only the first equality in (2.1). The proof of the first equality in (2.2) is based on symmetric arguments. The other equalities follow just by the definitions. So, take any $x \in \tau$ which is not the right endpoint of τ if it is right closed. Given $\epsilon > 0$, there is a $\delta > 0$ such that if $x < y < x + \delta$, then $\|f(y) - f(x+)\| < \epsilon$. It follows that $\|f(y-) - f(x+)\| \le \epsilon$, so $f_-(x+) = f(x+)$. The proof of Lemma 2.1 is complete. $\qquad\square$

It is clear that the maps $f \mapsto f_-$ and $f \mapsto f_+$ are linear on $\mathcal{R}(\tau; \mathbb{B})$. We will be using this fact without further mention. Also define

$$\left. \begin{aligned} (\Delta_-^+ f)(x) &:= f_+(x) - f_-(x) = f(x+) - f(x-) \\ (\Delta_- f)(x) &:= f(x) - f_-(x) = f(x) - f(x-) \\ (\Delta^+ f)(x) &:= f_+(x) - f(x) = f(x+) - f(x) \end{aligned} \right\} \tag{2.3}$$

except that if x is an endpoint of τ, we define the jump of f outside of τ to be zero.

The proof of the following statement is given in Hönig (1975, Corollary 3.2).

Lemma 2.2. *Given $\epsilon > 0$, for each regulated function f, there is at most a finite number of points $x \in \tau$ for which $\|\Delta_- f(x)\| > \epsilon$ or $\|\Delta^+ f(x)\| > \epsilon$.*

By Lemma 2.2, the set of all jumps of each regulated function f is countable. We will assume the jumps to be rearranged into a sequence $\{\xi_j : j \geq 1\}$ such that $\max(\|\Delta_- f(\xi_j)\|, \|\Delta^+ f(\xi_j)\|)$ is nonincreasing. For each $0 < p < \infty$, let

$$\mathfrak{S}_p(f) := \mathfrak{S}_p(f; \tau) := \Big\{ \sum_{j=1}^{\infty} (\|\Delta_- f(\xi_j)\|^p + \|\Delta^+ f(\xi_j)\|^p) \Big\}^{1/p}. \qquad (2.4)$$

2.2. Notation. Given an interval τ, a *partition* κ of τ is a finite increasing sequence of points of τ. Let x_0 be the left endpoint of τ. If τ is left closed we assume $x_0 \in \kappa$ and write $\kappa = \{x_0, x_1, \dots, x_n\}$. Otherwise, let $\kappa = \{x_1, x_2, \dots, x_n\}$. If τ is right closed we assume x_n is the right endpoint of τ. Let $Q = Q(\tau)$ denote the set of all partitions κ of τ. If all points of a partition κ are in a partition λ, we write $\kappa \subset \lambda$ and say λ is a *refinement* of κ. For each $i = 1, \dots, n$, define $\Delta_i^\kappa f := f(x_i) - f(x_{i-1})$ and $[\Delta_i^\kappa] := [x_{i-1}, x_i]$. Let $|\kappa|$ denote the *mesh* of the partition κ, i.e.

$$|\kappa| := \max\{x_i - x_{i-1} : i = 1, \dots, n\},$$

where x_n is the right endpoint of τ.

Let $0 < p < \infty$. A function $f : \tau \mapsto \mathbb{B}$ has *p-variation* defined by

$$v_p(f) := v_p(f; \tau) := \sup \Big\{ \sum_i \|\Delta_i^\kappa f\|^p : \kappa \in Q(\tau) \Big\}.$$

Also, let $\sum_\kappa \|\Delta^\kappa f\|^p := \sum_i \|\Delta_i^\kappa f\|^p$. Define

$$\mathcal{W}_p := \mathcal{W}_p(\tau; \mathbb{B}) := \{ f \in \mathcal{R}(\tau; \mathbb{B}) : v_p(f; \tau) < \infty \} \qquad (2.5)$$

and let $V_p(f) := V_p(f; \tau) := v_p(f; \tau)^{1/p}$, which is 0 if and only if f is a constant. In the case $p = \infty$ we define the ∞-variation seminorm of a function f to be the oscillation of f, i.e.

$$Osc(f) := Osc_\tau(f) := \sup\{\|f(x) - f(y)\| : x, y \in \tau\}.$$

One can also associate with the function f and $\delta > 0$, the quantity

$$V_p^{(\delta)}(f) := V_p^{(\delta)}(f; \tau) := \sup \Big\{ \Big(\sum_i \|\Delta_i^\kappa f\|^p \Big)^{1/p} : \kappa \in Q(\tau), \ |\kappa| \leq \delta \Big\}. \qquad (2.6)$$

Passing to the limit as $\delta \downarrow 0$ we get the p-th total variation of f as defined by Wiener (1924). We use also a related quantity defined by Love and Young (1938, p. 29) to be

$$V_p^*(f) := V_p^*(f; \tau) := \inf_{\lambda \in Q(\tau)} \sup \Big\{ \Big(\sum_\kappa \|\Delta^\kappa f\|^p \Big)^{1/p} : \lambda \subset \kappa \in Q(\tau) \Big\}. \qquad (2.7)$$

Lemma 2.3. *If $V_p^*(f; \tau) < \infty$ then $f \in W_p(\tau; \mathbb{B})$. Moreover, we have*

$$V_p^*(f)^p = \inf \left\{ \sum_{i=1}^n v_p(f; [x_{i-1}, x_i]): \{x_i: i = 0, \ldots, n\} \in Q(\tau) \right\}.$$

Proof. Let $\lambda = \{x_i: i = 0, \ldots, n\} \in Q(\tau)$ and let $A := \sum_{i=1}^n v_p(f; [x_{i-1}, x_i]) < \infty$. We claim that

$$B := \sup \left\{ \sum_\kappa \|\Delta^\kappa f\|^p: \lambda \subset \kappa \in Q(\tau) \right\} = A. \tag{2.8}$$

Since $\sum_\kappa \|\Delta^\kappa f\|^p \le A$ for each $\kappa \supset \lambda$, it follows that $B \le A$. To show $A \le B$ assume $A > B$. For each $i = 1, \ldots, n$, choose a partition $\kappa_i \in Q([x_{i-1}, x_i])$ such that

$$\sum_{\kappa_i} \|\Delta^{\kappa_i} f\|^p > (B/A) v_p(f; [x_{i-1}, x_i]).$$

Let κ be the union of the partitions κ_i. Then $\kappa \supset \lambda$ and

$$B \ge \sum_\kappa \|\Delta^\kappa f\|^p = \sum_{i=1}^n \sum_{\kappa_i} \|\Delta^{\kappa_i} f\|^p > (B/A) \sum_{i=1}^n v_p(f; [x_{i-1}, x_i]) = B.$$

The contradiction proves (2.8). To prove the lemma it suffices to show that, for each $a < c < b$,

$$v_p(f; [a, b]) \le (1 \vee 2^{p-1}) \big[v_p(f; [a, c]) + v_p(f; [c, b]) \big]. \tag{2.9}$$

Indeed, let $\kappa = \{x_i: i = 0, \ldots, n\} \in Q([a, b])$ and let $x_{j-1} < c \le x_j$ for some $j = 1, \ldots, n$. Then

$$\sum_{i=1}^n \|\Delta_i^\kappa f\|^p \le \sum_{i=1}^{j-1} \|\Delta_i^\kappa f\|^p + (1 \vee 2^{p-1}) \big[\|f(c) - f(x_{j-1})\|^p$$

$$+ \|f(x_j) - f(c)\|^p \big] + \sum_{j+1}^n \|\Delta_i^\kappa f\|^p \le (1 \vee 2^{p-1}) \big[v_p(f; [a, c]) + v_p(f; [c, b]) \big].$$

This implies (2.9) and the proof of Lemma 2.3 is complete. $\qquad\square$

The following is known for scalar-valued functions (Wiener, 1924) and the proof extends directly to Banach-valued functions.

Lemma 2.4. *Each function with finite p-variation, $0 < p < \infty$, is regulated.*

Therefore we can include in partitions also points $x\pm$. Let $\kappa^* = \{x_0, \ldots, x_n\}$ be such a partition of an interval τ with left endpoint a and right endpoint b, where $x_0 = a$ or $x_0 = a+$, if τ is left open, $x_n = b$ or $x_n = b-$, if τ is right open, and the remaining x_i may be at points x or $x\pm$. The set of partitions which may contain points $x\pm$ will be denoted by $Q^{\pm} = Q^{\pm}(\tau)$. Define $v_p(f;[y-,z]) := \|\Delta_- f(y)\|^p + v_p(f;[y,z])$ for $z \geq y$ and

$$v_p(f;[y+,z]) := \sup\Big\{\|f(x_1) - f(y+)\|^p + \sum_{i=2}^{n} \|f(x_i) - f(x_{i-1})\|^p\colon \{x_i\colon i=1,\dots,n\} \in Q((y,z])\Big\}$$

for $z > y$. Likewise define $v_p(f;[y,z+])$ for $z \geq y$ and $v_p(f;[y,z-])$ for $z > y$. Let

$$J_{\pm}(\kappa^*) := \{i\colon [x_{i-1}, x_i] = [x-, x] \text{ or } [x, x+] \text{ for some } x \in \tau\}. \tag{2.10}$$

If $|\kappa^*| < \delta$ then

$$\Big(\sum_{i=1}^{n} \|\Delta_i^{\kappa^*} f\|^p\Big)^{1/p} \leq V_p^{(\delta)}(f, \tau). \tag{2.11}$$

Indeed, for each $i = 0,\dots,n$, take a sequence $\{y_{mi}\colon m \geq 1\}$ of points of $[a,b]$ such that for $i \in J_{\pm}(\kappa^*)$, $y_{mi} \to x_i$ from the left if x_{i-1} is an $x-$ or $y_{mi} \to x_{i-1}$ from the right if x_i is an $x+$ and $y_{mn} \to x_n$ from the left if τ is right open, while $y_{mi} \equiv x_i$ for the rest of the indices. Then as $m \to \infty$,

$$\sum_{i=1}^{n} \|f(y_{mi}) - f(y_{m,i-1})\|^p \to \sum_{i=1}^{n} \|f(x_i) - f(x_{i-1})\|^p.$$

For m large enough, $\max_{1\leq i\leq n}(y_{mi} - y_{m,i-1}) < \delta$, and, therefore, (2.11) follows. The same argument implies (cf. (2.4) and (2.7))

$$\mathfrak{S}_p(f) \leq V_p^*(f) \leq V_p(f) \tag{2.12}$$

and

$$v_p(f;\tau) = \sup\Big\{\sum_{\kappa} \|\Delta^{\kappa} f\|^p\colon \kappa \in Q^{\pm}(\tau)\Big\} \tag{2.13}$$

for any interval τ which may be open or closed at either end or may have a point $x\pm$ at either end. Moreover, we have:

Lemma 2.5. *Let τ be a non-degenerate interval, open or closed at either end, and let f be a $\mathbb{B}$-valued function on τ. Then the following hold:*

(1) *If f has a right limit at $y \in \overline{\tau}$, which is not the right endpoint of τ, then for each $z > y$, $z \in \tau$,*

$$v_p(f;[y+,z]) = v_p(f;(y,z]).$$

(2) *If f has a left limit at $z \in \overline{\tau}$, which is not the left endpoint of τ, then for each $y < z$, $y \in \tau$,*

$$v_p(f;[y,z-]) = v_p(f;[y,z)).$$

Proof. It will be enough to prove the first statement since the second is symmetrical. Let y and z be as in statement (1). It is clear that

$$v_p(f; [y+, z]) \geq v_p(f; (y, z]).$$ (2.14)

To prove the reverse inequality assume $v_p(f; (y, z]) < \infty$. Let $\kappa := \{y+ = x_0 < x_1 < \dots, x_n = z\}$ and let $\epsilon > 0$. Since f has a right limit at y there exists $x \in (y, x_1)$ such that

$$\|f(x_1) - f(y+)\|^p \leq \epsilon + \|f(x_1) - f(x)\|^p.$$

Then $\{x, x_1, \dots, x_n\}$ is a partition of $(y, z]$ and

$$\sum_{i=1}^{n} \|\Delta_i^\kappa f\|^p \leq \epsilon + \|f(x_1) - f(x)\|^p + \sum_{i=2}^{n} \|\Delta_i^\kappa f\|^p \leq \epsilon + v_p(f; (y, z]).$$

Since κ and $\epsilon > 0$ are arbitrary this implies $v_p(f; [y+, z]) \leq v_p(f; (y, z])$ which together with (2.14) implies statement (2) and, hence, Lemma 2.5. $\square$

It follows from (2.11) for δ large enough that, for each $f \in \mathcal{R}([a, b]; \mathbb{B})$,

$$V_p(f_-^{(a)}; [a, b]) \leq V_p(f; [a, b)) \quad \text{and} \quad V_p(f_+^{(b)}; [a, b]) \leq V_p(f; (a, b]).$$ (2.15)

Here and later on we use the following notation: for any $f \in \mathcal{R}([a, b]; \mathbb{B})$ and point y, let

$$f_+^{(y)}(x) := \begin{cases} f(x+), & \text{for } x \neq y \\ f(y), & \text{for } x = y \end{cases} \quad \text{and} \quad f_-^{(y)}(x) := \begin{cases} f(x-), & \text{for } x \neq y \\ f(y), & \text{for } x = y. \end{cases}$$ (2.16)

Given a function f on $[a, b]$ with a left limit at b and a right limit at a, define a function $f^{(b-)}$ on $[a, b]$ by

$$f^{(b-)}(x) := \begin{cases} f(x), & \text{for } x \in [a, b) \\ f(b-), & \text{for } x = b \end{cases}$$ (2.17)

for $b > a$ and $f^{(a-)}(a) := f(a)$. Likewise define a function $f^{(a+)}$ on $[a, b]$ by

$$f^{(a+)}(x) := \begin{cases} f(a+), & \text{for } x = a \\ f(x), & \text{for } x \in (a, b] \end{cases}$$ (2.18)

for $a < b$ and $f^{(b+)}(b) := (b)$.

2.3. Quasinorms and the space $\mathcal{W}_p$. A function $\alpha(\cdot) \geq 0$ on a real vector space V will be called a *semi-quasinorm* if
(a) $\alpha(ru) = |r|\alpha(u)$ for all $r \in \mathbb{R}$ and $u \in V$;
(b) For some $c < \infty$, $\alpha(u + v) \leq c[\alpha(u) + \alpha(v)]$ for all $u, v \in V$.
    If in addition
(c) $\alpha(u) = 0$ if and only if $u = 0$,

then α will be called a *quasinorm* and (V, α) a *quasinormed space*, see e.g. Köthe (1969, §15.10).

For any semi-quasinorm $\alpha(\cdot)$ there is clearly a smallest $c = c(\alpha) \geq 0$ for which (b) holds. If for some $u \in V$, $\alpha(u) > 0$, then $c(\alpha) \geq 1$, and $c(\alpha) = 1$ if and only if α is a seminorm, or a norm if α is a quasinorm.

For $0 < q \leq 1$ a quasinorm α on V is called *q-subadditive* if

(d) $\alpha(u + v)^q \leq \alpha(u)^q + \alpha(v)^q$ for all $u, v \in V$.

By Köthe (1969, §15.10), every quasinorm is equivalent to a q-subadditive quasinorm for some q, $0 < q \leq 1$.

Recall the definition (2.5) of $\mathcal{W}_p = \mathcal{W}_p(\tau, \mathbb{B})$. Each function in $\mathcal{W}_p(\tau; \mathbb{B})$ belongs to $\ell_\infty(\tau; \mathbb{B}) := \{f: \tau \mapsto \mathbb{B}: \|f\|_\infty := \sup_{x \in \tau} \|f(x)\| < \infty\}$ (even if τ is unbounded). Let

$$V_{p,\infty}(f) := V_p(f) + \|f\|_\infty.$$

Then on $\mathcal{W}_p$, V_p is a semi-quasinorm and $V_{p,\infty}$ is a quasinorm: $f = 0$ iff $V_{p,\infty}(f) = 0$, $V_{p,\infty}(rf) = |r|V_{p,\infty}(f)$ for all $r \in \mathbb{R}$, and for all $f, g \in \mathcal{W}_p$,

$$V_p(f + g) \leq c_p[V_p(f) + V_p(g)], \tag{2.19}$$

and

$$V_{p,\infty}(f + g) \leq c_p[V_{p,\infty}(f) + V_{p,\infty}(g)], \tag{2.20}$$

where $c_p = 1 \vee 2^{(1/p)-1}$ and (2.20) follows from (2.19) since $c_p \geq 1$. To prove (2.19) it suffices to note that, by the Minkowski inequality if $p \geq 1$ or convexity of the function $t \mapsto t^{1/p}$ if $p \leq 1$, for any $u_i, v_i \in \mathbb{B}$, $i = 1, \ldots, n$, we have

$$\left(\sum_{i=1}^n \|u_i + v_i\|^p\right)^{1/p} \leq c_p\left[\left(\sum_{i=1}^n \|u_i\|^p\right)^{1/p} + \left(\sum_{i=1}^n \|v_i\|^p\right)^{1/p}\right]. \tag{2.21}$$

Let $\tilde{V}_{p,\infty}(f) := V_p(f) \vee \|f\|_\infty$. Then on $\mathcal{W}_p$, $\tilde{V}_{p,\infty}$ is a $1 \wedge p$-subadditive quasinorm. Moreover, $\tilde{V}_{p,\infty}(f) \leq V_{p,\infty}(f) \leq 2\tilde{V}_{p,\infty}(f)$ for all $f \in \mathcal{W}_p$. We then have:

Lemma 2.6. *For $0 < p < \infty$, $c(V_p) = c(V_{p,\infty}) = c(\tilde{V}_{p,\infty}) = c_p$.*

Proof. If $p \geq 1$ then V_p is a seminorm, $V_{p,\infty}$ and $\tilde{V}_{p,\infty}$ are norms, $c_p = 1$ and the lemma holds. If $p < 1$, we have $c(V_p) \leq c_p$, $c(\tilde{V}_{p,\infty}) \leq c_p$ by (2.19) and $c(V_{p,\infty}) \leq c_p$ by (2.20). Conversely, for $n = 1, 2, \cdots$, let $f_n(0) := 0$, $f_n(1/(4n)) = 1/n$, $f_n(1/(2n)) = f_n(1/n) = 0$, and let f_n be linear on each closed interval with endpoints where it was just defined. Let f_n be periodic of period $1/n$ on $[0,1]$ and 0 outside it. Let $g_n(x) \equiv f_n(x - [1/(2n)])$. Then $v_p(f_n) = v_p(g_n) = 2n^{1-p}$ and $v_p(f_n + g_n) = 4n^{1-p}$. It follows that $c(V_p) = c_p$. Letting $n \to \infty$, we also get $c(V_{p,\infty}) = c_p = (\tilde{V}_{p,\infty})$. $\quad\square$

For any quasinormed space (V, α), and $x \in V$, the sets $\{y \in V : \alpha(x - y) < \delta\}$ form a base for a system of neighborhoods of x. These neighborhoods define a unique topology $\mathcal{T} = \mathcal{T}(\alpha)$. It is easily verified that $(V, \mathcal{T}(\alpha))$ is a topological vector space. $\mathcal{T}(\alpha)$ is not necessarily locally convex, in fact $\mathcal{T}(V_{p,\infty})$ is not for $0 < p < 1$ (Proposition 2.17 below). $\mathcal{T}(\alpha)$ is locally bounded and thus is always metrizable by some metric d with the invariance property $d(x,y) = d(x-u,y-u)$ for all x, y, $u \in V$ (cf. §15.10 in Köthe, 1969).

The metric

$$d_{p,\infty}(f,g) := V_p^{p\wedge 1}(f-g) + \|f-g\|_\infty^{p\wedge 1}, \quad \forall f,g \in W_p,$$

defines $\mathcal{T}(V_{p,\infty})$.

If τ is a non-degenerate interval then the space $W_p(\tau; \mathbb{B})$, $0 < p < \infty$, is non-separable. This can be shown using an uncountable family of indicator functions of subintervals.

Lemma 2.7. $W_p(\tau; \mathbb{B})$, $0 < p < \infty$, is complete relative to $d_{p,\infty}$.

Proof. Let $\{f_k: k \geq 1\} \subset W_p(\tau, \mathbb{B})$ be a Cauchy sequence, i.e.

$$\lim_{k,m\to\infty} d_{p,\infty}(f_k, f_m) = 0. \tag{2.22}$$

Then $\lim_{k,m\to\infty} \|f_k - f_m\|_\infty = 0$. Since $\ell_\infty(\tau; \mathbb{B})$ is complete, there exists $f \in \ell_\infty(\tau; \mathbb{B})$ such that $\lim_{k\to\infty} \|f_k - f\|_\infty = 0$. To show that $f \in W_p$, by (2.22), there exists N such that for all $k, m \geq N$ and each partition $\kappa = \{x_i: i = 0, \ldots, n\}$ of τ, we have

$$\Big(\sum_{i=1}^n \|(f_k - f_m)(x_i) - (f_k - f_m)(x_{i-1})\|^p\Big)^{1/p} \leq V_{p,\infty}(f_k - f_m) \leq 1.$$

Letting $m \to \infty$ we get $\big(\sum_{i=1}^n \|(f_k - f)(x_i) - (f_k - f)(x_{i-1})\|^p\big)^{1/p} \leq 1$ for each partition κ and all $k \geq N$. So by (2.21), we have

$$\Big(\sum_{i=1}^n \|f(x_i) - f(x_{i-1})\|^p\Big)^{1/p} \leq c_p\Big[\Big(\sum_{i=1}^n \|\Delta_i^\kappa(f_N - f)\|^p\Big)^{1/p} + \Big(\sum_{i=1}^n \|\Delta_i^\kappa f_N\|^p\Big)^{1/p}\Big]$$

$$\leq c_p[1 + V_p(f_N)] < \infty.$$

The same argument except that 1 is replaced by an arbitrary $\epsilon > 0$ implies that $\lim_{n\to\infty} V_p(f_k - f) = 0$. This proves Lemma 2.7.   □

For $p \geq 1$, $V_{p,\infty}$ is a norm. So, the p-variation norm will be defined to be

$$\|f\|_{[p]} := \|f\|_{(p)} + \|f\|_\infty := V_p(f) + \|f\|_\infty.$$

By Lemma 2.7, W_p equipped with the p-variation norm is a Banach space.

The following is known for $p \geq 1$ (Krabbe, 1961a, 1961b).

Lemma 2.8. Let $0 < p < \infty$, let τ be any interval and f, $g \in W_p(\tau; \mathbb{B})$. Then $fg \in W_p(\tau; \mathbb{B})$ and

$$V_{p,\infty}(fg) \leq c_p V_{p,\infty}(f) V_{p,\infty}(g).$$

Thus for $1 \leq p < \infty$, $W_p(\tau; \mathbb{B})$ is a Banach algebra with pointwise multiplication.

Proof. For any $x, y \in \tau$,

$$\|(fg)(x) - (fg)(y)\| \leq \|(f(x) - f(y))g(x)\| + \|f(y)(g(x) - g(y))\|$$

$$\leq \|g\|_\infty \|f(x) - f(y)\| + \|f\|_\infty \|g(x) - g(y)\|.$$

So by (2.21),

$$V_p(fg) \leq c_p[\|g\|_\infty V_p(f) + \|f\|_\infty V_p(g)]$$

and the conclusions follow.   □

A quasinormed space (W, β) is called *complete* if it is complete for $\mathcal{T}(\beta)$ or equivalently for any invariant metric metrizing $\mathcal{T}(\beta)$. The uniform structure on W also has a base given by the vicinities

$$\{(x, y) \in W \times W : \beta(x - y) < r\}, \quad r > 0.$$

A *β-Cauchy sequence* $\{w_k\}$ is one such that $\lim_{j \to \infty} \sup_{k \geq j} \beta(w_k - w_j) = 0$. Thus, (W, β) is complete if and only if every β-Cauchy sequence converges.

Lemma 2.9. *Let (W, β) be a complete quasinormed space. If $f_j \in W$ for all $j = 1, 2, \ldots$, and $\sum_{j=1}^\infty c(\beta)^j \beta(f_j) < \infty$ then the series $\sum_{j=1}^\infty f_j$ converges in W to some f in W with $\beta(f) \leq \sum_{j=1}^\infty c(\beta)^j \beta(f_j)$.*

Proof. Let $w_k := \sum_{j=1}^k f_j$. Iterating (b) from the definition of a quasinorm gives

$$\beta(w_k) \leq c(\beta)[\beta(f_1) + \beta(\sum_{j=2}^k f_j)] \leq c(\beta)\beta(f_1) + c(\beta)^2 \beta(f_2) + \cdots + c(\beta)^k \beta(f_k).$$

We get likewise that, for each $k < n$, $\beta(w_n - w_k) \leq \sum_{j=k+1}^n c(\beta)^{j-k} \beta(f_j) \to 0$ uniformly in $n > k$ as $k \to \infty$. Thus $\{w_n\}$ is a β-Cauchy sequence and converges to some $f \in W$. Since $\beta(\cdot)$ is continuous, $\beta(f)$ is bounded as stated.   □

Proposition 2.10. $(\mathcal{W}_p, V_{p,\infty})$ *is complete.*

Proof. A $V_{p,\infty}$–Cauchy sequence $\{f_n\}$, being Cauchy for $\|\cdot\|_\infty$, converges uniformly to some function f. It is then easily seen that $V_{p,\infty}(f_n - f) \to 0$.   □

Next is a fact about $\mathcal{W}_p$ for $0 < p < 1$. It will follow that, for $0 < p < 1$, $\mathcal{W}_p$ is quite a small space: the only continuous functions in it are constants. Proposition 2.12 will give a characterization.

Theorem 2.11. *Let $\mathbb{B} = (\mathbb{B}, \|\cdot\|)$ be any Banach space, $a < b$, $0 < p < 1$ and $f \in \mathcal{W}_p([a, b]; \mathbb{B})$. Then, for $a < x \leq b$,*

$$f(x) = f(a) + \Delta^+ f(a) + \left[\sum_{a < y < x} \Delta_\pm f(y) \right] + \Delta_- f(x), \qquad (2.23)$$

where the series converges absolutely.

Proof. Since $p < 1$, (2.12) implies that the series in (2.23) converges absolutely. Let $g(a) := f(a)$ and for $a < x \leq b$ define $g(x)$ by the right side of (2.23). Then g is a regulated function and

$$V_p(g) \leq \mathfrak{S}_p(f). \qquad (2.24)$$

Indeed, let $\kappa = \{x_i : i = 0, \ldots, n\}$ be a partition of $[a,b]$. Since $p < 1$ it follows that

$$\sum_{i=1}^{n} \|\Delta_i^\kappa g\|^p \leq \sum_{i=1}^{n} \{\|\Delta_- f(x_i)\|^p + \sum_{x_{i-1} < y < x_i} (\|\Delta_- f(y)\|^p + \|\Delta_+ f(y)\|^p)$$
$$+ \|\Delta_+ f(x_{i-1})\|^p\} = \mathfrak{S}_p^p(f).$$

Since κ is an arbitrary partition this implies (2.24). Since the jumps of g are exactly those of f, $h := f - g$ is continuous. By (2.24), $g \in W_p$ and, hence, $h \in W_p$. It will be shown that a continuous function h in $W_p([a,b]; \mathbb{B})$, $0 < p < 1$, is a constant. Suppose not. Then we can assume $h(a) \neq h(b)$ and $\|h(a) - h(b)\| = 1$. Let $x_0 := a$, $x_1 := b$. There exists some $x_{1/2}$, $a < x_{1/2} < b$, such that $\|h(a) - h(x_{1/2})\| = 1/2$. For the least such $x_{1/2}$, $\|h(b) - h(x_{1/2})\| \geq 1/2$. Recursively, we can define $x(t) := x_t$ for each dyadic rational $t = j/2^k$ for $j = 0, 1, \ldots, 2^k$, such that

$$\left\|h(x(j/2^k)) - h(x((j-1)/2^k))\right\| \geq 1/2^k$$

for all $j = 1, \ldots, 2^k$. Thus $v_p(h) \geq 2^k/2^{kp} \to \infty$ as $k \to \infty$, a contradiction. So h is a constant and for $h = f - g$, with $h(a) = 0$, we get $f \equiv g$, proving Theorem 2.11.$\Box$

The following elaborates on Theorem 2.11.

Proposition 2.12. *If $0 < p < 1$ then $f \in W_p([a,b])$ if and only if there exist v_a, u_b, and u_y, v_y for $a < y < b$, all in $\mathbb{B}$, such that, for $a < x \leq b$,*

$$f(x) = f(a) + v_a + \sum_{a < y < x} (u_y + v_y) + u_x, \tag{2.25}$$

where each sum converges absolutely, and

$$\sum_{a < y < b} (\|u_y\|^p + \|v_y\|^p) < \infty. \tag{2.26}$$

Then

$$\Delta_+ f(x) = v_x \text{ for } a \leq x < b, \quad \Delta_- f(x) = u_x \text{ for } a < x \leq b \tag{2.27}$$

and

$$V_p^*(f) = V_p(f) = \mathfrak{S}_p(f) = \{\|v_a\|^p + \sum_{a < y < b} (\|u_y\|^p + \|v_y\|^p) + \|u_b\|^p\}^{1/p}. \tag{2.28}$$

Proof. "Only if" follows from Theorem 2.11 since the sum in (2.26) is $\leq \mathfrak{S}_p(f)^p \leq v_p(f) < \infty$ by (2.12).

To prove "if," (2.26) for $p < 1$ implies that it holds for $p = 1$, so for each x, the sum in (2.25) converges absolutely and f is well-defined. It follows easily that f is of bounded variation with

$$v_1(f) \leq \|v_a\| + \|u_b\| + \sum_{a < y < b} \|u_y\| + \|v_y\|.$$

Thus f is regulated. By (2.26) for $p = 1$, for each $x \in [a, b)$, we have $\| \sum_{x<y<z}(u_y + v_y) + u_z \| \to 0$ as $z \downarrow x$. Thus, the right limit of f at x exists and is equal to

$$f(x+) := f(a) + v_a + \sum_{a<y\leq x}(u_y + v_y).$$

Similarly, the left limit of f exists at any $x \in (a, b]$ and (2.27) holds. Then the third (last) equality in (2.28) holds. Since for $0 < p \leq 1$, $\| \sum w_i \|^p \leq \sum \|w_i\|^p$ for any summable $w_i \in \mathbb{B}$, it follows that $V_p(f) \leq \mathfrak{S}_p(f)$ in this case, so the other equalities in (2.28) hold by (2.12). The proof of Proposition 2.12 is complete. □

Next we define the set of f whose "local" p-variation comes only from jumps,

$$W_p^*(\tau; \mathbb{B}) := \{f \in W_p(\tau; \mathbb{B}): \mathfrak{S}_p(f) = V_p^*(f)\}. \tag{2.29}$$

By Proposition 2.12, $W_p^* = W_p$ for $p < 1$. For $p = 1$ we have:

Proposition 2.13. *For any function $f: [a, b] \mapsto \mathbb{B}$,*

(1) $V_1^*(f) = V_1(f)$.
(2) *We have $f \in W_1^*$ if and only if (2.23), or equivalently (2.27) and (2.25), and both (2.26) and (2.28) for $p = 1$ hold.*

Proof. For (1) we have $V_p^*(f) \leq V_p(f)$ always by (2.12). The converse inequality holds for $p = 1$ just from the triangle inequality for the norm in $\mathbb{B}$.

For (2), the "if" part, the proof of Proposition 2.12 applies since $p < 1$ is not needed in this direction. For "only if," since $f \in W_1$, it is regulated. Define u_y, v_y by (2.27), let $f_1(a) := f(a)$ and define $f_1(x)$ for $a < x \leq b$ by the right side of (2.25). Then also by the proof of Proposition 2.12, (2.28) with $p = 1$ holds for f_1. Clearly $\mathfrak{S}_1(f_1) = \mathfrak{S}_1(f)$ which by the assumption and part (1) also equal $V_1^*(f) = V_1(f)$. Now, $f - f_1$ is of bounded variation and has no jumps, so from the known decomposition of functions of bounded variation, $V_1(f) = V_1(f_1)+V_1(f-f_1)$. Thus $V_1(f - f_1) = 0$ and $f - f_1$ is a constant, 0 at a, so $f \equiv f_1$, and the rest follows, so Proposition 2.13 is proved. □

By Love and Young (1938, Theorems 18 and 19), if $p \geq 1$ then $f \in W_p^*([a,b]; \mathbb{R})$ if and only if f can be approximated in v_p by step-functions. Moreover, we have

Lemma 2.14. *If $1 \leq q < p < \infty$ then $W_q(\tau; \mathbb{B}) \subset W_p^*(\tau; \mathbb{B})$.*

The proof of Lemma 2.14 is based on the following known statement. We include it here for easier reference.

Lemma 2.15. *Let $f \in \mathcal{R}([a, b]; \mathbb{B})$, $a \leq c < d \leq b$ and $\|\Delta^+ f(x)\| \vee \|\Delta_- f(x)\| < \alpha$ for all $x \in (c, d)$. Then there exists a $\delta > 0$ such that*

$$\|f(x) - f(y)\| < 2\alpha \tag{2.30}$$

whenever $x, y \in (c, d)$ and $|x - y| < \delta$. The same statement is also true when the open interval (c, d) is replaced by any of the intervals $\tau = [c, d)$, $(c, d]$ or $[c, d]$ and, if $c \in \tau$, $\|\Delta^+ f(c)\| < \alpha$ and, if $d \in \tau$, $\|\Delta_- f(d)\| < \alpha$.

Remark. A variant of this statement is proved in Lebesgue (1973, p. 21) and attributed to R. Baire.

Proof. We will prove only the case when the interval (c, d) is open and note that the other cases can be treated similarly. Since f is regulated there exists $0 < \delta_1 < (d-c)/2$ such that (2.30) holds for all $x, y \in (c, c+2\delta_1] \cup [d-2\delta_1, d)$ with $|x-y| < 2\delta_1$. Thus, it suffices to find $0 < \delta \leq \delta_1$ such that (2.30) holds also for $x, y \in [c+\delta_1, d-\delta_1]$ with $|x-y| < \delta$. If not, then there exist $\{x_n, y_n \colon n \geq 1\} \subset [c+\delta_1, d-\delta_1]$ such that $0 < y_n - x_n < 1/n$ and $\|f(y_n) - f(x_n)\| \geq 2\alpha$. We can assume that $x_n \to z$ for some $z \in [c+\delta_1, d-\delta_1]$. Then there exists an infinite subsequence (n') such that one of the following three cases holds: $y_{n'} \leq z$ for all n'; or $x_{n'} \leq z < y_{n'}$ for all n'; or $z < x_{n'}$ for all n'. In each of these cases it follows that: either $f(z-)$ does not exist or $\|\Delta_- f(z)\| \geq 2\alpha$; either $\|\Delta_+ f(z)\| \geq 2\alpha$ or $\|\Delta_-^+ f(z)\| \geq 2\alpha$; $f(z+)$ does not exist, respectively. All these conclusions are impossible because of our assumptions and because $\|\Delta_-^+ f(z)\| \leq \|\Delta^+ f(z)\| + \|\Delta_- f(z)\| < 2\alpha$. This completes the proof of Lemma 2.15. □

Proof of Lemma 2.14. By (2.12), given $\epsilon > 0$, it suffices to find a partition λ of τ such that

$$\sup_{\kappa \supset \lambda} \sum_\kappa \|\Delta^\kappa f\|^p \leq \mathfrak{S}_p(f)^p + \epsilon. \tag{2.31}$$

To this end, by Lemma 2.2, there exists a finite set $\mu \subset \tau$ such that

$$\|\Delta_- f(x)\| \vee \|\Delta^+ f(x)\| < \frac{1}{2}\Big(\frac{\epsilon}{2v_q(f)}\Big)^{1/(p-q)} \tag{2.32}$$

for all $x \in \tau \setminus \mu$. Using Lemma 2.15 with α equal to the right side of (2.32), one can find a $\delta_1 > 0$ such that

$$\|f(z) - f(y)\| < \big(\epsilon/[2v_q(f)]\big)^{1/(p-q)} \tag{2.33}$$

whenever $[y, z] \subset \tau \setminus \mu$ and $z - y < \delta_1$. Let $\mu =: \{\xi_j \colon j = 1, \ldots, N\}$. Using Lemma 2.4 one can find a $\delta_2 > 0$ such that

$$\sum_{j=1}^N \big(\|f(\xi_j) - f(y_j)\|^p + \|f(z_j) - f(\xi_j)\|^p\big) \leq \mathfrak{S}_p(f)^p + \epsilon/2 \tag{2.34}$$

for all $y_j \in [\xi_j - \delta_2, \xi_j]$ with $y_1 = \xi_1$ if ξ_1 is the left endpoint of τ and for all $z_j \in [\xi_j, \xi_j + \delta_2]$ with $z_N = \xi_N$ if ξ_N is the right endpoint of τ.

Now take any partition λ of $[a, b]$ such that $\lambda \supset \mu$ and $|\lambda| < \delta_1 \wedge \delta_2$. Let $\kappa = \{x_i \colon i = 0, \ldots, n\}$ be any refinement of λ and let $J(\kappa) := \{i \colon x_i = \xi \text{ or } x_{i-1} = \xi$ for some $\xi \in \mu\}$. Then we have

$$\sum_\kappa \|\Delta^\kappa f\|^p = \sum_{i \in J(\kappa)} \|\Delta_i^\kappa f\|^p + \sum_{i \notin J(\kappa)} \|\Delta_i^\kappa f\|^p$$

by (2.34) $\qquad \leq \mathfrak{S}_p(f)^p + \epsilon/2 + v_q(f) \max_{i \notin J(\kappa)} \|\Delta_i^\kappa f\|^{p-q}$

by (2.33) $\qquad \leq \mathfrak{S}_p(f)^p + \epsilon,$

so (2.31) holds and the proof of Lemma 2.14 is complete. □

Lemma 2.16. *Let* $0 < p < \infty$. *If* $p < 1$ *let* $r = p$, *or if* $p \geq 1$ *let* $p < r < \infty$. *Let* τ *be a non-empty interval. Then for any* $f \in \mathcal{W}_p(\tau; \mathbb{B})$ *and* $\epsilon > 0$ *there exists a step function* ϕ *from* τ *into* $\mathbb{B}$ *such that* $V_{r,\infty}(f - \phi) < \epsilon$.

Proof. If $p = r < 1$, then by Theorem 2.11, f is a pure jump function. If ϕ is a step function with $\phi(c) = f(c)$ for some $c \in \tau$ and jumps $\Delta^+ \phi = \Delta^+ f$ and $\Delta_- \phi = \Delta_- f$ on large enough finite subsets of τ, where ϕ is constant on each interval between such jumps, then $V_{p,\infty}(f - \phi) < \epsilon$ by the proof of Theorem 2.11.

If $1 \leq p < r < \infty$ Love and Young (1938, Lemma 2) show for a function $f \in \mathcal{W}_p([a,b]; \mathbb{R})$ that there is a step function ϕ with $V_r(f - \phi) < \epsilon/2$. Their proof extends also to $\mathbb{B}$-valued functions. Taking $\phi(a) = f(a)$ as in their proof gives $V_{r,\infty}(f - \phi) < \epsilon$. The proof also adapts easily to the case where τ does not contain its left endpoint and a is replaced by a point $c \in \tau$. $\qquad \square$

We finish this subsection with another negative fact about the space $\mathcal{W}_p$ for $0 < p < 1$.

Proposition 2.17. *For* $0 < p < 1$ *and* $a < b$, $\mathcal{W}_p([a,b]; \mathbb{R})$ *with the metric* $d_{p,\infty}$ *is not locally convex.*

Proof. For any $\delta > 0$, each $m = 1, 2, \dots$ and $j = 1, \dots, m$, let

$$f_j(x) := \begin{cases} 0, & \text{if } 0 \leq x < j/m \\ \delta, & \text{if } j/m \leq x \leq 1. \end{cases}$$

Then $d_{p,\infty}(f_j, 0) = 2\delta^p$ for each $j = 1, \dots m$. Let $f := m^{-1} \sum_{j=1}^m f_j$. Then

$$d_{p,\infty}(f, 0) = m(\delta/m)^p + \delta^p = (m^{1-p} + 1)\delta^p \to \infty \quad \text{as } m \to \infty.$$

Thus, any neighborhood of zero in $\mathcal{W}_p$ for $d_{p,\infty}$ has a convex hull containing elements arbitrarily far from zero. $\qquad \square$

2.4. Auxiliary results. The main result of this subsection is Lemma 2.20 which is basic for the paper; it is the key to the proofs given for the main theorems. To prepare the proof of it we start with two other facts.

Lemma 2.18. *If* $f \in \mathcal{W}_p(\tau; \mathbb{B})$, $0 < p < \infty$, *and* $g \in \mathcal{R}(\tau; \mathbb{B})$ *then, for any* $\xi \in \bar{\tau}$ *which is not the right endpoint of* τ,

$$\lim_{y \downarrow \xi} v_p(f; (\xi, y]) = 0 \quad \text{and} \quad \lim_{y \downarrow \xi} Osc_{(\xi, y]}(g) = \lim_{y \downarrow \xi} Osc_{[\xi+, y]}(g) = 0.$$

Also, for any $\xi \in \bar{\tau}$ *which is not the left endpoint of* τ,

$$\lim_{y \uparrow \xi} v_p(f; [y, \xi)) = 0 \quad \text{and} \quad \lim_{y \uparrow \xi} Osc_{[y, \xi)}(g) = \lim_{y \uparrow \xi} Osc_{[y, \xi-]}(g) = 0.$$

Proof. It will be enough to prove the first statement for the function f because it follows for g just by the definition of regulated functions and the second statement is symmetrical. We have $v_p(f; (\xi, y]) \downarrow$ as $y \downarrow \xi$ to some limit. Suppose it is $\delta > 0$. Take $y > \xi$ such that $v_p(f, (\xi, y]) < 3\delta/2$. Choose $\xi < y_1 < \cdots < y_m < y$ such that $\sum_{j=2}^m \|f(y_j) - f(y_{j-1})\|^p > 3\delta/4$. Then choose $\xi < x_1 < \cdots < x_n < y_1$ such that $\sum_{i=2}^n \|f(x_i) - f(x_{i-1})\|^p > 3\delta/4$. Letting $x_{n+j} := y_j$ for $j = 1, \dots, m$ gives $\sum_{i=2}^{m+n} \|f(x_i) - f(x_{i-1})\|^p > 3\delta/2$, a contradiction. $\qquad \square$

The following statement is an extension of the previous one to closed intervals.

Lemma 2.19. *If $f \in \mathcal{W}_p(\tau; \mathbb{B})$, $0 < p < \infty$, and $g \in \mathcal{R}(\tau; \mathbb{B})$ then, for any $\xi \in \tau$ except for the right endpoint of τ,*

$$\lim_{y \downarrow \xi} v_p(f; [\xi, y]) = \|(\Delta^+ f)(\xi)\|^p =: v_p(f; \xi+) \quad and \quad \lim_{y \downarrow \xi} Osc_{[\xi, y]}(g) = \|(\Delta^+ f)(\xi)\|.$$

Also, for any $\xi \in \tau$ except for the left endpoint of τ,

$$\lim_{y \uparrow \xi} v_p(f; [y, \xi]) = \|(\Delta_- f)(\xi)\|^p =: v_p(f; \xi-) \quad and \quad \lim_{y \uparrow \xi} Osc_{[y, \xi]}(g) = \|(\Delta_- f)(\xi)\|.$$

Proof. As in the previous lemma, it will be enough to prove the first statement for the function f. First, $v_p(f; \xi+) \le v_p(f; [\xi, y])$ which decreases to some limit as $y \downarrow \xi$. Suppose the limit is $v_p(f; \xi+) + \delta$ for some $\delta > 0$. Hence one can find $z > \xi$, $z \in \tau$, such that

$$v_p(f; [\xi, z]) < v_p(f; \xi+) + 5\delta/4 \tag{2.35}$$

and, due to the existence of $f(\xi+)$, for all $y \in (\xi, z]$,

$$\|f(y) - f(\xi)\|^p < v_p(f; \xi+) + \delta/8. \tag{2.36}$$

Choose a partition $\{x_i : i = 0, \ldots, n\} \in Q([\xi, z])$ such that

$$\sum_{i=1}^{n} \|f(x_i) - f(x_{i-1})\|^p > v_p(f; \xi+) + 7\delta/8.$$

Then by (2.36), we have

$$\sum_{i=2}^{n} \|f(x_i) - f(x_{i-1})\|^p > 3\delta/4. \tag{2.37}$$

Doing the same for x_1 in place of z we get a partition $\{y_j : j = 1, \ldots, m\} \in Q((\xi, x_1])$ such that

$$\sum_{j=2}^{m} \|f(y_j) - f(y_{j-1})\|^p > 3\delta/4. \tag{2.38}$$

Due to the existence of $f(\xi+)$, again one can find $y_0 \in (\xi, y_1)$ such that

$$\|f(y_0) - f(\xi)\|^p > v_p(f; \xi+) - \delta/4. \tag{2.39}$$

Combining (2.37), (2.38) and (2.39), which are sums over nonoverlapping intervals, we get $v_p(f; [\xi, z]) > v_p(f, \xi+) + 5\delta/4$, contradicting (2.35). $\qquad\square$

Now, recalling the notation (2.10), we are ready to prove the main result of this subsection.

Lemma 2.20. *Let* $f \in \mathcal{W}_p([a,b]; \mathbb{B})$, $0 < p < \infty$. *Given a number* $c > 0$, *there exists a partition* $\kappa = \{x_i : i = 0, \ldots, n\} \in Q^{\pm}([a,b])$ *such that, for each* $i \notin J_{\pm}(\kappa)$, $v_p(f; [x_{i-1}, x_i]) \le c$, *and* $n + 1 = card(\kappa) \le 4 + 3v_p(f; [a,b])/c$.

Proof. Recursively, let $x_0 := a$. If x_i is an x or an $x+$ (not an $x-$) let $y_i := \sup\{y > x_i : y \le b, v_p(f; [x_i, y]) \le c\}$ if such y exist. If x_i is an $x+$ such y always exist by Lemmas 2.5 and 2.18. If such y exist, let $x_{i+1} := y_i$ if $v_p(f; [x_i, y_i]) \le c$, otherwise let $x_{i+1} := y_i-$. If x_i is an x and no such y exist let $x_{i+1} := x_i+$. If x_i is an $x-$ let $x_{i+1} := x$. The recursion stops if some $x_i = b$.

Claim: $v_p(f; [x_i, x_{i+3}]) \ge c$ whenever x_{i+3} is defined and $x_{i+3} < b$.

To prove the claim it will be enough to show that $v_p(f; [x_i, x_{i+2}]) \ge c$ when x_i is not an $x-$. If $x_{i+1} = y_i$ then either $x_{i+2} = y_i+$, and then

$$v_p(f; [x_{i+1}, x_{i+2}]) = \|f(y_i+) - f(y_i)\|^p \ge c$$

by Lemma 2.19, or $x_{i+2} > y_i+$, and then $v_p(f; [x_i, x_{i+2}]) > c$ by definition of y_i. If y_i is not defined then $x_{i+1} = x_i+$ and

$$v_p(f; [x_i, x_{i+1}]) = \|f(x_i+) - f(x_i)\|^p \ge c$$

by Lemma 2.19 again, proving the claim.

The claim implies that $v_p(f; [x_0, x_{3k}]) \ge kc$ if x_{3k} is defined and $< b$. This is a contradiction if $kc > v_p(f; [a,b])$. If k is the smallest integer $> v_p(f; [a,b])/c$, then either x_{3k} is not defined, and card$(\kappa) \le 3k$, or $x_{3k} = b$ and card$(\kappa) = 3k + 1$. So

$$\text{card}(\kappa) \le 3k + 1 \le 4 + 3v_p(f; [a,b])/c.$$

This completes the proof of Lemma 2.20.   □

The proof of the next statement is essentially the same as the proof of Lemma 2.20 and, hence, is omitted.

Lemma 2.21. *Let* $g \in \mathcal{R}([a,b]; \mathbb{B})$. *Given a number* $c > 0$, *there exists a partition* $\kappa = \{x_i : i = 0, \ldots, n\} \in Q^{\pm}([a,b])$ *such that, for each* $i \notin J_{\pm}(\kappa)$, $Osc_{[x_{i-1}, x_i]}(f) \le c$, *and*

$$n + 1 = card(\kappa) \le 4 + 3Osc_{[a,b]}(f)/c.$$

2.5. The space $\mathcal{W}_p^*$. Here we characterize the space $\mathcal{W}_p^*$, with $0 < p < \infty$, defined by (2.29). Recall that, by Proposition 2.12, $\mathcal{W}_p^* = \mathcal{W}_p$ for $0 < p < 1$ and, by Lemma 2.14, $\mathcal{W}_q \subset \mathcal{W}_p^*$ for $1 \le q < p < \infty$.

We start by showing that in the definition (2.7) of V_p^*, partitions may include points $x\pm$.

Lemma 2.22. *Let* $f \in \mathcal{W}_p(\tau; \mathbb{B})$ *with* $0 < p < \infty$. *Then*

$$V_p^*(f)^p = \inf_{\lambda \in Q^{\pm}} \sup \left\{ \sum_{\kappa} \|\Delta^{\kappa} f\|^p : \lambda \subset \kappa \in Q^{\pm} \right\}$$

$$= \inf \left\{ \sum_{i=1}^{n} v_p(f; [x_{i-1}, x_i]) : \{x_i : i = 0, \ldots, n\} \in Q^{\pm} \right\}.$$

Proof. Let $S(\kappa) := \sum_\kappa \|\Delta^\kappa f\|^p$. The argument used to prove (2.8) also yields

$$\sup\left\{S(\kappa) \colon \lambda \subset \kappa \in Q^\pm\right\} = \sum_{i=1}^n v_p(f; [x_{i-1}, x_i]) \tag{2.40}$$

for any partition $\lambda \in Q^\pm(\tau)$ such that the right side is finite. Moreover, the limiting argument used to show (2.11) implies that $\sup\{S(\kappa) \colon \lambda \subset \kappa \in Q\} = \sup\{S(\kappa) \colon \lambda \subset \kappa \in Q^\pm\}$ for each partition $\lambda \in Q(\tau)$. Hence we have $V_p^*(f)^p \geq \inf_{\lambda \in Q^\pm} \sup\{S(\kappa) \colon \lambda \subset \kappa \in Q^\pm\}$. To prove equality, assume to the contrary that $V_p^*(f)^p > \inf_{\lambda \in Q^\pm} \sup\{S(\kappa) \colon \lambda \subset \kappa \in Q^\pm\} + \epsilon$ for some $\epsilon > 0$. Then for some $\lambda \in Q^\pm$, we have $V_p^*(f)^p > \sup\{S(\kappa) \colon \lambda \subset \kappa \in Q^\pm\} + \epsilon$. By adding additional points to λ if necessary we can assume that $x \in \lambda$ if either $x-$ or $x+$ already belongs to λ. Moreover by (2.40), we can and do assume that $\lambda = \{a, a+, b\}$. Let $\lambda_h := \{a, a+h, b\}$. By Lemma 2.19 and by the above assumption, there exists $h > 0$ such that $v_p(f; [a, a+h]) < \|\Delta^+ f(a)\|^p + \epsilon/2$. Then by (2.8), we have

$$\|\Delta^+ f(a)\|^p + \epsilon/2 + v_p(f; [a+h, b]) > \sup\left\{S(\kappa) \colon \lambda_h \subset \kappa \in Q([a,b])\right\} \geq V_p^*(f)^p$$

$$> \sup\left\{S(\kappa) \colon \lambda \subset \kappa \in Q^\pm([a,b])\right\} + \epsilon = \|\Delta^+ f(a)\|^p + v_p(f; [a+, b]) + \epsilon$$

$$\geq \|\Delta^+ f(a)\|^p + v_p(f; [a+h, b]) + \epsilon.$$

This contradiction completes the proof of Lemma 2.22. $\qquad\square$

Using preceding statement one can give the following characterization of the elements of the class $\mathcal{W}_p^*$ defined by (2.29).

Proposition 2.23. *Let $f \in \mathcal{R}([a,b]; \mathbb{B})$ and $0 < p < \infty$. The following are equivalent:*

(1) *$f \in \mathcal{W}_p^*([a,b]; \mathbb{B})$;*

(2) *for every $\epsilon > 0$ there is a partition $\{z_j \colon j = 0, \dots, m\} \in Q([a,b])$ such that*

$$\sum_{j=1}^m v_p(f; [z_{j-1}+, z_j-]) < \epsilon; \tag{2.41}$$

(3) *$\mathfrak{S}_p(f; [a,b]) < \infty$ and for every $\epsilon > 0$ there is a partition $\lambda \in Q([a,b])$ such that*

$$\sum_{i=1}^n \|f(y_i-) - f(y_{i-1}+)\|^p < \epsilon \tag{2.42}$$

for each refinement $\{y_i \colon i = 0, \dots, n\} \in Q([a,b])$ of λ.

Remark. Since (2.41) implies $f \in \mathcal{W}_p([a,b]; \mathbb{B})$ the case $0 < p < 1$ of the above statement follows from Proposition 2.12.

The case $p > 1$ of the above statement can be compared with Theorem 1 of Love (1951). There p-th power generalization of absolute continuity, defining a class V^p with $p \geq 1$, is characterized as a continuous subclass of $\mathcal{W}_p^*$. Notice that a real-valued function $f \in V^p$ with $p > 1$ if and only if given $\epsilon > 0$ there is a $\delta > 0$ such that $\sum_j |f(\beta_j) - f(\alpha_j)|^p < \epsilon$ for all finite sets of non-overlapping intervals (α_j, β_j) such that $\max_j (\beta_j - \alpha_j) < \delta$.

Proof. (1) $\Rightarrow$ (2). Let $f \in \mathcal{W}_p^*([a,b]; \mathbb{B})$ and let $\epsilon > 0$. Since $\mathfrak{S}_p(f) < \infty$ there exists a finite subset $\mu \subset [a,b]$ such that

$$\sum_{x \in \mu} (\|\Delta^+ f(x)\|^p + \|\Delta_- f(x)\|^p) > \mathfrak{S}_p(f)^p - \epsilon/2. \tag{2.43}$$

By Lemma 2.22, there is a partition $\kappa = \{x_i : i = 0, \ldots, n\} \in Q^\pm([a,b])$ such that $\kappa \supset \mu$ and

$$\sum_{i=1}^{n} v_p(f; [x_{i-1}, x_i]) < V_p^*(f)^p + \epsilon/2. \tag{2.44}$$

Also we can and do assume that $a+, b- \in \kappa$ and each $x-, x, x+ \in \kappa$ whenever one of them already belongs to κ and x is not an endpoint of $[a,b]$. Let $\{z_j : j = 0, \ldots, m\}$ be the set of those points of κ which are not points $x\pm$. Since $v_p(f; [x, x+]) = \|\Delta^+ f(x)\|^p$ and $v_p(f; [x-, x]) = \|\Delta^- f(x)\|^p$, by (2.43) and (2.44) it then follows that

$$\sum_{j=1}^{m} v_p(f; [z_{j-1}+, z_j-]) \leq \sum_{i=1}^{n} v_p(f; [x_{i-1}, x_i]) - \sum_{j=1}^{m} (\|\Delta^+ f(z_{j-1})\|^p + \|\Delta^- f(z_j)\|^p)$$

$$< V_p^*(f)^p + \epsilon/2 - \mathfrak{S}_p(f)^p + \epsilon/2 = \epsilon.$$

Hence (2.41) holds.

   (2) $\Rightarrow$ (1). Using (2.9) one can show that $f \in \mathcal{W}_p([a,b]; \mathbb{B})$. Assume however that $f \notin \mathcal{W}_p^*([a,b]; \mathbb{B})$. Therefore, since (2.12) always holds, $V_p^*(f)^p - \mathfrak{S}_p(f)^p \geq C$ for some constant $C > 0$. Let $\lambda = \{z_j : j = 0, \ldots, m\} \in Q([a,b])$. Then by Lemma 2.22, we have

$$\sum_{j=1}^{m} (v_p(f; [z_{j-1}, z_{j-1}+]) + v_p(f; [z_{j-1}+, z_j-]) + v_p(f; [z_j-, z_j])) \geq V_p^*(f)^p.$$

It then follows that

$$\sum_{j=1}^{m} v_p(f; [z_{j-1}+, z_j-]) \geq V_p^*(f)^p - \sum_{j=1}^{m} (\|\Delta^+ f(z_{j-1})\|^p + \|\Delta_- f(z_j)\|^p)$$

$$\geq V_p^*(f)^p - \mathfrak{S}_p(f)^p \geq C > 0.$$

Thus (2.41) cannot hold because λ is arbitrary. This contradiction implies that $f \in \mathcal{W}_p^*([a,b]; \mathbb{B})$.

   (2) $\Rightarrow$ (3). Since (2.41) implies $f \in \mathcal{W}_p$, $\mathfrak{S}_p(f) < \infty$ by (2.12). Let $\epsilon > 0$ and let $\lambda = \{z_j : j = 0, \ldots, m\} \in Q([a,b])$ be such that (2.41) holds. Let $\{y_i : i = 0, \ldots, n\}$ be a refinement of λ and let the index $i = i(j)$ be such that $y_{i(j)} = z_j$ for $j = 0, \ldots, m$. Then we have

$$\sum_{i=1}^{n} \|f(y_i-) - f(y_{i-1}+)\|^p = \sum_{j=1}^{m} \sum_{i=i(j-1)+1}^{i(j)} \|f(y_i-) - f(y_{i-1}+)\|^p$$

$$\leq \sum_{j=1}^{m} v_p(f; [z_{j-1}+, z_j-]) < \epsilon.$$

Therefore (3) holds.

(3) $\Rightarrow$ (2). Let $\epsilon > 0$. Then there is a finite set $\nu \subset [a, b]$ such that

$$\sum_{x \in \mu} \left(\|\Delta^- f(x)\|^p + \|\Delta^+ f(x)\|^p \right) < \epsilon$$

for each finite subset μ of (a, b) with $\mu \cap \nu = \emptyset$. Moreover, there is $\lambda \in Q([a, b])$ such that (2.42) holds for each refinement $\{y_i : i = 0, \ldots, n\}$ of λ. Let $\{z_j : j = 0, \ldots, m\} := \mu \cup \lambda$ and let $\{z_{j-1}+ = x_0^j < x_1^j < \cdots < x_{n(j)}^j = z_j-\}$ be partitions of $[z_{j-1}+, z_j-]$ for each $j = 1, \ldots, m$ with $\{x_i^j : i = 1, \ldots, n(j) - 1\} \in Q((z_{j-1}, z_j))$. Then we have

$$\sum_{j=1}^{m} \sum_{i=1}^{n(j)} \|f(x_i^j) - f(x_{i-1}^j)\|^p \leq 4^{p-1} \sum_{j=1}^{m} \{ \|f(x_1^j-) - f(z_{j-1}+)\|^p + \|\Delta^- f(x_1^j)\|^p$$

$$+ \sum_{i=2}^{n(j)-1} \left[\|\Delta^+ f(x_{i-1}^j)\|^p + \|f(x_i^j-) - f(x_{i-1}^j+)\|^p + \|\Delta^- f(x_i^j)\|^p \right]$$

$$+ \|\Delta^+ f(x_{n(j)-1}^j)\|^p + \|f(z_j-) - f(x_{n(j)-1}^j+)\|^p \} < 4^p \epsilon/2.$$

Since all partitions $\{x_i^j : i = 0, \ldots, n(j)\}$ are arbitrary, it follows that

$$\sum_{j=1}^{m} v_p(f; [z_{j-1}+, z_j-]) \leq 4^p \epsilon/2.$$

Since $\epsilon > 0$ is arbitrary, this proves (2). The proof of Proposition 2.23 is complete. $\square$

Statement (2) of the preceding Proposition can be strengthened as follows:

Corollary 2.24. *If $f \in \mathcal{W}_p^*([a, b]; \mathbb{B})$ with $0 < p < \infty$ then, for every $\epsilon > 0$, there is a partition $\lambda \in Q([a, b])$ such that (2.41) holds for each refinement $\{z_j : j = 0, \ldots, m\} \in Q([a, b])$ of λ.*

Proof. This is a consequence of statement (2) of Proposition 2.23, the fact that

$$v_p(f; [u+, c-]) + v_p(f; [c+, d-]) \leq v_p(f; [u+, d-])$$

for any $a \leq u < c < d \leq b$, and induction. $\square$

3. STIELTJES AND YOUNG INTEGRALS

We assume from here on that $\mathbb{B} = (\mathbb{B}, \| \cdot \|)$ is a Banach algebra. Since $\mathbb{B}$ may not be commutative we will consider integrals

$$\int_a^b g \, dh \, f \tag{3.1}$$

of two $\mathbb{B}$-valued integrands f and g with respect to a $\mathbb{B}$-valued function h, all of them defined on an interval $[a, b]$. Derivatives of the product integral (Theorem 5.16 below) will involve integrals of the form (3.1). We will always assume that $\mathbb{B}$ contains an identity $\mathbb{I}$. This will be needed for product integrals in later sections. If a Banach algebra does not have an identity, then one can be adjoined (e. g. Segal and Kunze, 1968, p. 216). Setting $g \equiv \mathbb{I}$ or $f \equiv \mathbb{I}$, we get integrals

$$\int dh \, f := \int \mathbb{I} \, dh \, f, \quad \int g \, dh := \int g \, dh \, \mathbb{I}, \quad \text{or} \quad \int dh := \int \mathbb{I} \, dh \, \mathbb{I}.$$

3.1. Moore-Pollard-Stieltjes integrals. Let g, h, f be functions on an interval τ (which may be open or closed at either end) with values in $\mathbb{B}$. Let $\kappa = \{x_0, \ldots, x_n\}$ be a partition of τ. A *Riemann-Stieltjes sum for g, h, f based on κ* will be any sum

$$\sum_{i=1}^n g(y_i) \big[h(x_i) - h(x_{i-1}) \big] f(y_i)$$

where $x_{i-1} \le y_i \le x_i$ for $i = 1, \ldots, n$.

Let $\tau = [a, b]$. We will say that the integral (3.1) exists in the *Riemann-Stieltjes sense* and equals C, if the limit

$$\lim_\kappa \sum_{i=2}^n g(y_i) \left(\Delta_i^\kappa h \right) f(y_i) = C \tag{3.2}$$

exists in $\mathbb{B}$ as the mesh $|\kappa| \to 0$ for any $y_i \in [x_{i-1}, x_i]$, $i = 1, \ldots, n$.

Let τ be any one of the intervals $[a, b]$, $[a, b)$, $(a, b]$ or (a, b). Let $\mathcal{S}(\kappa, \tau, g, h, f)$ denote the set of all Riemann-Stieltjes sums for g, h, f based on partitions of τ which are refinements of κ. For a set $\mathcal{S} \subset \mathbb{B}$, $T \in \mathbb{B}$ and $\epsilon > 0$ let $\|\mathcal{S} - T\| \le \epsilon$ mean that $\|S - T\| \le \epsilon$ for all $S \in \mathcal{S}$. The *Moore-Pollard-Stieltjes* integral

$$(MPS) \int_\tau g \, dh \, f$$

is defined as the $T \in \mathbb{B}$, if it exists, such that for every $\epsilon > 0$ there is a partition κ of τ such that $\|\mathcal{S}(\kappa, \tau, g, h, f) - T\| \le \epsilon$. Pollard (1923) gave a detailed treatment of this integral. E. H. Moore (1915) had treated limits based on refinements. The books of Gochman (1958) and Hildebrandt (1963, Secs. 9–18) also contain detailed expositions of the (MPS) integral.

If the functions g, h, f are defined on τ equal to $(a, b]$, $[a, b)$ or (a, b) , then (MPS) integrals will be defined over intervals $[a+, b]$, $[a, b-]$ or $[a+, b-]$ as before, except for the following: if the left endpoint is $a+$, then we require that $h(a+)$ is defined, in other words h has a right limit at a, and in Riemann-Stieltjes sums for partitions $\kappa = \{x_0, \ldots, x_n\}$ with $x_0 = a+$, we take terms $g(y_1)[h(x_1) - h(a+)]f(y_1)$ where $a < y_1 \leq x_1$. Likewise if the right endpoint is $b-$, then we require that $h(b-)$ is defined and in Riemann-Stieltjes sums for partitions $\kappa = \{x_0, \ldots, x_n\}$ with $x_n = b-$, we take terms $g(y_n)[h(b-) - h(x_{n-1})]f(y_n)$ where $x_{n-1} \leq y_n < b$. We write

$$(MPS) \int_c^d g\, dh\, f := (MPS) \int_{[c,d]} g\, dh\, f$$

where c may be an $a+$ and/or d a $b-$.

We start by describing (MPS) integrals over half open intervals. Recalling the notation (2.17) and (2.18) we have:

Lemma 3.1. *Let* $g, h, f \in \ell_\infty([a,b]; \mathbb{B})$.

(1) *If* h *has a left limit at* b *then*

$$(MPS) \int_{[a,b)} g\, dh\, f = (MPS) \int_a^{b-} g\, dh\, f = (MPS) \int_a^b g\, dh^{(b-)}\, f, \qquad (3.3)$$

where each integral exists if and only if both the others exist.

(2) *If* h *has a right limit at* a *then*

$$(MPS) \int_{(a,b]} g\, dh\, f = (MPS) \int_{a+}^b g\, dh\, f = (MPS) \int_a^b g\, dh^{(a+)}\, f, \qquad (3.4)$$

where each integral exists if and only if both the others exist.

Proof. We will prove part (1) only and note that a proof of (2) is based on symmetric arguments. First assume $T_1 := (MPS) \int_{[a,b)} g\, dh\, f$ exists. Given $\epsilon > 0$, take a partition $\kappa = \{x_0, \ldots, x_{n-1}\}$ of $[a,b)$ such that $\|S(\kappa, [a,b), g, h, f) - T_1\| \leq \epsilon$. We can assume x_{n-1} is close enough to b so that

$$Osc_{[x_{n-1},b)}(h)\|g\|_\infty\|f\|_\infty < \epsilon. \qquad (3.5)$$

Let ζ be the partition $\{x_0, \ldots, x_n\}$ of $[a,b-]$ where $x_n := b-$. Then for any refinement $\zeta' = \{y_0, \ldots, y_m\}$ of ζ, we have $y_m = b-$, and $\kappa' := \{y_0, \ldots, y_{m-1}\}$ is a partition of $[a,b)$ and a refinement of κ. For any $z \in [y_{m-1}, b-]$, $\|g(z)(h(b-) - h(y_{m-1}))f(z)\| < \epsilon$ by (3.5). Thus $\|S(\zeta, [a,b-], g, h, f) - T_1\| < 2\epsilon$. So $T_2 := (MPS) \int_a^{b-} g\, dh\, f$ exists and equals T_1.

Next, suppose T_2 exists. Given $\epsilon > 0$, take a partition $\zeta = \{x_0, \ldots, x_n\}$ of $[a,b-]$ such that $\|S(\zeta, [a,b-], g, h, f) - T_2\| \leq \epsilon$. We can assume that $x_n = b-$ and (3.5) holds. Let $\zeta_b = \{z_0, \ldots, z_n\}$ where $z_i = x_i$ for $i = 0, \ldots, n-1$ and $z_n = b$. Then ζ_b is a partition of $[a,b]$ and it is clear that $\|S(\zeta_b, [a,b], g, h^{(b-)}, f) - T_2\| \leq \epsilon$, so $T_3 := (MPS) \int_a^b g\, dh^{(b-)} f$ exists and equals T_2.

Now assume T_3 exists. Given $\epsilon > 0$, take a partition $\zeta_b = \{x_0, \ldots, x_n\}$ of $[a,b]$ such that $\|S(\zeta_b, [a,b], g, h^{(b-)}, f) - T_3\| \leq \epsilon$, where we can assume $x_n = b$ and that (3.5) holds. Then $\kappa = \{x_0, \ldots, x_{n-1}\}$ is a partition of $[a,b)$ such that $\|S(\kappa, [a,b), g, h, f) - T_3\| < 2\epsilon$, so T_1 exists and equals T_3. Lemma 3.1 is proved. $\qquad\square$

The next statement describes relations between

$$(MPS) \int_a^b g\, dh\, f \qquad (3.6)$$

and the integrals in (3.3) and (3.4).

Lemma 3.2. Let $g, h, f \in \ell_\infty([a, b]; \mathbb{B})$.

(1) If h has a left limit at b and (3.6) exists then the integrals in (3.3) exist and

$$(MPS) \int_a^b g \, dh \, f = (MPS) \int_a^{b-} g \, dh \, f + [g(\Delta_- h)f](b). \qquad (3.7)$$

(2) If h is left continuous at b then $(MPS) \int_a^{b-} g \, dh \, f$ exists if and only if (3.6) exists and both are equal.

(3) If h has a right limit at a and (3.6) exists then the integrals in (3.4) exist and

$$(MPS) \int_a^b g \, dh \, f = (MPS) \int_{a+}^b g \, dh \, f + [g(\Delta^+ h)f](a). \qquad (3.8)$$

(4) If h is right continuous at a then $(MPS) \int_{a+}^b g \, dh \, f$ exists if and only if (3.6) exists and both are equal.

Proof. We will prove parts (1) and (2) only and note that proofs of (3) and (4) are symmetrical. If $h(b-) = h(b)$, then $[g(\Delta_- h)f](b) = 0$, $h^{(b-)} \equiv h$ and part (1) holds by Lemma 3.1. So suppose $(\Delta_- h)(b) \neq 0$. Then f and g must be left-continuous at b. As $x \uparrow b$, for any $y \in [x, b]$,

$$g(y)[h(b) - h(x)]f(y) \to [g(\Delta_- h)f](b).$$

Given $\epsilon > 0$, let $\zeta = \{x_0, \dots, x_n\}$ be a partition of $[a, b]$ and $\|S(\zeta, [a, b], g, h, f) - (MPS) \int_a^b g \, dh \, f\| \leq \epsilon$. We can assume x_{n-1} is close enough to b so that whenever $x_{n-1} \leq x \leq y \leq b$, we have

$$\|g(y)[h(b) - h(x))]f(y) - [g(\Delta_- h)f](b)\| < \epsilon.$$

Let $\kappa = \{x_0, \dots, x_{n-1}\}$. Then κ is a partition of $[a, b)$ and

$$\|S(\kappa, [a, b), g, h, f) - [(MPS) \int_a^b g \, dh \, f - [g(\Delta_- h)f](b)]\| < 2\epsilon,$$

so part (1) holds by part (1) of Lemma 3.1.

To prove part (2) we note that the Riemann-Stieltjes sums for the integrals (3.6) and (3.3) corresponding to the partitions $\{x_0, \dots, x_n = b\}$ and $\{x_0, \dots, x_n = b-\}$ either may be taken the same or may differ at most by

$$g(b)[h(b) - h(x_{n-1})]f(b) - g(y_n)[h(b-) - h(x_{n-1})]f(y_n) = g(y_n)[h(b) - h(b-)]f(b)$$

$$+[g(b) - g(y_n)][h(b) - h(x_{n-1})]f(b) + g(y_n)[h(b-) - h(x_{n-1})][f(b) - f(y_n)],$$

where $x_{n-1} \leq y_n < b$. Since $g, f \in \ell_\infty([a, b]; \mathbb{B})$ and $h(b) = h(b-)$ the right side tends to zero as $x_{n-1} \uparrow b$. This proves part (2) and the proof of Lemma 3.2 is complete. □

Gochman (1958, p. 131) defined the (MPS) integral over a half open interval $[a, b)$ as the limit as $x \uparrow b$ of the (MPS) integrals over closed intervals $[a, x]$. The next fact shows a possibly surprising distinction between the (MPS) integral over a half open interval as defined above and as defined by Gochman.

Proposition 3.3. *There are $g, h \in C([0,1])$ such that $\lim_{y\uparrow 1}(MPS)\int_0^y g\,dh$ exists, but $(MPS)\int_0^{1-} g\,dh$ does not.*

Proof. Let $y_m = 1 - 1/m$ for $m = 1, 2, \ldots$. For each m let $g(y_{4m-2}) = g(y_{4m}) = 0$, $g(y_{4m-1}) = m^{-1/2}$, $g(0) = g(1) = 0$, and let g be linear and continuous on intervals between adjacent points where it has been defined. Let $h(y_{4m-3}) = 0$, $h(y_{4m-2}) = h(y_{4m}) = m^{-1/2}$, $h(1) = 0$, and let h also be linear between points where it has been defined. Then g and h are both continuous. Since h is constant on each interval where g is non-zero, we have for $0 < y < 1$ that $(MPS)\int_0^y g\,dh = 0$, taking a partition containing y and all points $y_m < y$.

But for any partition $\kappa = \{x_0, \ldots, x_n\}$ of $[0, 1-]$, there exist arbitrarily large Riemann-Stieltjes sums for g, h based on refinements of κ, as follows. Let $z = x_{n-1}$. Let m_0 be the smallest m such that $y_{4m-3} > z$. Form a partition κ' by adjoining to κ all points y_{4m-3}, y_{4m} for $m = m_0, m_0+1, \ldots, m_0+N$, and form Riemann-Stieltjes sums containing terms $g(y_{4m-1})[h(y_{4m}) - h(y_{4m-3})] = 1/m$ for all such m, and 0 terms $g(y_{4m})[h(y_{4m+1}) - h(y_{4m})]$. As $N \to \infty$ these sums become arbitrarily large, as claimed. $\square$

3.2. L. C. Young integrals.

Two variants of L. C. Young integrals called (Y_1) and (Y_2) integrals will be defined for regulated functions g, h, f with values in a Banach algebra $\mathbb{B} = (\mathbb{B}, \|\cdot\|)$. It will be shown that each of them exists if and only if the other does and then the two are equal. The integral will then be called the (CY) integral. Then two other variants called the (LY) and (RY) integrals will be defined which will usually equal each other but differ from the (CY) integral if the integrator h and an integrand g or f have jumps in common.

To begin with we recall some notation. Let $(\mathbb{B}, \|\cdot\|)$ be a Banach space and τ a set. Let $\mathcal{F}$ be the system of finite subsets λ of τ directed by inclusion. Suppose that for each $x \in \tau$ we have $u_x \in \mathbb{B}$. Let $S_\lambda := \sum_{x\in\lambda} u_x$ for each $\lambda \in \mathcal{F}$. We say that the sum $\sum_{x\in\tau} u_x$ converges *unconditionally* to $S := S_\tau \in \mathbb{B}$ if the net $\{S_\lambda: \lambda \in \mathcal{F}\}$ converges to S. Clearly, $\sum_{x\in\tau} u_x$ converges unconditionally (to some limit) if and only if, for every $\epsilon > 0$, there is a $\mu \in \mathcal{F}$ such that for each $\nu \in \mathcal{F}$ disjoint from μ, $\|S_\nu\| \le \epsilon$. It follows that for any $\nu \subset \tau$ disjoint from μ we have $\|S_\nu\| \le \epsilon$ and, for any $\rho \subset \tau$, the sum $\sum_{x\in\rho} u_x$ also converges unconditionally. Clearly, unconditional convergence of the sum implies that $u_x = 0$ except for at most countably many $x \in \tau$. The sum $\sum_{x\in\tau} u_x$ is said to converge *absolutely* if $\sum_{x\in\tau} \|u_x\| < \infty$. Absolute convergence implies unconditional convergence. It is well known and easily seen that in $\mathbb{R}$, and thus in any finite-dimensional Banach space, absolute and unconditional convergence are equivalent. In an infinite dimensional Banach space, unconditional convergence never implies absolute convergence (Dvoretzky and Rogers, 1950).

Recalling the notation (2.16) and setting $\sum_\tau F := \sum_{x\in\tau} F(x)$ for a set τ and a function F we have:

Definition 3.4. Let $g, h, f \in \mathcal{R}([a,b]; \mathbb{B})$. If $a < b$ we define the *Young* (Y_1) integral

$$(Y_1)\int_a^b g\,dh\,f := (MPS)\int_a^b g_+^{(b)}\,dh_-^{(a)}\,f_+^{(b)} + [g(\Delta^+ h)f - g_+(\Delta^+ h)f_+](a)$$

$$+ \sum_{(a,b)} [g(\Delta_-^+ h)f - g_+(\Delta_-^+ h)f_+] + [g(\Delta_- h)f](b)$$

if the (MPS) integral exists and the sum converges unconditionally in $\mathbb{B}$. If $a = b$ we define the (Y_1) integral as 0. Similarly, if $a < b$ we define the *Young* (Y_2) integral

$$(Y_2) \int_a^b g\,dh\,f := (MPS) \int_a^b g_-^{(a)}\,dh_+^{(b)}\,f_-^{(a)} + [g(\Delta^+ h)f](a)$$
$$+ \sum_{(a,b)} [g(\Delta_-^\pm h)f - g_-(\Delta^\pm h)f_-] + [g(\Delta_- h)f - g_-(\Delta_- h)f_-](b)$$

also if the (MPS) integral exists and the sum converges unconditionally in $\mathbb{B}$. If $a = b$ the (Y_2) integral is also defined as 0.

These definitions are the same as those given in Dudley (1992) for real-valued functions except that absolute convergence of the sums has been replaced by unconditional convergence (which, as noted above, is equivalent in the real-valued case) and, in view of possible non-commutativity, one integrand g has been replaced by two, g and f. Young (1936, p. 263) defined an extension of the Stieltjes integral for $g, h \in \mathcal{R}([a,b]; \mathbb{C})$ by

$$(Y_0) \int_a^b g\,dh := (MPS) \int_a^b g_+\,dh_- + \sum_{a \le x \le b} [g(x) - g(x+)](\Delta_-^\pm h)(x)$$

if the (MPS) integral exists and the sum converges absolutely. However, the value of the integral (Y_0) depends on $h(a-)$, $g(b+)$ and $h(b+)$. So, it can't be a limit of sums $\sum_{i=1}^n g(y_i)[h(x_i) - h(x_{i-1})]$, where $a = x_0 < x_1 < \cdots < x_n = b$ and $x_{i-1} \le y_i \le x_i$ for $i = 1, \ldots, n$ (cf. footnote on p. 266 of Young, 1936), as the (Y_1) and (Y_2) integrals are (Proposition 3.10 below). However, if we redefine the functions g_+ and g_- on endpoints as in footnote 6 p. 583 in Young (1938), then the (Y_0) integral would coincide with the (Y_1) integral. In retrospect and on an account of footnote 9 p. 585, again from Young (1938), one may believe that L. C. Young had these definitions in mind already in his 1936 paper.

We continue by proving the following relation between the (Y_1) and (Y_2) integrals:

Theorem 3.5. *For any regulated functions g, h, f from $[a, b]$ into $\mathbb{B}$,*

$$(Y_1) \int_a^b g\,dh\,f = -(Y_2) \int_{-b}^{-a} g(-x)\,dh(-x)\,f(-x) \qquad (3.9)$$

and

$$(Y_2) \int_a^b g\,dh\,f = -(Y_1) \int_{-b}^{-a} g(-x)\,dh(-x)\,f(-x), \qquad (3.10)$$

where each integral on the left exists if and only if the one on the right does.

Proof. Assuming that the (Y_1) integral on the left side of (3.9) exists we will prove that the (Y_2) integral on the right side exists and the equality (3.9) holds. Proofs of the converse implication and of (3.10) are based on symmetric arguments and therefore are omitted. Let the (Y_1) integral exist and let $\overline{g}(x) := g(-x)$, $\overline{h}(x) := h(-x)$, $\overline{f}(x) := f(-x)$. To show the existence of the integral

$$(MPS) \int_{-b}^{-a} \overline{g}_-^{(-b)}\,d\overline{h}_+^{(-a)}\,\overline{f}_-^{(-b)} \qquad (3.11)$$

take any Riemann-Stieltjes sum S for $\overline{g}_-^{(-b)}$, $\overline{h}_+^{(-a)}$, $\overline{f}_-^{(-b)}$ based on a partition of the interval $[-b, -a]$. Let

$$S = \sum_{i=1}^{n} \overline{g}_-^{(-b)}(y_i) \left[\overline{h}_+^{(-a)}(x_i) - \overline{h}_+^{(-a)}(x_{i-1}) \right] \overline{f}_-^{(-b)}(y_i),$$

where $\{x_i: i = 0, \ldots, n\}$ is a partition of $[-b, -a]$ and $y_i \in [x_{i-1}, x_i]$ for all $i = 1, \ldots, n$. Let $\overline{x}_j := -x_{n-j}$ for $j = 0, \ldots, n$ and $\overline{y}_j := -y_{n+1-j}$ for $j = 1, \ldots, n$. Then $\{\overline{x}_j: j = 0, \ldots, n\}$ is a partition of $[a, b]$ and $\overline{y}_j \in [\overline{x}_{j-1}, \overline{x}_j]$ for all $j = 1, \ldots, n$. So

$$-S = \sum_{j=1}^{n} g_+^{(b)}(\overline{y}_j) \left[h_-^{(a)}(\overline{x}_j) - h_-^{(a)}(\overline{x}_{j-1}) \right] f_+^{(b)}(\overline{y}_j)$$

is a Riemann-Stieltjes sum for $g_+^{(b)}$, $h_-^{(a)}$, $f_+^{(b)}$ based on the partition $\{\overline{x}_j: j = 0, \ldots, n\}$ of $[a, b]$. Since the integral $I := (MPS) \int_a^b g_+^{(b)} dh_-^{(a)} f_+^{(b)}$ exists, it follows that the integral (3.11) exists and equals $-I$.

Similarly one can show that the sum $-\sum_{(-b,-a)} \left[\overline{g}(\Delta_-^+ \overline{h})\overline{f} - \overline{g}_-(\Delta_-^+ \overline{h})\overline{f}_- \right]$ converges unconditionally to the sum $\sum_{(a,b)} \left[g(\Delta_-^+ h)f - g_+(\Delta_-^+ h)f_+ \right]$, and

$$-\left[\overline{g}(\Delta^+ \overline{h})\overline{f} \right](-b) - \left[\overline{g}(\Delta_- \overline{h})\overline{f} - \overline{g}_-(\Delta_- \overline{h})\overline{f}_- \right](-a)$$

$$= \left[g(\Delta_- h)f \right](b) + \left[g(\Delta^+ h)f - g_+(\Delta^+ h)f_+ \right](a).$$

This proves that the (Y_2) integral on the right side of (3.9) exists and the equality (3.9) holds. The proof of Theorem 3.5 is complete. $\qquad\square$

The next statement is an extension of Theorem 3.2 of Dudley (1992) to integrals of two $\mathbb{B}$-valued integrands with respect to a $\mathbb{B}$-valued function. Since the proof is essentially the same we omit it.

Theorem 3.6. *Let* $g, h, f \in \mathcal{R}([a, b]; \mathbb{B})$. *Then if both of the integrals*

$$(Y_1) \int_a^b g\, dh\, f \quad and \quad (Y_2) \int_a^b g\, dh\, f \qquad (3.12)$$

exist, they are equal.

Next is the main result of this subsection.

Theorem 3.7. *Let* $g, h, f \in \mathcal{R}([a, b]; \mathbb{B})$. *If one of the integrals in* (3.12) *exists then so does the other.*

Proof. It will be proved that if the (Y_1) integral exists then so does the (Y_2) integral. The converse statement then will follow from Theorem 3.5. First it will be shown that

$$\sum_{(a,b)} \left[g(\Delta_-^+ h)f - g_-(\Delta_-^+ h)f_- \right] \quad \text{converges unconditionally.} \qquad (3.13)$$

Suppose not. Then for some $\epsilon > 0$, for every finite set $\mu \subset (a, b)$, there is a finite set $\nu \subset (a, b)$, disjoint from μ, such that

$$\left\| \sum_{\nu} \left[g(\Delta_-^+ h)f - g_-(\Delta_-^+ h)f_- \right] \right\| > \epsilon. \tag{3.14}$$

Take a partition $\kappa = \{x_i : i = 0, \ldots, n\}$ of $[a, b]$ such that for every Riemann-Stieltjes sum S for $g_+^{(b)}$, $h_-^{(a)}$, $f_+^{(b)}$ based on a refinement of κ,

$$\left\| S - (MPS) \int_a^b g_+^{(b)} dh_-^{(a)} f_+^{(b)} \right\| < \epsilon/4. \tag{3.15}$$

Since $\sum_{(a,b)} \left[g(\Delta_-^+ h)f - g_+(\Delta_-^+ h)f_+ \right]$ converges unconditionally, there is a finite set $\mu \subset (a, b)$ such that for every finite set $\nu \subset (a, b)$ disjoint from μ,

$$\left\| \sum_{\nu} \left[g(\Delta_-^+ h)f - g_+(\Delta_-^+ h)f_+ \right] \right\| < \epsilon/4. \tag{3.16}$$

We can assume that $\mu \subset \kappa$. Take a finite set $\nu \subset (a, b)$, disjoint from κ, such that (3.14) holds. Then by (3.16), we have

$$\left\| \sum_{\nu} \left[g_+(\Delta_-^+ h)f_+ - g_-(\Delta_-^+ h)f_- \right] \right\| > 3\epsilon/4.$$

Let $\nu = \{y_j : j = 1, \ldots, m\}$. We can choose $s_j < y_j < t_j$ for $j = 1, \ldots, m$ such that $t_{j-1} < s_j$ for $j = 2, \ldots, m$, and for the interval (x_{i-1}, x_i) containing y_j we have $x_{i-1} < s_j < y_j < t_j < x_i$, such that

$$\left\| \sum_{j=1}^m \left\{ g(y_j+) \left[h(t_j-) - h(s_j-) \right] f(y_j+) \right. \right.$$
$$\left. \left. - g(s_j+) \left[h(t_j-) - h(s_j-) \right] f(s_j+) \right\} \right\| > \epsilon/2. \tag{3.17}$$

Let λ be the partition including κ and containing all the points $s_j, t_j, j = 1, \ldots, k$. Then (3.17) implies that there exists two Riemann-Stieltjes sums for $g_+^{(b)}$, $h_-^{(a)}$, $f_+^{(b)}$, both based on the refinement λ of κ, and more than $\epsilon/2$ apart (for intervals of λ other than $[s_j, t_j]$, evaluate g_+ and f_+ at the same point for the two sums). This contradicts (3.15), so (3.13) is proved.

It remains to prove existence of

$$(MPS) \int_a^b g_-^{(a)} dh_+^{(b)} f_-^{(a)}. \tag{3.18}$$

Given $\epsilon > 0$, take a partition $\kappa = \{x_i : i = 0, \ldots, n\}$ of $[a, b]$ such that for every Riemann-Stieltjes sum S for $g_+^{(b)}$, $h_-^{(a)}$, $f_+^{(b)}$ based on a refinement of κ, we have

$$\left\| S - (MPS) \int_a^b g_+^{(b)} dh_-^{(a)} f_+^{(b)} \right\| < \epsilon. \tag{3.19}$$

We can assume κ is chosen large enough so that for every finite set $\nu \subset (a, b)$ disjoint from κ, we have

$$\left\| \sum_\nu \left[g(\Delta_-^\pm h) f - g_+(\Delta_-^\pm h) f_+ \right] \right\| < \epsilon \tag{3.20}$$

and by (3.13),

$$\left\| \sum_\nu \left[g(\Delta_-^+ h) f - g_-(\Delta_-^+ h) f_- \right] \right\| < \epsilon. \tag{3.21}$$

Since g, h and f are regulated, they are bounded and for each $i = 1, \ldots, n$, we can choose $u_i < x_i$ close enough to x_i so that

$$Osc_{(u_i, x_i)}(g) \leq \epsilon / \left[4n \|h\|_\infty \|f\|_\infty \right], \tag{3.22}$$

$$Osc_{(u_i, x_i)}(h) \leq \epsilon / \left[2n \|g\|_\infty \|f\|_\infty \right], \tag{3.23}$$

and

$$Osc_{(u_i, x_i)}(f) \leq \epsilon / \left[4n \|h\|_\infty \|g\|_\infty \right]. \tag{3.24}$$

Likewise, for each $i = 0, \ldots, n-1$, choose $v_i > x_i$ close enough to x_i so that (3.22), (3.23) and (3.24) hold with (x_i, v_i) in place of (u_i, x_i). We can assume that $v_{i-1} < u_i$ for $i = 1, \ldots, n$. Let $\lambda = \{ y_k \colon k = 0, \ldots, l \}$ be a partition of $[a, b]$ including κ and containing u_i and v_{i-1} for $i = 1, \ldots, n$. For each $k = 1, \ldots, l$, let $\zeta_k := (y_{k-1} + y_k)/2$. Let κ' be the partition $\lambda \cup \{ \zeta_k \colon k = 1, \ldots, l \}$. Let $\mu := \{ z_j \colon j = 0, \ldots, m \}$ be any refinement of κ'. For each $j = 1, \ldots, m$, suppose $z_{j-1} \leq \xi_j \leq z_j$. For each $k = 1, \ldots, l$, let $\lambda(k) := \{ j = 1, \ldots, m \colon z_j \in (y_{k-1}, y_k] \}$. For each $j = 1, \ldots, m$, let $k(j) := k$ such that $j \in \lambda(k)$. Let

$$W := \sum_{j=1}^m T_j := \sum_{j=1}^m \left\{ g_-(\zeta_{k(j)}) \left[h_+^{(b)}(z_j) - h_+(z_{j-1}) \right] f_-(\zeta_{k(j)}) \right.$$
$$\left. - g_-^{(a)}(\xi_j) \left[h_+^{(b)}(z_j) - h_+(z_{j-1}) \right] f_-^{(a)}(\xi_j) \right\}.$$

Then W is the difference between a particular Riemann-Stieltjes sum for $g_-^{(a)}$, $h_+^{(b)}$, $f_-^{(a)}$ based on λ, and an arbitrary such sum based on μ. We aim to find a small upper bound for $\|W\|$.

By the choice of κ', for each $j = 1, \ldots, m$, at most one of z_{j-1} and z_j is in κ. Let $\nu(1) := \{ j = 1, \ldots, m \colon z_{j-1} \in \kappa \}$. Then for $j \in \nu(1)$, by (3.23) with (x_i, v_i) for $x_i = z_{j-1}$ in place of (u_i, x_i), since $z_{j-1} = y_{k(j)-1} < z_j \leq \zeta_{k(j)} = (x_i + v_i)/2 < v_i$, it follows that

$$\sum_{j \in \nu(1)} \|T_j\| \leq \sum_{j \in \nu(1)} 2\|h_+(z_j) - h_+(z_{j-1})\| \|g\|_\infty \|f\|_\infty \leq \epsilon. \tag{3.25}$$

Let $\nu(2) := \{ j = 1, \ldots, m \colon z_j \in \kappa \}$. Then for $j \in \nu(2)$, by (3.22) and (3.24) and since in this case $z_j = y_{k(j)} > z_{j-1} \geq \zeta_{k(j)} > y_{k(j)-1} = u_i$,

$$\sum_{j \in \nu(2)} \|T_j\| \leq \sum_{j \in \nu(2)} \left\{ 2\|g_-(\zeta_{k(j)}) - g_-^{(a)}(\xi_j)\| \|h\|_\infty \|f\|_\infty \right.$$
$$\left. + 2\|g\|_\infty \|h\|_\infty \|f_-(\zeta_{k(j)}) - f_-(\xi_j)\| \right\} \leq \epsilon. \tag{3.26}$$

Let $\nu(3)$ be the set of all $j = 1, \ldots, m$ such that neither z_{j-1} nor z_j is in κ. Then $\{1, \ldots, m\} = \nu(1) \cup \nu(2) \cup \nu(3)$. For each $j = 2, \ldots, m$, take $t_{j-1} \in (z_{j-1}, z_j)$ if $z_{j-1} \notin \kappa$ and let $t_{j-1} := z_{j-1}$ if $z_{j-1} \in \kappa$. Then $\{t_0 := a\} \cup \{t_j : j = 1, \ldots, m-1\} \cup \{t_m := b\}$ is a partition of $[a, b]$ and a refinement of κ. By (3.19), we have

$$\Big\| \sum_{j=1}^{m} \big\{ g_+(\beta_{k(j)}) \big[h_-(t_j) - h_-^{(a)}(t_{j-1}) \big] f_+(\beta_{k(j)})$$

$$-g_+^{(b)}(\alpha_j) \big[h_-(t_j) - h_-^{(a)}(t_{j-1}) \big] f_+^{(b)}(\alpha_j) \big\} \Big\| < 2\epsilon \qquad (3.27)$$

whenever $t_{j-1} \leq \alpha_j \leq t_j$ and $t_{r(k(j)-1)} < \beta_{k(j)} < t_{r(k(j))}$ for $j = 1, \ldots, m$ where $r(k) := r$ such that $z_r = y_k$. Letting $t_j \downarrow z_j$ for each j such that $z_j \notin \kappa$ in (3.27), we get $\|M_1 + M_2 + M_3\| \leq 2\epsilon$ where

$$M_1 := \sum_{j \in \nu(1)} \big\{ g_+(\beta_{k(j)}) \big[h_+(z_j) - h_-^{(a)}(z_{j-1}) \big] f_+(\beta_{k(j)})$$

$$-g_+(\alpha_j) \big[h_+(z_j) - h_-^{(a)}(z_{j-1}) \big] f_+(\alpha_j) \big\},$$

$$M_2 := \sum_{j \in \nu(2)} \big\{ g_+(\beta_{k(j)}) \big[h_-(z_j) - h_+(z_{j-1}) \big] f_+(\beta_{k(j)})$$

$$-g_+(\alpha_j) \big[h_-(z_j) - h_+(z_{j-1}) \big] f_+(\alpha_j) \big\},$$

$$M_3 := \sum_{j \in \nu(3)} \big\{ g_+(\beta_{k(j)}) \big[h_+(z_j) - h_+(z_{j-1}) \big] f_+(\beta_{k(j)})$$

$$-g_+(\alpha_j) \big[h_+(z_j) - h_+(z_{j-1}) \big] f_+(\alpha_j) \big\},$$

where now we can take any $\alpha_j \in [z_{j-1}, z_j]$, clearly if $\alpha_j > z_{j-1}$, while for z_{j-1} we can let $\alpha_j \downarrow z_{j-1}$ and note that only the functions f_+ and g_+ are evaluated at α_j. Also, we can take $\beta_{k(j)} \in (y_{k(j)-1}, y_{k(j)})$ for $j = 1, \ldots, m$. Then by the proofs of (3.25) and (3.26), it follows that

$$\|M_1\| \leq \sum_{j \in \nu(1)} \big\{ 2\|g_+(\beta_{k(j)}) - g_+(\alpha_j)\| \|h\|_\infty \|f\|_\infty$$

$$+ 2\|g\|_\infty \|h\|_\infty \|f_+(\beta_{k(j)}) - f_+(\alpha_j)\| \big\} \leq \epsilon. \qquad (3.28)$$

Similarly as for (3.28), by (3.23) one can conclude that $\|M_2\| \leq \epsilon$. Thus

$$\|M_3\| \leq 4\epsilon. \qquad (3.29)$$

We can let $\beta_k \uparrow \zeta_k$ for each k in (3.29) and get

$$\Big\| \sum_{j \in \nu(3)} \big\{ g_-(\zeta_{k(j)}) \big[h_+(z_j) - h_+(z_{j-1}) \big] f_-(\zeta_{k(j)})$$

$$-g_+(\alpha_j) \big[h_+(z_j) - h_+(z_{j-1}) \big] f_+(\alpha_j) \big\} \Big\| \leq 4\epsilon. \qquad (3.30)$$

Let $\nu(3,1) := \{j \in \nu(3): \xi_j > z_{j-1}\}$ and $\nu(3,2) := \{j \in \nu(3): \xi_j = z_{j-1}\}$. Then $\nu(3) = \nu(3,1) \cup \nu(3,2)$. For $j \in \nu(3,1)$, letting $\alpha_j \uparrow \xi_j$ and, for $j \in \nu(3,2)$, taking $\alpha_j := z_{j-1} = \xi_j$ in (3.30) we get

$$\Big\| \sum_{j \in \nu(3,1)} T_j + \sum_{j \in \nu(3,2)} \{g(\zeta_{k(j)}-)[\Delta_j^\mu h_+]f(\zeta_{k(j)}-)$$
$$-g(z_{j-1}+)[\Delta_j^\mu h_+]f(z_{j-1}+)\}\Big\| \leq 4\epsilon. \qquad (3.31)$$

We get from (3.20) and (3.21)

$$\Big\| \sum_{j \in \nu(3,2)} \{g(z_{j-1}+)[\Delta_-^\pm h(z_{j-1})]f(z_{j-1}+)$$
$$-g(z_{j-1}-)[\Delta_-^\pm h(z_{j-1})]f(z_{j-1}-)\}\Big\| < 2\epsilon. \qquad (3.32)$$

Let $C := \{j \in \nu(3,2): j \text{ odd}\}$ and $D := \{j \in \nu(3,2): j \text{ even}\}$. Take $s_0 := a$, $s_m := b$ and, for each $j = 2, \ldots, m$, let $s_{j-1} = z_{j-1}$ if $z_{j-1} \in \kappa$, otherwise let $s_{j-1} \in (z_{j-1}, z_j)$ if $j \notin C$ and then let $s_{j-1} \in (s_{j-2}, z_{j-1})$ if $j \in C$. Then $\{s_j: j = 0, \ldots, m\}$ is a partition of $[a,b]$ and a refinement of κ. Suppose $s_{j-1} \leq \alpha_j \leq \beta_j \leq s_j$ for $j = 1, \ldots, m$. Then by (3.19),

$$\Big\| \sum_{j=1}^m \{g_+^{(b)}(\beta_j)[h_-(s_j) - h_-^{(a)}(s_{j-1})]f_+^{(b)}(\beta_j)$$
$$-g_+^{(b)}(\alpha_j)[h_-(s_j) - h_-^{(a)}(s_{j-1})]f_+^{(b)}(\alpha_j)\}\Big\| < 2\epsilon. \qquad (3.33)$$

First, let $\alpha_j = \beta_j$ for $j \notin C$. Thus the sum in (3.33) can be restricted to C. For $j \in C$, let $s_j \downarrow z_j$, $s_{j-1} \uparrow z_{j-1}$, $\beta_j \downarrow z_{j-1}$ and $\alpha_j \uparrow z_{j-1}$. We obtain

$$\Big\| \sum_{j \in C} \{g_+(z_{j-1})[h(z_j+) - h(z_{j-1}-)]f_+(z_{j-1})$$
$$-g_-(z_{j-1})[h(z_j+) - h(z_{j-1}-)]f_-(z_{j-1})\}\Big\| \leq 2\epsilon. \qquad (3.34)$$

Interchanging "even" and "odd", we get (3.34) with D in place of C. Then, summing, we get (3.34) with C replaced by $\nu(3,2)$ and 2ϵ by 4ϵ. Combining this with (3.32) gives

$$\Big\| \sum_{j \in \nu(3,2)} \{g_+(z_{j-1})[\Delta_j^\mu h_+]f_+(z_{j-1}) - g_-(z_{j-1})[\Delta_j^\mu h_+]f_-(z_{j-1})\}\Big\| < 6\epsilon.$$

Then combining with (3.31) and recalling $\xi_j = z_{j-1}$, $j \in \nu(3,2)$, gives $\|\sum_{j \in \nu(3)} T_j\|$ $< 10\epsilon$. It follows from (3.25) and (3.26) that $\|W\| \leq 12\epsilon$. Thus, any two Riemann-Stieltjes sums for $g_-^{(a)}$, $h_+^{(b)}$, $f_-^{(a)}$ based on partitions which are refinements of κ' are at most 24ϵ apart. So the integral (3.18) exists and, by (3.13), $(Y_2) \int_a^b g\, dh\, f$ exists. □

The next statement is a consequence of Theorems 3.7 and 3.6.

Corollary 3.8. *Let* $g, h, f \in \mathcal{R}([a, b]; \mathbb{B})$. *If one of the integrals in* (3.12) *exists then so does the other and they are equal.*

Since the (Y_1) and (Y_2) integrals have been shown to coincide for regulated functions g, h, f, if either is defined, we now define

$$(CY) \int_a^b g \, dh \, f := (Y_1) \int_a^b g \, dh \, f = (Y_2) \int_a^b g \, dh \, f \qquad (3.35)$$

and call this the *central Young integral*. A main reason for using the adjective "central" is that the value of the (CY) integral does not depend on the value at a discontinuity point of the integrator h, as is proved by the following statement. It is shown below (cf. Definition 3.11 and the discussion following it) that modifications of the (CY) integral are possible which make the new integrals dependent on the left jumps and right jumps of the integrator h.

If h is defined on $[a, b]$, call the value of $h(x)$ *special* if $h(x-) \neq h(x) \neq h(x+)$ and $a < x < b$. In an integral $\int_a^b g \, dh \, f$ call g and f the *integrands* and h the *integrator*. Let

$$h_+^{(a,b)} := \begin{cases} h(x+), & \text{for } x \in (a, b) \\ h(x), & \text{for } x = a \text{ or } b. \end{cases}$$

Define $h_-^{(a,b)}$ analogously. The following shows that special values of the integrator do not affect the central Young integral:

Theorem 3.9. *If* $(CY) \int_a^b g \, dh \, f$ *exists, then so do*

$$(CY) \int_a^b g \, dh_+^{(a,b)} f \qquad and \qquad (CY) \int_a^b g \, dh_-^{(a,b)} f$$

and all three integrals have the same value.

Proof. For $h_+^{(a,b)}$ we apply the (Y_2) definition. We have $\Delta^+_- h_+^{(a,b)}(x) = \Delta^+_- h(x)$, $a < x < b$, and

$$h_+^{(a,b)}(b) - h_+^{(a,b)}(b-) = h(b) - h(b-).$$

Also, $(h_+^{(a,b)})_+^{(b)} \equiv h_+^{(b)}$, so we need to show

$$(MPS) \int_a^b g_-^{(a)} dh_+^{(b)} f_-^{(a)} + [g\Delta^+ h f](a) = (MPS) \int_a^b g_-^{(a)} dh_+^{(b)} f_-^{(a)} + [g\Delta^+ h_+^{(a,b)} f](a),$$

which is true. The integral for $h_-^{(a,b)}$ can be treated symmetrically. $\square$

Special values of the integrands at points x do influence the integral if the integrator h has $\Delta^+_- h(x) \neq 0$.

Next we show that a (CY) integral is also a limit of Riemann-Stieltjes sums.

Proposition 3.10. *Let $g, h, f \in \mathcal{R}([a,b]; \mathbb{B})$ and let the central Young integral (3.35) exist. Then, given $\epsilon > 0$ and a partition κ_0 of $[a,b]$, there is a refinement $\nu = \{z_i : i = 0, \dots, n\}$ of κ_0 and there exist $y_i \in [z_{i-1}, z_i]$ for $i = 1, \dots, n$ such that*

$$\left\| (CY) \int_a^b g \, dh \, f - \sum_{i=1}^n g(y_i)[h(z_i) - h(z_{i-1})] f(y_i) \right\| < \epsilon.$$

Proof. We prove the proposition for the (Y_1) integral, which is sufficient by Corollary 3.8. Let $\epsilon > 0$. $(MPS) \int_a^b g_+^{(b)} dh_-^{(a)} f_+^{(b)}$ exists, so there is a partition κ of $[a,b]$ such that, in the notation given after (3.2),

$$\left\| (MPS) \int_a^b g_+^{(b)} dh_-^{(a)} f_+^{(b)} - S(\kappa, [a,b], g, h, f) \right\| < \epsilon/3. \tag{3.36}$$

Take a finite set $\mu \subset (a,b)$ such that

$$R_\mu := \sup_{\lambda \subset (a,b) \setminus \mu} \left\| \sum_\lambda \left[g(\Delta_-^+ h) f - g_+ (\Delta_-^+ h) f_+ \right] \right\| < \epsilon/3. \tag{3.37}$$

Let $\kappa_0 \cup \kappa \cup \mu = \{x_i : i = 0, \dots, m\}$, $a = x_0 < x_1 < \cdots < x_m = b$. Since g, f are regulated, they are bounded, and since h is regulated, there exist V_{i-1} and U_i for $i = 1, \dots, m$ such that

$$\sum_{i=1}^m \left[Osc_{(x_{i-1}, V_{i-1}]}(h) + Osc_{[U_i, x_i)}(h) \right] \|g\|_\infty \|f\|_\infty < \epsilon/6. \tag{3.38}$$

Take $v_{i-1} \in (x_{i-1}, V_{i-1})$ and $u_i \in (U_i, x_i)$ at which h is continuous, and such that $v_{i-1} < u_i$ for all $i = 1, \dots, m$. Let $\nu = \{z_j : j = 0, 1, \dots, n\} := \kappa_0 \cup \kappa \cup \mu \cup \bigcup_{i=1}^m \{v_{i-1}, u_i\}$. If $z_j = x_i$ for some i, let $y_j = x_i$. If $z_j = v_{i-1}$ for some i, let $y_j = x_{i-1}$. If $z_j = u_i$ for some i, choose y_j such that $v_{i-1} < y_j < u_i$ and g, f are continuous at y_j. Then by (3.36),

$$\left\| (MPS) \int_a^b g_+^{(b)} dh_-^{(a)} f_+^{(b)} - \sum_{j=1}^n g_+^{(b)}(y_j)[h_-^{(a)}(z_j) - h_-^{(a)}(z_{j-1})] f_+^{(b)}(y_j) \right\| < \epsilon/3.$$

$$\tag{3.39}$$

For $j = 1, \dots, n$ let

$$D_j := g(y_j)[h(z_j) - h(z_{j-1})] f(y_j) - g_+^{(b)}(y_j)[h_-^{(a)}(z_j) - h_-^{(a)}(z_{j-1})] f_+^{(b)}(y_j).$$

If $z_j = x_i$ for some $i = 1, \dots, m-1$ then

$$D_j = g(x_i)[h(x_i) - h(u_i)] f(x_i) - g_+(x_i)[h_-(x_i) - h(u_i)] f_+(x_i).$$

If $z_j = v_i$ for some $i = 1, \dots, m-1$ then

$$D_j = g(x_i)[h(v_i) - h(x_i)] f(x_i) - g_+(x_i)[h(v_i) - h_-(x_i)] f_+(x_i).$$

The sum of the latter two D_j's for the same i is

$$T_i := g(x_i)[h(v_i) - h(u_i)]f(x_i) - g_+(x_i)[h(v_i) - h(u_i)]f_+(x_i).$$

Let

$$T_0 := D_1 = g(a)[h(v_0) - h(a)]f(a) - g_+(a)[h(v_0) - h(a)]f_+(a),$$

$$T_m := D_n = g(b)[h(b) - h(u_m)]f(b) - g(b)[h_-(b) - h(u_m)]f(b)$$
$$= g(b)[h(b) - h_-(b)]f(b).$$

Then by (3.38),

$$\|T_0 - [g(\Delta^+h)f - g_+(\Delta^+h)f_+](a)\|$$

$$+ \left\| \sum_{i=1}^{m-1} (T_i - [g(\Delta_-^+h)f - g_+(\Delta_-^+h)f_+](x_i)) \right\| \le \epsilon/3. \qquad (3.40)$$

If $z_j = u_i$ for some $i = 1, \dots, m$, so that $z_{j-1} = v_{i-1}$, then

$$D_j = g(y_j)[h(u_i) - h(v_{i-1})]f(y_j) - g_+(y_j)[h_-(u_i) - h_-(v_{i-1})]f_+(y_j) = 0,$$

since g, f are continuous at y_j and h is continuous at u_i and v_{i-1}. So $\sum_{j=1}^{n} D_j = \sum_{i=0}^{m} T_i$. Then by (3.37), (3.39) and (3.40), we have

$$\left\| (CY) \int_a^b g\, dh\, f - \sum_{j=1}^{n} g(y_j)[h(z_j) - h(z_{j-1})]f(y_j) \right\|$$

$$\le \left\| (MPS) \int_a^b g_+^{(b)} dh_-^{(a)} f_+^{(b)} - \sum_{j=1}^{n} g_+^{(b)}(y_j)[h_-^{(a)}(z_j) - h_-^{(a)}(z_{j-1})]f_+^{(b)}(y_j) \right\|$$

$$+ R_\mu + \left\| \sum_{j=1}^{n} D_j - \sum_{i=1}^{m-1} [g(\Delta_-^+h)f - g_+(\Delta_-^+h)f_+](x_i) \right.$$

$$\left. - [g(\Delta^+h)f - g_+(\Delta^+h)f_+](a) - [g(\Delta_-)f](b) \right\| < \epsilon/3 + \epsilon/3 + \epsilon/3 = \epsilon.$$

Since $\epsilon > 0$ is arbitrary this proves Proposition 3.10. $\qquad \square$

We finish this subsection by defining two integrals of Stieltjes type. They will allow concise formulations of several of the results of this paper. Recalling the notations (2.16), (2.17) and (2.18) we state:

Definition 3.11. Let $g, h, f \in \mathcal{R}([a, b]; \mathbb{B})$. We define the *left Young* (LY) integral

$$(LY) \int_a^b g\, dh\, f := (CY) \int_a^b g^{(a+)}\, dh\, f_-^{(a)} + \sum_{(a,b)} [g_+(\Delta^+h)f - g(\Delta^+h)f_-]$$

if the (CY) integral exists and the sum converges unconditionally in $\mathbb{B}$. Similarly, we define the *right Young* (RY) integral

$$(RY) \int_a^b g\,dh\,f := (CY) \int_a^b g_+^{(b)}\,dh\,f^{(b-)} - \sum_{(a,b)} [g_+(\Delta_-h)f - g(\Delta_-h)f_-]$$

if the (CY) integral exists and the sum converges unconditionally in $\mathbb{B}$. The sum in both integrals over an empty interval (a, b) is defined as 0.

We notice that the order of the integrands and the integrator in the (LY) and (RY) integrals is important. Since the role of both integrands is non-symmetric, in general we have

$$(LY) \int_a^b g\,dh\,f \neq (LY) \int_a^b (gf)\,dh$$

even if the Banach algebra $\mathbb{B}$ is commutative. The same remark applies to the right Young integral. Propositions 3.13 and 3.14 below show that the (LY), (RY) and (CY) integrals are different in general. However we have the following:

Theorem 3.12. *The (LY) and (RY) integrals $\int_a^b g\,dh\,f$ are equal whenever both are defined.*

Proof. Let the (LY) and (RY) integrals $\int_a^b g\,dh\,f$ exist. By Definition 3.11 and the (Y_1) definition of the (CY) integral (Definition 3.4), we have

$$(RY) \int_a^b g\,dh\,f = (MPS) \int_a^b g_+^{(b)}\,dh_-^{(a)}\,f_+^{(b-)} - [g_+(\Delta^+h)(\Delta^+f)](a)$$
$$+ \sum_{(a,b)} [g(\Delta_-h)f_- - g_+(\Delta_-h)f_+ - g_+(\Delta^+h)(\Delta^+f)]$$
$$+ [g(\Delta_-h)f_-](b). \tag{3.41}$$

The above sum converges unconditionally since it results from adding two unconditionally convergent sums and writing $\Delta^\pm h = \Delta^+h + \Delta_-h$. The analogous expression of the (LY) integral using the (Y_1) definition of the (CY) integral gives a value identical to the right side of (3.41). This proves Theorem 3.12. $\square$

Remark. We used the (Y_1) definition of (CY) integral in the proof above. If instead the (Y_2) integral is used then similarly we get that

$$(LY) \int_a^b g\,dh\,f = (MPS) \int_a^b g_-^{(a+)}\,dh_+^{(b)}\,f_-^{(a)} + [g_+(\Delta^+h)f](a)$$

$$+ \sum_{(a,b)} [g_+(\Delta^+h)f - g_-(\Delta^+h)f_- + (\Delta_-g)(\Delta_-h)f_-] + [(\Delta_-g)(\Delta_-h)f_-](b). \tag{3.42}$$

Similarly, the (RY) integral has the same value whenever it is defined.

We write integrands $\mathbb{I}$ in the next statement and its proof, as we did not earlier, because the rôle of the functions g and f in the definitions of $(LY) \int_a^b g\,dh\,f$ and $(RY) \int_a^b g\,dh\,f$ is not symmetric.

Proposition 3.13. *For any regulated functions* f, h *on* $[a, b]$, *if* $(RY) \int_a^b \mathbb{1} \, dh \, f$ *exists then* $(LY) \int_a^b \mathbb{1} \, dh \, f$ *also exists (and has the same value), but the converse can fail for* $f, h \in \mathcal{W}_p([a, b], \mathbb{R})$ *for* $p > 2$.

Similarly for any regulated functions g, h, *if* $(LY) \int_a^b g \, dh \, \mathbb{1}$ *exists then so does* $(RY) \int_a^b g \, dh \, \mathbb{1}$ *(and has the same value) but not conversely for* $g, h \in \mathcal{W}_p$, $p > 2$.

Proof. If $(RY) \int_a^b \mathbb{1} \, dh \, f$ exists then by the definition the sum $-\sum_{(a,b)}(\Delta_- h)(\Delta_- f)$ converges (unconditionally) and from the (Y_2) definition of (CY) integral, the integral $(MPS) \int_a^b \mathbb{1} \, dh_+^{(b)} f_-^{(a)}$ exists and the sum $\sum_{(a,b)}(\Delta_-^+ h)(\Delta_- f)$ converges. Adding the two sums gives that $\sum_{(a,b)}(\Delta^+ h)(\Delta_- f)$ converges, which is the sum in the definition of (LY) integral. The sum in the (Y_2) form of $(CY) \int_a^b \mathbb{1} \, dh \, f_-^{(a)}$ is 0 and the (MPS) integral is the same, so the (LY) integral exists.

To see that the converse can fail let $a = 0$ and $b = 1$. Let $f(1/(2n)) = h_-(1/(2n)) = n^{-1/2}$ and $f(1/(2n-1)) = h(1/(2n-1)) = 0$ for $n = 1, 2, \cdots$. Also let f be linear and $h \equiv 0$ on each interval $[1/(2n), 1/(2n-1)]$ while h is linear and $f \equiv 0$ on $[1/(2n+1), (1/(2n))-]$. Let $f(0) := h(0) := 0$. It's easy to check that f and $h \in \mathcal{W}_p$ for any $p > 2$.

To show that $(LY) \int_0^1 \mathbb{1} \, dh \, f$ exists, note that since h is right continuous, the sum in the definition of (LY) integral is 0. The $\sum_{(0,1)}$ in the (Y_2) definition of $(CY) \int_0^1 \mathbb{1} \, dh \, f^{(0)}$ is also 0. The (MPS) integral reduces to the sum of values of f_- times jumps of h at the points $1/(2n)$, which is also 0, by taking partitions $\{0, 1/n, 1/(n-1), \cdots, 1\}$. But the definition of (RY) integral includes $\sum_{(0,1)}(\Delta_- h)(\Delta_- f)$ which diverges.

The statements about $\int g \, dh \, \mathbb{1}$ can be proved symmetrically. The proof of Proposition 3.13 is complete. $\qquad \square$

The next statement relates the (LY) and (RY) integrals with the (CY) integral.

Proposition 3.14. *The value of the* (CY) *integral* $\int_a^b g \, dh \, f$ *can differ from the value of the corresponding* (LY) *and* (RY) *integrals.*

Proof. Let $g, f \in \mathcal{R}([a, b]; \mathbb{R})$ and let h be a real-valued function on $[a, b]$ with at most three different values, being constant on $[a, c)$ and on $(c, b]$ for some $a < c < b$. Then

$$(LY) \int_a^b g \, dh \, f = (RY) \int_a^b g \, dh \, f = [g(\Delta_- h)f_- + g_+(\Delta^+ h)f](c) \qquad (3.43)$$

and $(CY) \int_a^b g \, dh \, f = [g(\Delta_-^+ h)f](c)$. This implies Proposition 3.14. $\qquad \square$

The (LY) and (RY) integrals with two integrands appear in the Duhamel formula (cf. Theorem 5.5) and the differentiation of indefinite product integral operators (Theorem 5.16). These integrals thus seem exactly suited for those purposes. Special cases of these integrals with one integrand are used to define an analogue of Peano series (cf. (4.74) and (4.75)), in linear integral equations (cf. Definition 5.19) and to define the logarithmic operator (cf. Definition 6.2). Namely, we *define* $(LY) \int_a^b dh \, f := (LY) \int_a^b \mathbb{1} \, dh \, f$ and $(RY) \int_a^b g \, dh := (RY) \int_a^b g \, dh \, \mathbb{1}$. Note: we do not define $(LY)\int_a^b g\, dh$ or $(RY)\int dh\, f$ since these are never used below and could cause confusion because of the non-symmetric definitions of (LY) and (RY) integrals.

Then, respectively by (3.42) and (3.41), we have

$$(LY)\int_a^b dh\, f = (MPS)\int_a^b dh_+^{(b)}\, f_-^{(a)} + [(\Delta^+ h)f](a) + \sum_{(a,b)} \Delta^+ h \Delta_- f \qquad (3.44)$$

and

$$(RY)\int_a^b g\, dh = (MPS)\int_a^b g_+^{(b)}\, dh_-^{(a)} - \sum_{(a,b)} \Delta^+ g \Delta_- h + [g(\Delta_- h)](b).$$

These two integrals with the same integrands $g \equiv f$ can have different values, even when $\mathbb{B}$ is commutative. Indeed, if f and h are real-valued functions as in the proof of Proposition 3.14 then, by (3.43), we have

$$(LY)\int_a^b dh\, f = [(\Delta_- h)f_- + (\Delta^+ h)f](c)$$

and

$$(RY)\int_a^b f\, dh = [f(\Delta_- h) + f_+(\Delta^+ h)](c).$$

It is worth mentioning that the values of the (LY) and (RY) integrals in this example are the same as the values of the Left Cauchy and Right Cauchy integrals, respectively. Recall (cf. also Section II.19.2 in Hildebrandt, 1963) that (3.1) exists as the Left Cauchy or Right Cauchy integral if it exists as the (MPS) integral except that in the definition of the Riemann-Stieltjes sum, for each $i = 1, \ldots, n$, $y_i = x_{i-1}$ or $y_i = x_i$, respectively.

3.3. Refinement-Young-Stieltjes integrals.

The results of the present subsection are not used in the rest of the paper. We will define the refinement-Young-Stieltjes integral and show that whenever it exists, and so whenever the (MPS) integral exists, for regulated functions, then the (CY) integral also exists and the integrals have the same value.

The refinement-Young-Stieltjes integral, or (RYS) integral of $\mathbb{B}$-valued functions g and f with respect to a regulated $\mathbb{B}$-valued function h is defined by

$$(RYS)\int_a^b g\, dh\, f := \lim_{(\kappa)} \sum_{i=1}^n \Psi([\Delta_i^\kappa])$$

if it exists, where the limit is taken under refinements of partitions of $[a, b]$ and the interval function Ψ is defined by

$$\Psi([c,d]) := g(c)\Delta^+ h(c)f(c) + g(y)[h(d-) - h(c+)]f(y) + g(d)\Delta_- h(d)f(d)$$

for $y \in (c, d)$. The proof of the next proposition is the same as the proof of this statement for real-valued functions given by Theorem II.19.3.3 in Hildebrandt (1963). Therefore we omit it.

Proposition 3.15. *Let $h \in \mathcal{R}([a,b]; \mathbb{B})$ and let $g, f \in \ell_\infty([a,b]; \mathbb{B})$. The integral $(RYS) \int g \, dh \, f$ exists whenever the integral $(MPS) \int g \, dh \, f$ exists and they are equal.*

The integral $(MPS) \int_a^b g \, dh$ fails to exist if g and h have jumps on the same side of the same point in (a,b) (cf. Corollary II.10.6 in Hildebrandt, 1963). The integral $(Y_1) \int_a^b g \, dh$ avoids this problem by taking $\int g_+ dh_-$ and adding a special sum over points where both g and h have jumps. The next fact shows that even when h is continuous, $(Y_1) \int_a^b g \, dh$ may exist when $(RYS) \int_a^b g \, dh$ and, hence, $(MPS) \int_a^b g \, dh$ do not.

Proposition 3.16. *Let g be a regulated, real-valued function on $[0,1]$ such that $g(x) = 0$ except for x in a countable set S. Let h be a regulated real-valued function on $[0,1]$, continuous on S. Then*
(a) $(Y_1) \int_0^1 g \, dh$ exists and is 0, but
(b) There exist such g, h, where h is continuous on $[0,1]$ and S is a sequence converging to 0, for which $(RYS) \int_0^1 g \, dh$ does not exist.

Proof. For (a), since g is regulated and 0 on a dense set, $g_+^{(1)}(x) = g_-^{(0)}(x) = 0$ for $0 \le x \le 1$. Since h is continuous on S, $(Y_1) \int_0^1 g \, dh = (MPS) \int_0^1 g_+^{(1)} dh_-^{(0)}$ exists and is 0.

For (b), let $g(x) := m^{-1/2}$ if $x = 1/(3m)$ for $m = 1, 2, \cdots$, and $g(x) := 0$ otherwise. Let $h(1/(3m+1)) := 0$ and $h(1/(3m-1)) := m^{-1/2}$ for $m = 1, 2, \cdots$. Let $h(0) := h(1) := 0$. Let h be "linear in between," i.e. on each closed interval where h is so far defined only at the endpoints, let h be linear. Let $\kappa = \{x_i : i = 0, \ldots, n\}$ be any partition of $[0,1]$. Take the smallest m such that $y_m := 1/(3m+1) \le x_1$. Consider refinements λ of κ containing y_m. If $\lambda_1 := \kappa \cup \{y_m\}$ then the contribution to any R-S (Riemann-Stieltjes) sum for λ_1 coming from $[0, y_m]$ is 0. For $N > m$ let

$$\lambda_N := \kappa \cup \{1/(3k+1), 1/(3k-1) : k = m+1, \ldots, N\}.$$

We form a R-S sum for λ_N by letting it be the same as one for λ_1 on $[y_m, 1]$, and by evaluating g at $1/(3k)$ for $k = m+1, \ldots, N$. Thus the part of our R-S sum for λ_N coming from $[0, y_m]$ is

$$\sum_{k=m+1}^{N} g\left(\frac{1}{3k}\right) \left[h\left(\frac{1}{3k-1}\right) - h\left(\frac{1}{3k+1}\right)\right] + 0 \cdot \left[h\left(\frac{1}{3k-2}\right) - h\left(\frac{1}{3k-1}\right)\right]$$

$$= \sum_{k=m+1}^{N} k^{-1/2} k^{-1/2} \to \infty$$

as $N \to \infty$. Thus two R-S sums for $\int_0^1 g \, dh$, both based on refinements of κ, differ by an arbitrarily large amount. Since h is continuous, sums as in the definition of (RYS) integral are ordinary Riemann-Stieltjes sums. So $(RYS) \int_0^1 g \, dh$ does not exist. $\qquad \square$

In the example in (b) of the above Proposition 3.16, the following three statements are equivalent: $g \in \mathcal{W}_r([0,1])$; $h \in \mathcal{W}_r([0,1])$; $r > 2$. Thus existence of $(Y_1) \int_0^1 g \, dh$, although it holds, does not follow from the Love-Young inequality (L. C. Young, 1936; cf. also (3.94) below) which requires $g \in \mathcal{W}_p$, $h \in \mathcal{W}_q$, $\frac{1}{p} + \frac{1}{q} > 1$.

The next statement relates the (RYS) integral with the (Y_1) integral.

Proposition 3.17. *Let* $g, h, f \in \mathcal{R}([a,b]; \mathbb{B})$. *If* $(Y) \int_a^b g\,dh\,f$ *exists then so does* $(Y_1) \int_a^b g\,dh\,f$ *and the two are equal.*

Proof. We will prove that if $(RYS) \int_a^b g\,dh\,f$ exists then, first,

$$\sum_{(a,b)} [g_+(\Delta_-^+ h)f_+ - g(\Delta_-^+ h)f] \quad \text{converges unconditionally in } \mathbb{B}, \tag{3.45}$$

second,

$$(MPS) \int_a^b g_+^{(b)}\,dh_-^{(a)} f_+^{(b)} \quad \text{exists} \tag{3.46}$$

and, finally, that the (RYS) and (Y_1) integrals coincide.

For any $\epsilon > 0$, since $(RYS) \int_a^b g\,dh\,f$ exists there is a partition $\kappa := \{x_i : i = 0, \dots, n\}$ of $[a,b]$ such that, for any refinement $\lambda := \{y_j : j = 0, \dots, m\}$ of κ,

$$\Big\| \sum_{j=1}^m \{ [g(\Delta^+ h)f](y_{j-1}) + g(\xi_j)[h(y_j-) - h(y_{j-1}+)]f(\xi_j) + [g(\Delta_- h)f](y_j) \}$$

$$- (RYS) \int_a^b g\,dh\,f \Big\| < \epsilon \tag{3.47}$$

for all $\xi_j \in (y_{j-1}, y_j)$, $j = 1, \dots, m$. This implies also the bound

$$\Big\| \sum_{j=1}^m \{ g(\beta_j)[h(y_j-) - h(y_{j-1}+)]f(\beta_j)$$

$$- g(\alpha_j)[h(y_j-) - h(y_{j-1}+)]f(\alpha_j) \} \Big\| < 2\epsilon \tag{3.48}$$

whenever $y_{j-1} < \alpha_j \le \beta_j < y_j$ for $j = 1, \dots, m$.

To prove (3.45), suppose it fails. Then for some $\epsilon_0 > 0$, for any finite $\nu \subset (a,b)$ there is a finite $\mu \subset (a,b) \setminus \nu$ with $\| \sum_\mu [g_+(\Delta_-^+ h)f_+ - g(\Delta_-^+ h)f] \| > \epsilon_0$. Take κ such that (3.47) and (3.48) hold for $\epsilon = \epsilon_0/2$. Let $\nu := \kappa \cap (a,b)$ and take the above $\mu =: \{t_k : k = 1, \dots, r\}$. Consider a refinement $\lambda = \{y_j : j = 0, \dots, m\}$ of κ such that λ contains no points of μ, but between any two different points of $\kappa \cup \mu$ there are at least two points of λ. Define $j(k)$ so that $y_{j(k)-1} < t_k < y_{j(k)}$, $k = 1, \dots, r$. Then $y_{j(k)} \notin \nu$ and $y_{j(k)-1} \notin \nu$. Let

$$\alpha_{j(k)} = t_k, \quad \beta_{j(k)} \downarrow t_k, \quad y_{j(k)-1} \uparrow t_k \text{ and } y_{j(k)} \downarrow t_k,$$

as is possible by the "at least two points" condition. Let $\alpha_j = \beta_j$ if $j \ne j(k)$ for all k. Then the norm in (3.48) converges to

$$\Big\| \sum_\mu [g_+(\Delta_-^+ h)f_+ - g(\Delta_-^+ h)f] \Big\| \le \epsilon_0,$$

a contradiction. Hence (3.45) holds.

To prove (3.46), given $\epsilon > 0$, by (3.45), we can assume $\kappa = \{x_i : i = 0, 1, \ldots, n\}$ is chosen large enough so that for any refinement λ of κ, (3.47) and (3.48) hold and

$$\left\| [g_+(\Delta^+ h)f_+ - g(\Delta^+ h)f](a) + \sum_{\lambda \setminus \{a,b\}} [g_+(\Delta_-^{\pm} h)f_+ - g(\Delta_-^{\pm} h)f] \right.$$
$$\left. - [g(\Delta_- h)f](b) - S(g,h,f) \right\| < \epsilon, \quad (3.49)$$

where

$$S(g,h,f) := [g_+(\Delta^+ h)f_+ - g(\Delta^+ h)f](a)$$
$$+ \sum_{(a,b)} [g_+(\Delta_-^{\pm} h)f_+ - g(\Delta_-^{\pm} h)f] - [g(\Delta_- h)f](b).$$

Since g, h and f are regulated, they are bounded and, for each $i = 1, \ldots, n$, one can choose $u_i \in (x_{i-1}, x_i)$ close enough to x_i so that

$$Osc_{(u_i, x_i)}(h) \leq \epsilon / [2n \|g\|_\infty \|f\|_\infty]. \quad (3.50)$$

(If $f \equiv 0$ or $g \equiv 0$, the conclusions hold.) Let $t_i := (u_i + x_i)/2$. Likewise, for each $i = 0, \ldots, n-1$, choose $v_i \in (x_i, x_{i+1})$ close enough to x_i so that

$$Osc_{(x_i, v_i)}(g) \leq \epsilon / [4n \|h\|_\infty \|f\|_\infty], \quad Osc_{(x_i, v_i)}(f) \leq \epsilon / [4n \|h\|_\infty \|g\|_\infty]. \quad (3.51)$$

Let $w_i := (x_i + v_i)/2$. Let κ' be the partition $\kappa \cup \{w_{i-1}, t_i : i = 1, \ldots, n\}$. Let $\lambda := \{y_j : j = 0, \ldots, m\}$ be any refinement of κ'. We aim to find a small upper bound for the norm of

$$R(\xi, \lambda) - (RYS) \int_a^b g \, dh \, f - S(g, h, f),$$

where ξ is any vector $(\xi_1 \ldots, \xi_m)$ with $\xi_j \in [y_{j-1}, y_j]$ for all $j = 1, \ldots, m$,

$$R(\xi, \lambda) := \sum_{j=1}^m r_j(\xi_j), \quad \text{and} \quad r_j(w) := g_+^{(b)}(w)[h_-(y_j) - h_-^{(a)}(y_{j-1})]f_+^{(b)}(w) \quad (3.52)$$

for any $j = 1, \ldots, m$ and real w. Then for each $j = 1, \ldots, m$, we have

$$r_j(\xi_j) = [g_+^{(b)}(\xi_j)(\Delta_j^\lambda h^{(a)})f_+^{(b)}(\xi_j) - g_+(y_{j-1})(\Delta_j^\lambda h^{(a)})f_+(y_{j-1})]$$

$$+ \{g(y_{j-1}+)[h(y_j-) - h(y_{j-1}+)]f(y_{j-1}+) + [g(h_+ - h_-^{(a)})f](y_{j-1})\}$$

$$+ [g_+(h_+ - h_-^{(a)})f_+ - g(h_+ - h_-^{(a)})f](y_{j-1})$$
$$= t(\xi_j; y_{j-1}, y_j) + w(y_{j-1}, y_j) + s(y_{j-1}), \quad (3.53)$$

where

$$t(\eta; u, v) := g_+^{(b)}(\eta)[h_-(v) - h_-^{(a)}(u)]f_+^{(b)}(\eta) - g_+(u)[h_-(v) - h_-^{(a)}(u)]f_+(u), \quad (3.54)$$

$$w(u, v) := g(u+)[h(v-) - h(u+)]f(u+) + [g(h_+ - h_-^{(a)})f](u) \quad (3.55)$$

and

$$s(u) := \left[g_+(h_+ - h_-^{(a)})f_+ - g(h_+ - h_-^{(a)})f\right](u). \tag{3.56}$$

Then, summing (3.53) over j, adding and subtracting $[g(\Delta_- h)f](b)$, we get

$$R(\xi, \lambda) = \sum_{j=1}^{m} t(\xi_j; y_{j-1}, y_j) + \left\{ \sum_{j=1}^{m} w(y_{j-1}, y_j) + [g(\Delta_- h)f](b) \right\}$$

$$+ \left\{ \sum_{j=1}^{m} s(y_{j-1}) - [g(\Delta_- h)f](b) \right\} =: R_1 + R_2 + R_3. \tag{3.57}$$

Letting in (3.47) $\xi_j \downarrow y_{j-1}$ for each $j = 1, \ldots, m$ it follows that

$$\left\| R_2 - (RYS) \int_a^b g \, dh \, f \right\| \leq \epsilon. \tag{3.58}$$

By (3.49), we have

$$\| R_3 - S(g, h, f) \| < \epsilon. \tag{3.59}$$

Now consider ξ such that $\xi_j \in [y_{j-1}, y_j)$ for all $j = 1, \ldots, m$. To bound $\|R_1\|$, define a mapping $j(\cdot)$ by $\{0, \ldots, n\} \ni i \mapsto j(i) \in \{0, \ldots, m\}$ such that $y_{j(i)} = x_i$ for all i. By (3.51), we have

$$\left\| \sum_{i=0}^{n-1} \left[g_+(\xi_{j(i)+1}) \left[h_-(y_{j(i)+1}) - h_-^{(a)}(x_i) \right] f_+(\xi_{j(i)+1}) \right. \right.$$

$$\left. \left. - g_+(x_i) \left[h_-(y_{j(i)+1}) - h_-^{(a)}(x_i) \right] f_+(x_i) \right] \right\| < \epsilon. \tag{3.60}$$

Let $M := \{2, \ldots, m\} \setminus \{j(i) + 1 : i = 0, \ldots, n-1\}$, $M_o := \{j \in M : j \text{ odd}\}$ and $M_e := \{j \in M : j \text{ even}\}$. Take $z_j \in (\xi_j, y_j)$ for each $j = 1, \ldots, m$ and replace each $y_j \in \lambda \setminus \kappa$ with j odd by z_j. Then by (3.48) with $\alpha_j = \beta_j$ for all $j \notin M_e$, we have

$$\left\| \sum_{j \in M_e} \left\{ g(\beta_j) \left[h(y_j-) - h(z_{j-1}+) \right] f(\beta_j) \right. \right.$$

$$\left. \left. - g(\alpha_j) \left[h(y_j-) - h(z_{j-1}+) \right] f(\alpha_j) \right\} \right\| < 2\epsilon. \tag{3.61}$$

Interchanging "even" and "odd", we get (3.61) with M_o in place of M_e. Then summing, we get (3.61) with M_e replaced by M and 2ϵ by 4ϵ. For all $j \in M$, let $z_{j-1} \uparrow y_{j-1}$, $\alpha_j \downarrow y_{j-1}$ and $\beta_j \downarrow \xi_j$. Combining the resulting bound with (3.60) gives $\|R_1\| \leq 5\epsilon$. Therefore, by (3.58) and (3.59), we get

$$\left\| R(\xi, \lambda) - (RYS) \int_a^b g \, dh \, f - S(g, h, f) \right\| \leq 7\epsilon \tag{3.62}$$

whenever $\xi_j \in [y_{j-1}, y_j)$ for all $j = 1, \ldots, m$. Note that (3.60) and (3.61) remain true if any terms of the sums are omitted. Thus if V is any sum of some of the terms of R_1, then $\|V\| \leq 5\epsilon$.

Next consider an arbitrary ξ with possible values $\xi_j = y_j$ for any $j = 1, \ldots, m$. For all ξ_j in ξ such that $\xi_j = y_j$ with j odd, replace ξ_j by y_{j-1} and denote the resulting vector by ξ^o. Define also ξ^e in the same way, interchanging "odd" and "even". Let ξ^m be ξ with all $\xi_j = y_j$ replaced by y_{j-1}. Reshuffling Riemann-Stieltjes sums we get

$$R(\xi, \lambda) + R(\xi^m, \lambda) = R(\xi^o, \lambda) + R(\xi^e, \lambda).$$

Then, adding and subtracting $R(\xi^m, \lambda)$ and $(RYS) \int_a^b g \, dh \, f + S(g, h, f)$ we get

$$\left\| R(\xi, \lambda) - (RYS) \int_a^b g \, dh \, f - S(g, h, f) \right\|$$

$$\leq \left\| R(\xi^o, \lambda) - (RYS) \int_a^b g \, dh \, f - S(g, h, f) \right\|$$

$$+ \left\| R(\xi^e, \lambda) - (RYS) \int_a^b g \, dh \, f - S(g, h, f) \right\|$$

$$+ \left\| R(\xi^m, \lambda) - (RYS) \int_a^b g \, dh \, f - S(g, h, f) \right\|. \tag{3.63}$$

Since, in ξ^m, we have $\xi_j < y_j$ for all $j = 1, \ldots, m$, by (3.62), the last term in (3.63) is $\leq 7\epsilon$. Therefore it suffices to consider vectors ξ such that $\xi_{j+1} < y_{j+1}$ or $j = m$ whenever $\xi_j = y_j$. Let J be the set of $j = 1, \ldots, m - 1$ for which $\xi_j = y_j = x_i$ for some $i = 1, \ldots, n - 1$, i.e. $j = j(i)$ for some $i = 1, \ldots, n - 1$. Let $K := \{j \notin J: 1 \leq j < m, \xi_j = y_j\}$ and let $L := \{j = 1, \ldots, m-1: \{j-1, j\} \cap (J \cup K) = \emptyset\}$. $I(m) := \emptyset$ if $m \in K + 1$ and $I(m) := \{m\}$ otherwise. Then $\{1, \ldots, m\}$ is the disjoint union of J, $J + 1$, K, $K + 1$, L and $I(m)$, where $J + 1$ and $K + 1$ are $\{j + 1: j \in J\}$ and $\{j + 1: j \in K\}$, respectively.

Next we will represent $R(\xi, \lambda)$ as a sum of three sums, different from those in (3.57). Using the notations (3.52), (3.54), (3.55) and (3.56), for $j \in J \cup K$ we have by (3.53)

$$r_j(y_j) + r_{j+1}(\xi_{j+1}) =$$

$$= t(\xi_{j+1}; y_j, y_{j+1}) + t(y_j; y_{j-1}, y_j) + w(y_j, y_{j+1}) + w(y_{j-1}, y_j) + s(y_j) + s(y_{j-1}) \tag{3.64}$$

$$= t(\xi_{j+1}; y_j, y_{j+1}) + t(y_j; y_{j-1}, y_{j+1}) + w(y_{j-1}, y_{j+1}) + s(y_{j-1}). \tag{3.65}$$

Let

$$R_1' := A_{J1} + A_{J2} + A_{K1} + A_{K2} + A_L + T_m$$

$$:= \sum_{j \in J} t(\xi_{j+1}; y_j, y_{j+1}) + \sum_{j \in J} t(y_j; y_{j-1}, y_j) + \sum_{j \in K} t(\xi_{j+1}; y_j, y_{j+1})$$

$$+ \sum_{j \in K} t(y_j; y_{j-1}, y_{j+1}) + \sum_{j \in L} t(\xi_j; y_{j-1}, y_j) + t(\xi_m; y_{m-1}, b),$$

$$R_2' := \sum_{j \in J} \{w(y_j, y_{j+1}) + w(y_{j-1}, y_j)\} + \sum_{j \in K} w(y_{j-1}, y_{j+1}) + \sum_{j \in L \cup I(m)} w(y_{j-1}, y_j)$$

$$+ [g(\Delta_- h) f](b)$$

and

$$R'_3 := \sum_{j \in J} \{s(y_j) + s(y_{j-1})\} + \sum_{j \in K} s(y_{j-1}) + \sum_{j \in L \cup I(m)} s(y_{j-1}) - [g(\Delta_- h)f](b).$$

Then using (3.64) for $j \in J$, (3.65) for $j \in K$ and (3.53) for $j \in L \cup I(m)$ we have from (3.52)

$$R(\xi, \lambda) = \sum_{j \in J} [r_j(y_j) + r_{j+1}(\xi_{j+1})] + \sum_{j \in K} [r_j(y_j) + r_{j+1}(\xi_{j+1})] + \sum_{j \in L \cup I(m)} r_j(\xi_j)$$

$$= \sum_{j \in J} [t(\xi_{j+1}; y_j, y_{j+1}) + t(y_j; y_{j-1}, y_j) + w(y_j, y_{j+1}) + w(y_{j-1}, y_j) + s(y_j) + s(y_{j-1})]$$

$$+ \sum_{j \in K} [t(\xi_{j+1}; y_j, y_{j+1}) + t(y_j; y_{j-1}, y_{j+1}) + w(y_{j-1}, y_{j+1}) + s(y_{j-1})]$$

$$+ \sum_{j \in L \cup I(m)} [t(\xi_j; y_{j-1}, y_j) + w(y_{j-1}, y_j) + s(y_{j-1})]$$

$$+ [g(\Delta_- h)f](b) - [g(\Delta_- h)f](b) = R'_1 + R'_2 + R'_3.$$

Let $\lambda' := \lambda \setminus \{y_j: j \in K\}$. Note that R'_k for $k = 2, 3$ are the same as R_k except that they are based on the partition λ'. Since $\lambda' \supset \kappa$ we have that (3.58) and (3.59) hold for R'_k in place of R_k. Now, $A_{J1} + A_{K1} + A_L$ is a sum of terms of R_1, in all of which $\xi_j < y_j$, so $\|A_{J1} + A_{K1} + A_L\| \leq 5\epsilon$. A_{K2} is the sum of terms of R_1 for the partition $\lambda' =: \{z_l: l = 0, \dots, L\}$ over values of l such that for some $j \in K$, $z_{l-1} = y_{j-1}$ and $z_l = y_{j+1}$. Here $y_j < y_{j+1}$, so $\|A_{K2}\| \leq 5\epsilon$. Next, by (3.50) it follows that $\|A_{J2} + T_m\| \leq \epsilon$. So $\|R'_1\| \leq 11\epsilon$. Thus by (3.58) and (3.59) for R'_k, the first two norms in (3.63) are each $< 13\epsilon$. Hence, by (3.63), for an arbitrary vector ξ, (3.62) holds with 7ϵ replaced by 33ϵ. Since ϵ is an arbitrary positive number this implies that (3.46) holds and, hence, the (Y_1) integral exists.

To show that the (RYS) and (Y_1) integrals coincide let $\kappa = \{x_i: i = 0, \dots, n\}$ be a partition of $[a, b]$ and let $y_i \in (x_{i-1}, x_i)$ for $i = 1, \dots, n$. Then we have

$$\sum_{i=1}^{n} \Psi([\Delta_i^\kappa]) = \sum_{i=1}^{n} g(y_i)(\Delta_i^\kappa h_-^{(a)}) f(y_i) + [g(\Delta^+ h)f](a) - g(y_1)(\Delta^+ h)(a)f(y_1)$$

$$+ \sum_{i=1}^{n-1} \{[g(\Delta_-^+ h)f](x_i) - g(y_{i+1})(\Delta_-^+ h)(x_i)f(y_{i+1})\} + [g(\Delta_- h)f](b).$$

Letting $y_i \downarrow x_{i-1}$ for all $i = 1, \dots, n$ the right side of this identity approaches the following expression:

$$\sum_{i=1}^{n} g_+(x_{i-1})(\Delta_i^\kappa h_-^{(a)}) f_+(x_{i-1}) + [g(\Delta^+ h)f - g_+(\Delta^+ h)f_+](a)$$

$$+ \sum_{i=1}^{n-1} [g(\Delta_-^+ h)f - g_+(\Delta_-^+ h)f_+](x_i) + [g(\Delta_- h)f](b).$$

By the definition (1.4) of (Y_1) integral, refining the partition κ, this expression can be made arbitrarily close to $(Y_1) \int_a^b g\, dh\, f$. The proof of Proposition 3.17 is complete. $\square$

The following is an immediate consequence of Propositions 3.15 and 3.17.

Corollary 3.18. *Let $g, h, f \in \mathcal{R}([a,b]; \mathbb{B})$. If $(MPS) \int_a^b g\, dh\, f$ exists then so does $(Y_1) \int_a^b g\, dh\, f$ and the two are equal.*

Next we will show that the converse to Proposition 3.17 holds if an additional condition on functions g, h and f holds.

Proposition 3.19. *Let $h \in \mathcal{R}([a,b]; \mathbb{B})$ and let $g, f\colon [a,b] \mapsto \mathbb{B}$. If $(RYS) \int_a^b g\, dh\, f$ exists then (A) for each $\epsilon > 0$, there exists a partition λ of $[a,b]$ such that*

$$\left\| \sum_{i=1}^n \{ g(\xi_i)[h(x_i-) - h(x_{i-1}+)]f(\xi_i) - g(\eta_i)[h(x_i-) - h(x_{i-1}+)]f(\eta_i) \} \right\| < \epsilon \quad (3.66)$$

for any refinement $\kappa = \{x_i\colon i = 0, \ldots, n\}$ of λ and for all $x_{i-1} < \eta_i \le \xi_i < x_i$, $i = 1, \ldots, n$.

(B) If $f, g, h \in \mathcal{R}([a,b]; \mathbb{B})$, (A) holds and $(Y_1) \int_a^b g\, dh\, f$ exists, then so does $(RYS) \int_a^b g\, dh\, f$, and the two integrals are equal.

Proof. If $(RYS) \int_a^b g\, dh\, f$ exists then subtracting two sums defining the (RYS) integral and based on the same partition, we get that (A) holds.

Under the assumptions of (B), let $\kappa = \{x_i\colon i = 0, \ldots, n\}$ be a partition of $[a,b]$ and let $x_{i-1} < \eta_i < \xi_i < x_i$ for $i = 1, \ldots, n$. Then any sum for defining the (RYS) integral

$$[g(\Delta_- h)f](b) + \sum_{i=1}^n \{ g(\xi_i)[h(x_i-) - h(x_{i-1}+)]f(\xi_i) + [g(h_+ - h_-^{(a)})f](x_{i-1}) \}$$

$$= \sum_{i=1}^n g_+(x_{i-1})[h_-(x_i) - h_-^{(a)}(x_{i-1})]f_+(x_{i-1}) \quad (3.67)$$

$$+ \sum_{i=1}^n [g(h_+ - h_-^{(a)})f - g_+(h_+ - h_-^{(a)})f_+](x_{i-1}) + [g(\Delta_- h)f](b)$$

$$+ \sum_{i=1}^n \{ g(\xi_i)[h(x_i-) - h(x_{i-1}+)]f(\xi_i) - g(x_{i-1}+)[h(x_i-) - h(x_{i-1}+)]f(x_{i-1}+) \}.$$

For each $i = 1, \ldots, n$, letting $\eta_i \downarrow x_{i-1}$ in (3.66) we get that the last sum in (3.67) can be made arbitrarily small for sufficiently refined partitions. The other terms on the right side of (3.67) approach the (Y_1) integral as partitions are refined also. Therefore the (RYS) integral exists and equals the (Y_1) integral. The proof of Proposition 3.19 is complete. $\qquad\square$

The next statement provides a simple sufficient condition for the coincidence of the (RYS) and (Y_1) integrals. For notational convenience we let $\mathcal{W}_\infty([a,b]; \mathbb{B}) = \mathcal{W}_\infty^*([a,b]; \mathbb{B}) := \mathcal{R}([a,b]; \mathbb{B})$.

Corollary 3.20. *Let $1/p + 1/p' = 1$ with $1 \le p < \infty$ and let either $g, f \in \mathcal{W}_{p'}([a,b]; \mathbb{B})$ and $h \in \mathcal{W}_p^*([a,b]; \mathbb{B})$ or $g, f \in \mathcal{W}_{p'}^*([a,b]; \mathbb{B})$ and $h \in \mathcal{W}_p([a,b]; \mathbb{B})$. If* *one of the integrals* $(Y_1) \int_a^b g \, dh \, f$ *and* $(RYS) \int_a^b g \, dh \, f$ *exists then so does the other, and the two are equal.*

Remark. The integrals in Corollary 3.20 need not exist under the stated assumptions, or even for $g, f \in W_{p'}^*$ and $h \in W_p^*$. This follows from Theorem 4.21 of Leśniewicz and Orlicz (1973).

Proof. We prove the statement in the case when $h \in W_p^*([a,b]; \mathbb{B})$ for $1 < p < \infty$ and when $h \in W_1([a,b]; \mathbb{B})$ only because the proof in the other cases is similar. By Propositions 3.17 and 3.19, it is enough to show that (3.66) holds. Let $h \in W_p^*([a,b]; \mathbb{B})$ for $1 < p < \infty$ and let $\epsilon > 0$. Then there exists a partition $\lambda = \{z_j : j = 0, \ldots, m\}$ satisfying statement (2) of Proposition 2.23. Let $\kappa = \{x_i : i = 0, \ldots, n\}$ be any refinement of λ and $x_{i-1} < \eta_i < \xi_i < x_i$ for $i = 1, \ldots, n$. Then applying Hölder's inequality, we get

$$\Big\| \sum_{i=1}^n \{ g(\xi_i)[h(x_i-) - h(x_{i-1}+)] f(\xi_i) - g(\eta_i)[h(x_j-) - h(x_{i-1}+)] f(\eta_i) \} \Big\|$$

$$\leq \|f\|_\infty \Big(\sum_{i=1}^n \|g(\xi_j) - g(\eta_i)\|^{p'} \Big)^{1/p'} \Big(\sum_{i=1}^n \|h(x_i-) - h(x_{i-1}+)\|^p \Big)^{1/p}$$

$$+ \|g\|_\infty \Big(\sum_{i=1}^n \|f(\xi_j) - f(\eta_i)\|^{p'} \Big)^{1/p'} \Big(\sum_{i=1}^n \|h(x_i-) - h(x_{i-1}+)\|^p \Big)^{1/p}$$

$$\leq [\|f\|_\infty V_{p'}(g) + \|g\|_\infty V_{p'}(f)] \Big(\sum_{j=1}^m v_p(h; [z_{j-1}+, z_j-]) \Big)^{1/p} \leq \epsilon^{1/p} V_{p',\infty}(g) V_{p',\infty}(f).$$

Since $\epsilon > 0$ is arbitrary (3.66) holds. If $p = 1$ then, since f and g are regulated, we choose a partition $\lambda = \{z_j : j = 0, \ldots, m\}$ of $[a, b]$ such that the oscilation of f and g over each open interval (z_{j-1}, z_j), $j = 1, \ldots, m$, is less than any given $\epsilon > 0$. Therefore the bound $\epsilon v_1(h)(\|g\|_\infty + \|f\|_\infty)$ follows instead. The proof of Corollary 3.20 is over. $\qquad \square$

3.4. Properties.

Here standard properties of integrals are extended to the integrals defined so far. We start with linearity.

Proposition 3.21. *Let* $g, h, f, g_j, h_j, f_j \in \mathcal{R}([a, b]; \mathbb{B})$ *for* $j = 1, 2$ *and let* $u_j, v_j \in \mathbb{B}$ *for* $j = 1, 2$. *Let* $\int$ *denote any of the integrals* $(MPS) \int$, $(CY) \int$, $(LY) \int$ *or* $(RY) \int$.

(1) *If* $\int_a^b g_1 \, dh \, f$ *and* $\int_a^b g_2 \, dh \, f$ *both exist, so does* $\int_a^b (u_1 g_1 + u_2 g_2) \, dh \, f$ *and*

$$\int_a^b (u_1 g_1 + u_2 g_2) \, dh \, f = u_1 \int_a^b g_1 \, dh \, f + u_2 \int_a^b g_2 \, dh \, f.$$

(2) *If* $\int_a^b g u_1 dh_1 v_1 f$ *and* $\int_a^b g u_2 dh_2 v_2 f$ *both exist, so does* $\int_a^b g d(u_1 h_1 v_1 + u_2 h_2 v_2) f$ *and*

$$\int_a^b g \, d(u_1 h_1 v_1 + u_2 h_2 v_2) \, f = \int_a^b g u_1 \, dh_1 \, v_1 f + \int_a^b g u_2 \, dh_2 \, v_2 f.$$

(3) *If $\int_a^b g\,dh\,f_1$ and $\int_a^b g\,dh\,f_2$ both exist, so does $\int_a^b g\,dh\,(f_1 u_1 + f_2 u_2)$ and*

$$\int_a^b g\,dh\,(f_1 u_1 + f_2 u_2) = \left(\int_a^b g\,dh\,f_1 \right) u_1 + \left(\int_a^b g\,dh\,f_2 \right) u_2.$$

Proof. The proof for (MPS) integrals repeats the arguments of Pollard (1923, pp. 77-79, 91) used by him to prove (simpler forms of) these properties when functions are scalar-valued. The cases when $\int$ denotes other integrals follow just by applying what is true for (MPS) integrals and by using linearity properties of unconditionally convergent series.   □

The next statement shows that (MPS) and (CY) integrals are additive as interval functions. So are (LY) and (RY) integrals, by Proposition 3.25 below.

Proposition 3.22. *Let $g, h, f \in \mathcal{R}([a,b]; \mathbb{B})$ and let $a \le c \le b$. Suppose $\int$ denotes either $(MPS)\int$ or $(CY)\int$. Then $\int_a^b g\,dh\,f$ exists if and only if both $\int_a^c g\,dh\,f$ and $\int_c^b g\,dh\,f$ exist, and then*

$$\int_a^b g\,dh\,f = \int_a^c g\,dh\,f + \int_c^b g\,dh\,f. \tag{3.68}$$

Proof. The proof for (MPS) integrals is the same as Pollard's (1923, pp. 91-94) proof of this property for scalar-valued functions.

Let us consider the case when $\int$ is the central Young integral $(CY)\int$. We prove the "only if" part because a proof of the "if" part is similar. If $(Y_1)\int_a^b g\,dh\,f$ exists, then so does $(MPS)\int_a^b g_+^{(b)}\,dh_-^{(a)}\,f_+^{(b)}$ and

$$\sum_{(a,b)} \left[g(\Delta_-^+ h)f - g_+(\Delta_-^+ h)f_+ \right]$$

converges unconditionally. It follows that the sums of the same terms over (a,c) and (c,b) both converge unconditionally, and by what is true for (MPS) integrals,

$$(MPS)\int_a^c g_+\,dh_-^{(a)}\,f_+ \quad \text{and} \quad (MPS)\int_c^b g_+^{(b)}\,dh_-\,f_+^{(b)}$$

both exist, and

$$(MPS)\int_a^b g_+^{(b)}\,dh_-^{(a)}\,f_+^{(b)} = (MPS)\int_a^c g_+\,dh_-^{(a)}\,f_+ + (MPS)\int_c^b g_+^{(b)}\,dh_-\,f_+^{(b)}. \tag{3.69}$$

Since $h_-^{(a)}$ is left-continuous at c the integral $(MPS)\int_a^c \chi_{\{c\}}\,dh_-^{(a)}\,f_+$ exists and equals zero. Thus, by (1) of Proposition 3.21, the integral $(MPS)\int_a^c g_+^{(c)}\,dh_-^{(a)}\,f_+$ exists and

$$(MPS)\int_a^c g_+^{(c)}\,dh_-^{(a)}\,f_+ = (MPS)\int_a^c g_+\,dh_-^{(a)}\,f_+$$

$$+[g(c) - g(c+)](MPS) \int_a^c \chi_{\{c\}} dh_-^{(a)} f_+ = (MPS) \int_a^c g_+ dh_-^{(a)} f_+. \qquad (3.70)$$

Left-continuity of $h_-^{(a)}$ at c also implies that $(MPS) \int_a^c g_+^{(c)} dh_-^{(a)} \chi_{\{c\}}$ exists and equals zero. Therefore, by (3.70) and by (3) of Proposition 3.21, one can conclude that the integral $(MPS) \int_a^c g_+^{(c)} dh_-^{(a)} f_+^{(c)}$ exists and that

$$(MPS) \int_a^c g_+^{(c)} dh_-^{(a)} f_+^{(c)} = (MPS) \int_a^c g_+^{(c)} dh_-^{(a)} f_+$$

$$+(MPS) \int_a^c g_+^{(c)} dh_-^{(a)} \chi_{\{c\}}[f(c) - f(c+)] = (MPS) \int_a^c g_+ dh_-^{(a)} f_+. \qquad (3.71)$$

Next, since $g_+^{(b)}$ and $f_+^{(b)}$ are right-continuous at c the integral

$$(MPS) \int_c^b g_+^{(b)} (h_-^{(c)} - h_-)(c) d\chi_{\{c\}} f_+^{(b)}$$

exists and equals $[g_+(h_-^{(c)} - h_-)f_+](c)$. Therefore, by (2) of Proposition 3.21, the integral $(MPS) \int_c^b g_+^{(b)} dh_-^{(c)} f_+^{(b)}$ exists and

$$(MPS) \int_c^b g_+^{(b)} dh_-^{(c)} f_+^{(b)} = (MPS) \int_c^b g_+^{(b)} dh_- f_+^{(b)} \qquad (3.72)$$

$$+(MPS) \int_c^b g_+^{(b)} (h_-^{(c)} - h_-)(c) d\chi_{\{c\}} f_+^{(b)} = (MPS) \int_c^b g_+^{(b)} dh_- f_+^{(b)} - [g_+ \Delta_- h f_+](c).$$

It follows that the desired (MPS) integrals exist and we can plug (3.71) and (3.72) into (3.69), so that

$$\begin{aligned}
(CY) \int_a^b g\, dh\, f &= (MPS) \int_a^c g_+^{(c)} dh_-^{(a)} f_+^{(c)} + [g(\Delta^+ h)f - g_+(\Delta^+ h)f_+](a) \\
&\quad + \sum_{(a,c]} [g(\Delta_-^+ h)f - g_+(\Delta_-^+ h)f_+] \\
&\quad + [g_+(\Delta_- h)f_+](c) + (MPS) \int_c^b g_+^{(b)} dh_-^{(c)} f_+^{(b)} \\
&\quad + \sum_{(c,b)} [g(\Delta_-^+ h)f - g_+(\Delta_-^+ h)f_+] + [g(\Delta_- h)f](b) \\
&= (CY) \int_a^c g\, dh\, f + (CY) \int_c^b g\, dh\, f
\end{aligned}$$

and we arrive at (3.68) for the central Young integral. The proof of Proposition 3.22 is complete.   □

Next we show that indefinite (MPS) and (CY) integrals are regulated functions.

Lemma 3.23. *Let* $g, h, f \in \mathcal{R}([a, b]; \mathbb{B})$ *and let* $\int_a^b g \, dh \, f$ *exist, where* $\int$ *denotes either* $(MPS)\int$ *or* $(CY)\int$. *Then the indefinite integrals*

$$\Psi_a := \{\Psi_a(y) := \int_a^y g \, dh \, f : y \in [a, b]\} \in \mathcal{R}([a, b]; \mathbb{B}) \qquad (3.73)$$

and

$$\Psi^b := \{\Psi^b(y) := \int_y^b g \, dh \, f : y \in [a, b]\} \in \mathcal{R}([a, b]; \mathbb{B}). \qquad (3.74)$$

Moreover,

$$(\Delta_- \Psi_a)(y) = -(\Delta_- \Psi^b)(y) = [g(\Delta_- h)f](y) =: \int_{y-}^y g \, dh \, f \qquad (3.75)$$

for any $y \in (a, b]$ *and*

$$(\Delta^+ \Psi_a)(y) = -(\Delta^+ \Psi^b)(y) = [g(\Delta^+ h)f](y) =: \int_y^{y+} g \, dh \, f \qquad (3.76)$$

for any $y \in [a, b)$.

Proof. By Proposition 3.22, Ψ_a and Ψ^b are regulated whenever either one of them is regulated. Moreover, the first equalities in (3.75) and (3.76) each hold. Therefore it suffices to prove the proposition for the indefinite integral Ψ_a only. We start by proving (3.75) for the (MPS) integral. Let $y \in (a, b]$. Given $\epsilon > 0$, take a partition $\kappa_0 = \{x_i : i = 0, \ldots, n\}$ of $[a, y]$ such that for every refinement κ of κ_0, and every Riemann-Stieltjes sum S^κ for g, h, f,

$$\left\| (MPS) \int_a^y g \, dh \, f - S^\kappa \right\| < \epsilon/3. \qquad (3.77)$$

We can assume x_{n-1} is chosen so that $\|g(y)[h(x_{n-1}) - h(y-)]f(y)\| < \epsilon/3$. Let T be any Riemann-Stieltjes sum for g, h, f, based on any partition of $[x_{n-1}, y]$. Then

$$\left\| T - g(y)[h(y) - h(x_{n-1})]f(y) \right\| \leq 2\epsilon/3,$$

since otherwise, adjoining T and $g(y)(h(y) - h(x_{n-1}))f(y)$ separately to any Riemann-Stieltjes sum for g, h, f based on $\{x_i : i = 0, 1, \ldots, n-1\}$, we get two Riemann-Stieltjes sums for g, h, f based on refinements of κ_0 which differ by more than $2\epsilon/3$, which is impossible by (3.77). Thus

$$\left\| (MPS) \int_{x_{n-1}}^y g \, dh \, f - [g(\Delta_- h)f](y) \right\|$$

$$\leq \left\| (MPS) \int_{x_{n-1}}^y g \, dh \, f - g(y)[h(y) - h(x_{n-1})]f(y) \right\|$$

$$+ \|g(y)[h(x_{n-1}) - h(y-)]f(y)\| < 2\epsilon/3 + \epsilon/3 = \epsilon.$$

Here we can let $x_{n-1} \uparrow y$ and $\epsilon \downarrow 0$, so (3.75) is proved for the (MPS) integral. The proof of (3.76) in this case uses symmetric arguments and, hence, is omitted.

To prove (3.75) for the central Young integral, by definition of (Y_1) integral, one may write

$$(CY) \int_z^y g \, dh \, f - [g(\Delta_- h)f](y) = (MPS) \int_z^y g_+^{(y)} dh_-^{(z)} f_+^{(y)}$$

$$+ [g(\Delta^+ h)f - g_+(\Delta^+ h)f_+](z) + \sum_{(z,y)} [g(\Delta_-^\pm h)f - g_+(\Delta_-^\pm h)f_+] \quad (3.78)$$

for $a \leq z < y$. Since $h_-^{(z)}$ is left-continuous at y, by the first part of the proof the (MPS) integral approaches zero as $z \uparrow y$. Since the other terms on the right side of (3.78) also tend to zero (3.75) holds for the central Young integral.

To prove (3.76), by (2.1), we note that $\Delta^+ h_-^{(y)}(y) = \Delta^+ h(y)$. Therefore, by the definition we have

$$(CY) \int_y^z g \, dh \, f - [g(\Delta^+ h)f](y) = (MPS) \int_y^z g_+^{(z)} dh_-^{(y)} f_+^{(z)}$$

$$- [g_+(\Delta^+ h_-^{(y)})f_+](y) + \sum_{(y,z)} [g(\Delta_-^\pm h)f - g_+(\Delta_-^\pm h)f_+] + [g(\Delta_- h)f](z), \quad (3.79)$$

for $y < z \leq b$. Then (3.76) for the (MPS) integral implies that the right side of (3.79) approaches zero as $z \downarrow y$. Thus (3.76) holds for the central Young integral. The proof of Lemma 3.23 is complete. $\qquad \square$

By Lemma 3.23, the left-continuous modification of the indefinite integral Ψ_a is given by

$$\Psi_a(y-) = \int_a^y g \, dh \, f - \int_{y-}^y g \, dh \, f, \quad (3.80)$$

for any $y \in (a, b]$, and the right-continuous modification of the indefinite integral Ψ^b is given by

$$\Psi^b(y+) = \int_y^b g \, dh \, f - \int_y^{y+} g \, dh \, f, \quad (3.81)$$

for any $y \in [a, b)$. By Lemma 3.2, the right sides of (3.80) and (3.81) for the Moore-Pollard-Stieltjes integrals are equal to

$$(MPS) \int_a^{y-} g \, dh \, f \quad \text{and} \quad (MPS) \int_{y+}^b g \, dh \, f,$$

respectively. For the central Young integral we *define*, for each $a < y \leq b$,

$$(CY) \int_a^{y-} g \, dh \, f := (MPS) \int_a^{y-} g_+ \, dh_-^{(a)} f_+ + [g(\Delta^+ h)f - g_+(\Delta^+ h)f_+](a)$$

$$+ \sum_{(a,y)} [g(\Delta_-^\pm h)f - g_+(\Delta_-^\pm h)f_+] \quad (3.82)$$

if the (MPS) integral exists and the sum converges unconditionally in $\mathbb{B}$ and, for all $a \leq y < b$,

$$(CY) \int_{y+}^{b} g \, dh \, f := (MPS) \int_{y+}^{b} g_- \, dh_+^{(b)} \, f_- + [g(\Delta_-h)f - g_-(\Delta_-h)f_-](b)$$
$$+ \sum_{(y,b)} [g(\Delta_-^+h)f - g_-(\Delta_-^+h)f_-] \qquad (3.83)$$

if the (MPS) integral exists and the sum converges unconditionally in $\mathbb{B}$. With such definitions, Lemma 3.1 shows that the integrals in (3.82) and (3.83) do not depend on the values of f, g and h at y.

The next statement partially extends Lemmas 3.1 and 3.2 and justifies definitions (3.82) and (3.83) for the central Young integral.

Lemma 3.24. *Let* $g, h, f \in \mathcal{R}([a,b]; \mathbb{B})$.

(1) *If* $a < y \leq b$ *then* (3.82) *exists if and only if* $(CY) \int_a^y g \, dh \, f$ *does and*

$$(CY) \int_a^{y-} g \, dh \, f = (CY) \int_a^y g \, dh \, f - [g(\Delta_-h)f](y)$$
$$= C\Psi_a(y-) = (CY) \int_a^y g \, dh^{(y-)} f, \qquad (3.84)$$

where $C\Psi_a$ *denotes the indefinite integral* (3.73) *for the* (CY) *integral.*

(2) *If* $a \leq y < b$ *then* (3.83) *exists if and only if* $(CY) \int_y^b g \, dh \, f$ *does and*

$$(CY) \int_{y+}^{b} g \, dh \, f = (CY) \int_y^b g \, dh \, f - [g(\Delta^+h)f](y)$$
$$= C\Psi^b(y+) = (CY) \int_y^b g \, dh^{(y+)} f, \qquad (3.85)$$

where $C\Psi^b$ *denotes the indefinite integral* (3.74) *for the* (CY) *integral.*

Proof. We prove part (1) only since a proof of (2) is similar. Let $a < y \leq b$. By Lemma 3.1, the (MPS) integral in (3.82) does not depend on the values of functions at y, so it coincides with $(MPS) \int_a^{y-} g_+^{(y)} \, dh_-^{(a)} \, f_+^{(y)}$. Since $h^{(a)}$ is left continuous at y, by part (2) of Lemma 3.2, the latter integral exists if and only if $(MPS) \int_a^y g_+^{(y)} \, dh_-^{(a)} \, f_+^{(y)}$ does and the two are equal. Hence, by (3.82) and by the definition of (Y_1) integral, we have

$$(CY) \int_a^{y-} g \, dh \, f = (MPS) \int_a^y g_+^{(y)} \, dh_-^{(a)} \, f_+^{(y)} + [g(\Delta^+h)f - g_+(\Delta^+h)f_+]$$

$$+ \sum_{(a,y)} [g(\Delta_-^+h)f - g_+(\Delta_-^+h)f_+] = (CY) \int_a^y g \, dh \, f - [g(\Delta_-h)f](y)$$

This proves the first equality in (3.84). The second one follows from (3.75). To prove the last one we note that, by (2.16) and (2.2), for $y \in (a, b]$, $\left(h^{(y-)}\right)^{(a)}_- = h^{(a)}_-$ on $[a, y]$ and $\left(\Delta_- h^{(y-)}\right)(y) = h^{(y-)}(y) - h^{(y-)}(y-) = 0$. Hence, we get

$$(CY)\int_a^y g\, dh^{(y-)} f = (MPS)\int_a^y g^{(b)}_+ d(h^{(y-)})^{(a)}_- f^{(b)}_+ + [g(\Delta^+ h)f - g_+(\Delta^+ h)f_+](a)$$

$$+ \sum_{(a,y)} [g(\Delta^{\pm}_- h)f - g_+(\Delta^{\pm}_- h)f_+]$$

$$= (CY)\int_a^y g\, dh\, f - [g(\Delta_- h)f](y),$$

so (3.84) holds. The proof of Lemma 3.24 is complete. $\qquad\square$

Next is an extension of the additivity property (3.68) to the left Young and right Young integrals.

Proposition 3.25. *Let* $g, h, f \in \mathcal{R}([a, b]; \mathbb{B})$ *and let* $a \le c \le b$. *Suppose* $\int$ *denotes either* $(LY)\int$ *or* $(RY)\int$. *Then the conclusion of Proposition 3.22 holds.*

Proof. We give a proof for the (LY) integral only because a proof for the (RY) integral is similar. We can assume that $a < c < b$. Let $(LY)\int_a^b g\, dh\, f$ exist. Then so does $(CY)\int_a^b g^{(a+)}\, dh\, f^{(a)}_-$ and the sum

$$S(a, b) := \sum_{(a,b)} [g_+(\Delta^+ h)f - g(\Delta^+ h)f_-]$$

converges unconditionally in $\mathbb{B}$. It follows that the sums of the same terms over (a, c) and (c, b) both converge unconditionally and

$$S(a, b) = S(a, c) + [g_+(\Delta^+ h)f - g(\Delta^+ h)f_-](c) + S(c, b). \tag{3.86}$$

By Proposition 3.22 for (CY) integrals, the integrals

$$(CY)\int_a^c g^{(a+)}\, dh\, f^{(a)}_- \quad \text{and} \quad (CY)\int_c^b g\, dh\, f_-$$

exist. Hence, by part (2) of Lemma 3.24, so does $(CY)\int_{c+}^b g\, dh\, f_-$. Using (3.68) for (CY) integrals and (3.85) we have

$$(CY)\int_a^b g^{(a+)}\, dh\, f^{(a)}_- = (CY)\int_a^c g^{(a+)}\, dh\, f^{(a)}_- \tag{3.87}$$

$$+ [g(\Delta^+ h)f_-](c) + (CY)\int_{c+}^b g\, dh\, f_-.$$

By (3.83) and Lemma 3.1, the last integral over $[c+, b]$ does not depend on the values of g, h or f at the point c. Since $g = g^{(c+)}$ and $f_- = f^{(c)}_-$ over $[c+, b]$, adding (3.86) and (3.87) and using (3.85) again we get

$$(LY)\int_a^b g\, dh\, f = (CY)\int_a^c g^{(a+)}\, dh\, f^{(a)}_- + \sum_{(a,c)} [g_+(\Delta^+ h)f - g(\Delta^+)f_-]$$

$$+(CY) \int_{c+}^{b} g^{(c+)} \, dh \, f_{-}^{(c)} + [g_{+}(\Delta^{+}h)f](c) + \sum_{(c,b)} [g_{+}(\Delta^{+}h)f - g(\Delta^{+})f_{-}]$$

$$= (LY) \int_{a}^{c} g \, dh \, f + (LY) \int_{c}^{b} g \, dh \, f.$$

This proves the "only if" part of Proposition 3.25.

To prove the "if" part assume that $(LY) \int_{a}^{c} g \, dh \, f$ and $(LY) \int_{c}^{b} g \, dh \, f$ exist. Therefore

$$(CY) \int_{a}^{c} g^{(a+)} \, dh \, f_{-}^{(a)} \quad \text{and} \quad (CY) \int_{c}^{b} g^{(c+)} \, dh \, f_{-}^{(c)}$$

exist and the sum on the left side of (3.86) converges unconditionally in $\mathbb{B}$. By Lemmas 3.24 and 3.2, we have that the following three integrals exist and

$$(CY) \int_{c}^{b} g^{(c+)} \, dh \, f_{-}^{(c)} = (CY) \int_{c+}^{b} g \, dh \, f_{-} + [g_{+}(\Delta^{+}h)f](c) \qquad (3.88)$$

$$= (CY) \int_{c}^{b} g \, dh \, f_{-} + [g_{+}(\Delta^{+}h)f - g(\Delta^{+}h)f_{-}](c).$$

Hence, by Proposition 3.22 for (CY) integrals, $(CY) \int_{a}^{b} g^{(a+)} \, dh \, f_{-}^{(a)}$ exists and

$$(CY) \int_{a}^{b} g^{(a+)} \, dh \, f_{-}^{(a)} = (CY) \int_{a}^{c} g^{(a+)} \, dh \, f_{-}^{(a)} + (CY) \int_{c}^{b} g \, dh \, f_{-}$$

$$\text{by (3.88)} \quad = (CY) \int_{a}^{c} g^{(a+)} \, dh \, f_{-}^{(a)} + (CY) \int_{c}^{b} g^{(c+)} \, dh \, f_{-}^{(c)}$$

$$- [g_{+}(\Delta^{+}h)f - g(\Delta^{+}h)f_{-}](c).$$

Therefore $(LY) \int_{a}^{b} g \, dh \, f$ exists and invoking (3.86) we have

$$(LY) \int_{a}^{c} g \, dh \, f + (LY) \int_{c}^{b} g \, dh \, f = (CY) \int_{a}^{c} g^{(a+)} \, dh \, f_{-}^{(a)}$$

$$+(CY) \int_{c}^{b} g^{(c+)} \, dh \, f_{-}^{(c)} + \left(\sum_{(a,c)} + \sum_{(c,b)} \right) [g_{+}(\Delta^{+}h)f - g\Delta^{+}h)f_{-}]$$

$$= (CY) \int_{a}^{b} g^{(a+)} \, dh \, f_{-}^{(a)} + \sum_{(a,b)} [g_{+}(\Delta^{+}h)f - g\Delta^{+}h)f_{-}] = (LY) \int_{a}^{b} g \, dh \, f.$$

The proof of Proposition 3.25 is complete.   □

We finish this subsection by extending Lemma 3.23 to the left Young and right Young integrals. Denote by $L\Psi_{a}$ the indefinite (LY) integral (3.73) and by $R\Psi^{b}$ the indefinite (RY) integral (3.74).

Lemma 3.26. *Let $g, h, f \in \mathcal{R}([a,b]; \mathbb{B})$ and let $\int_a^b g \, dh \, f$ exist where $\int$ denotes either $(LY) \int$ or $(RY) \int$. Then $L\Psi_a$ and $R\Psi^b$ are regulated functions. Moreover,*

$$(\Delta^+ L\Psi_a)(y) = [g_+(\Delta^+ h)f](y) = -(\Delta^+ R\Psi^b)(y) \tag{3.89}$$

for any $y \in [a, b)$ and

$$(\Delta_- L\Psi_a)(y) = [g(\Delta_- h)f_-](y) = -(\Delta_- R\Psi^b)(y) \tag{3.90}$$

for any $y \in (a, b]$.

Proof. We prove the lemma for $L\Psi_a$ only and note that a proof for $R\Psi^b$ is similar. For each $y \in [a, b]$, let

$$S_a(y) := \sum_{(a,y)} [g_+(\Delta^+ h)f - g(\Delta^+ h)f_-].$$

First we show that, for each $y \in (a, b)$,

$$(\Delta^+ S_a)(y) = [g_+(\Delta^+ h)f - g(\Delta^+ h)f_-](y). \tag{3.91}$$

Let $\epsilon > 0$. Since the sum in the definition of (LY) integral converges unconditionally in $\mathbb{B}$ there exists a finite set $D(\epsilon) \subset (a, b)$ such that

$$\left\| \sum_{\mu} [g_+(\Delta^+ h)f - g(\Delta^+ h)f_-] \right\| < \epsilon$$

for any set $\mu \subset (a, b)$ disjoint from $D(\epsilon)$. Then for each $y < z < b$ such that $(y, z) \subset (a, b) \setminus D(\epsilon)$ we have

$$\left\| S_a(z) - S_a(y) - [g_+(\Delta^+ h)f - g(\Delta^+ h)f_-](y) \right\|$$
$$= \left\| \sum_{(y,z)} [g_+(\Delta^+ h)f - g(\Delta^+ h)f_-] \right\| < \epsilon.$$

Since $\epsilon > 0$ is arbitrary this implies (3.91). The same argument implies also that

$$(\Delta^+ S_a)(a) = 0 \quad \text{and} \quad (\Delta_- S_a)(y) = 0, \quad \forall y \in (a, b]. \tag{3.92}$$

To prove the first equality in (3.89) let $a < y < b$. Then by Lemma 3.23 for (CY) integrals and by (3.91), it follows that

$$(\Delta^+ L\Psi_a)(y) = [g(\Delta^+ h)f_-](y) + [g_+(\Delta^+ h)f - g(\Delta^+ h)f_-](y) = [g_+(\Delta^+ h)f](y).$$

If $y = a$ then by Lemma 3.23 again and by the first equality in (3.92), it follows that

$$(\Delta^+ L\Psi_a)(a) = [g^{(a+)}(\Delta^+ h)f_-^{(a)}](a) + 0 = [g_+(\Delta^+ h)f](a).$$

This proves the first equality in (3.89). To prove the first equality in (3.90), by Lemma 3.23 for (CY) integrals and by the second equality in (3.92), we get for each $a < y \le b$

$$(\Delta_- L\Psi_a)(y) = [g(\Delta_- h)f_-](y).$$

The proof of Lemma 3.26 is complete. $\qquad\qquad\qquad\qquad\qquad\qquad\qquad\qquad\quad\square$

3.5. Existence. This subsection contains sufficient conditions for existence of Moore-Pollard-Stieltjes and Young integrals. The following statement is an extension of a theorem on Stieltjes integrability due to E. R. Love and L. C. Young, see Young (1936, pp. 251, 264), who gives credit to Love, and Love (1993, Theorem 21).

Theorem 3.27. *Let* $h \in \mathcal{W}_p([a, b]; \mathbb{B})$ *and* $f, g \in \mathcal{W}_q([a, b]; \mathbb{B})$, *with* $p, q > 0$, $1/p + 1/q > 1$. *Then the integral* (3.1) *exists*

    (1) *in the Riemann-Stieltjes sense if the pairs* h, g *and* h, f *have no common discontinuities,*

    (2) *in the* (MPS) *sense if the pairs* h, g *and* h, f *have no common discontinuities on the right and no common discontinuities on the left,*

    (3) *in the* (CY) *sense always.*

In whichever of the three senses the integral (3.1) *is taken, the inequality*

$$\Big\| \int_a^b g \, dh \, f - g(x_0)[h(b) - h(a)]f(x_0) \Big\|$$

$$\leq c_q(1 + \zeta(1/p + 1/q))V_p(h)\big(V_q(f)\|g\|_\infty + \|f\|_\infty V_q(g)\big) \tag{3.93}$$

holds for any $x_0 \in [a, b]$, *where* ζ *is the Riemann zeta function* $\zeta(s) := \sum_{n \geq 1} n^{-s}$ *and* $c_q = 1 \vee 2^{(1/q)-1}$. *Also*

$$\Big\| \int_a^b g \, dh \, f \Big\| \leq c_q(1 + \zeta(1/p + 1/q))V_p(h)V_{q,\infty}(g)V_{q,\infty}(f). \tag{3.94}$$

We call (3.93) the Love-Young inequality. Young (1936, p. 258) gave an example showing that the condition $1/p + 1/q > 1$ in the preceding theorem can't be replaced by the weaker condition $1/p + 1/q = 1$ in general. Leśniewicz and Orlicz (1973, Theorem 4.21) used this example to prove that the stated condition and the later extension of Young (1938, Theorem 5.1) are best possible.

Before giving the proof of Theorem 3.27, we have two corollaries.

Corollary 3.28. *Let* $h \in \mathcal{W}_p([a, b]; \mathbb{B})$ *and* $f, g \in \mathcal{W}_q([a, b]; \mathbb{B})$, *with* $p, q > 0$, $1/p + 1/q > 1$. *Then the integrals* $(LY) \int_a^b g \, dh \, f$ *and* $(RY) \int_a^b g \, dh \, f$ *exist and are equal, with the sums in Definition 3.11 converging absolutely, and for both we have*

$$\Big\| \int_a^b g \, dh \, f \Big\| \leq c_q(1 + \zeta(1/p + 1/q))V_p(h)V_{q,\infty}(g)V_{q,\infty}(f)$$

$$+ \mathfrak{S}_p(h; (a, b))\big[\|f\|_\infty \mathfrak{S}_q(g; (a, b)) + \|g\|_\infty \mathfrak{S}_q(f; (a, b))\big].$$

Proof. Apply (3.94) and Hölder's inequality in the form of Young (1936, p. 252):

$$\sum_{i=1}^n |u_i v_i| \leq \Big(\sum_{i=1}^n |u_i|^p\Big)^{1/p} \Big(\sum_{i=1}^n |v_i|^q\Big)^{1/q} \tag{3.95}$$

valid for any $p, q > 0$ such that $1/p + 1/q \geq 1$. The inequality extends directly to norms in $\mathbb{B}$ in place of absolute values. We then have

$$\sum_{(a,b)} \big\| g_+(\Delta^+ h)f - g(\Delta^+ h)f_- \big\|$$

$$\leq \|f\|_\infty \sum_{(a,b)} \left[\|\Delta^+ g\|\,\|\Delta^+ h\|\right] + \|g\|_\infty \sum_{(a,b)} \left[\|\Delta^+ h\|\,\|\Delta_- f\|\right]$$

$$\leq \mathfrak{S}_p(h;(a,b))\left[\|f\|_\infty \mathfrak{S}_q(g;(a,b)) + \|g\|_\infty \mathfrak{S}_q(f;(a,b))\right].$$

We note that by part (1) of Lemma 2.5, $V_q(g^{(a+)}) = V_q(g;[a+,b]) \leq V_q(g)$ and, by the first inequality in (2.15), $V_q(f_-^{(a)}) \leq V_q(f)$. Therefore, by part (3) of Theorem 3.27, the (LY) integral exists. Moreover, by (3.93), it follows that

$$\|(CY)\int_a^b g^{(a+)}\,dh\,f_-^{(a)}\| \leq c_q(1 + \zeta(1/p + 1/q))V_p(h)V_{q,\infty}(g)V_{q,\infty}(f).$$

Similar arguments apply to the (RY) integral which then equals the (LY) integral by Theorem 3.12. □

The proof of the second corollary is a straightforward adaptation of the proof of (3.94) and therefore is omitted.

Corollary 3.29. *Let $h \in \mathcal{W}_p([a,b];\mathbb{B})$ and $f \in \mathcal{W}_q([a,b];\mathbb{B})$, with $p,q > 0$, $1/p + 1/q > 1$. Then the integral*

$$\int_a^b dh\,f = \int_a^b \mathbb{1}\,dh\,f \quad (or \quad \int_a^b f\,dh = \int_a^b f\,dh\mathbb{1}) \tag{3.96}$$

in the Riemann-Stieltjes, (MPS) or (CY) sense, whichever exists, satisfies

$$\|\int_a^b dh\,f - [h(b) - h(a)]f(x_0)\| \leq (1 + \zeta(1/p + 1/q))V_p(h)V_q(f) \tag{3.97}$$

for any $x_0 \in [a,b]$.

For the proof of Theorem 3.27 we need two auxiliary statements. First we will extend a Hölder type inequality of E. R. Love and L. C. Young (see Young, 1936). Given a partition $\kappa = \{x_i: i = 1,\dots,n\}$ of $[c,d]$ and functions $g,h,f : [c,d] \mapsto \mathbb{B}$, define

$$S_{p,q}^\kappa(g,h,f) := \sup_{\kappa' \subset \kappa}\left\{\left(\sum_j \|\Delta_j^{\kappa'} h\|^p\right)^{1/p}\left[\left(\sum_j \|\Delta_j^{\kappa'} f\|^q\right)^{1/q}\max_i \|g(x_i)\|\right.\right.$$
$$\left.\left. + \left(\sum_j \|\Delta_j^{\kappa'} g\|^q\right)^{1/q}\max_i \|f(x_i)\|\right]\right\}.$$

We note that the inequality

$$S_{p,q}^\kappa(g,h,f) \leq V_p(h;[c,d])\left[V_q(f;[c,d])\|g\|_\infty + V_q(g;[c,d])\|f\|_\infty\right] \tag{3.98}$$

holds for all partitions κ of $[c,d]$.

Lemma 3.30. *Let κ be a partition of an interval $[c,d]$ and let $p,q > 0$, $1/p+1/q > 1$. Then, for each integer $n \geq 1$ and for all $m = 1,\ldots,n$,*

$$\left\| \left(- \sum_{1 \leq i < j \leq m} + \sum_{m < j \leq i \leq n} \right) \Delta_j^\kappa \big(g(\cdot)(\Delta_i^\kappa h)f(\cdot)\big) \right\| \leq c_q \big(1 + \zeta(1/p + 1/q)\big) \mathcal{S}_{p,q}^\kappa(g,h,f),$$

where $c_q = 1 \vee 2^{(1/q)-1}$. One can drop the constant c_q if $f \equiv \mathbb{1}$ or $g \equiv \mathbb{1}$.

Remark. Lemma 3.30 is a complement to inequality (5.1) of Young (1936, p. 254) which is the case $m = 0$ where one of the functions f and g is $\equiv \mathbb{1}$ and all functions are complex-valued. Such an inequality for $m \geq 1$ seems needed to justify some of Young's statements as well as for later developments in this paper.

Proof. For any partition $\kappa' \subset \kappa$ containing x_m, where $\kappa = \{x_0, x_1, \ldots, x_n\}$, $\kappa' = \{y_0, y_1, \ldots, y_r\}$ and $y_k = x_m$ for some $k = 1, \ldots, r-1$, let

$$S_L(\kappa') := \sum_{1 \leq i < j \leq k} \Delta_j^{\kappa'}\big(g(\cdot)(\Delta_i^{\kappa'} h)f(\cdot)\big), \quad S_R(\kappa') := \sum_{k < j \leq i \leq r} \Delta_j^{\kappa'}\big(g(\cdot)(\Delta_i^{\kappa'} h)f(\cdot)\big).$$

For an $l = 1, \ldots, m-1, m+1, \ldots, n-1$, let $\kappa(l)$ denote the partition $\kappa \setminus \{x_l\}$. Then, for each $l = 1, \ldots, m-1$, we have

$$S_L(\kappa(l)) = \sum_{j=2}^{m-1} \Delta_j^{\kappa(l)}\Big(g(\cdot)\Big(\sum_{i=1}^{j-1}\Delta_i^{\kappa(l)}h\Big)f(\cdot)\Big)$$

$$= \sum_{j=2}^{l-1} \Delta_j^\kappa\Big(g(\cdot)\Big(\sum_{i=1}^{j-1}\Delta_i^\kappa h\Big)f(\cdot)\Big) + \Delta_l^\kappa\Big(g(\cdot)\Big(\sum_{i=1}^{l-1}\Delta_i^\kappa h\Big)f(\cdot)\Big)$$

$$+ \Delta_{l+1}^\kappa\Big(g(\cdot)\Big(\sum_{i=1}^{l-1}\Delta_i^\kappa h\Big)f(\cdot)\Big) + \sum_{j=l+1}^{m-1} \Delta_{j+1}^\kappa\Big(g(\cdot)\Big(\sum_{i=1}^{j}\Delta_i^\kappa h\Big)f(\cdot)\Big)$$

$$= -\Delta_{l+1}^\kappa\big(g(\cdot)(\Delta_l^\kappa h)f(\cdot)\big) + \sum_{j=2}^{m} \Delta_j^\kappa\Big(g(\cdot)\Big(\sum_{i=1}^{j-1}\Delta_i^\kappa h\Big)f(\cdot)\Big)$$

$$= -\Delta_{l+1}^\kappa\big(g(\cdot)(\Delta_l^\kappa h)f(\cdot)\big) + S_L(\kappa), \qquad (3.99)$$

and, for each $l = m+1, \ldots, n-1$, we have

$$S_R(\kappa(l)) = \sum_{j=m+1}^{n-1} \Delta_j^{\kappa(l)}\Big(g(\cdot)\Big(\sum_{i=j}^{n-1}\Delta_i^{\kappa(l)}h\Big)f(\cdot)\Big)$$

$$= \sum_{j=m+1}^{l-1} \Delta_j^\kappa\Big(g(\cdot)\Big(\sum_{i=j}^{n}\Delta_i^\kappa h\Big)f(\cdot)\Big) + \Delta_l^\kappa\Big(g(\cdot)\Big(\sum_{i=l}^{n}\Delta_i^\kappa h\Big)f(\cdot)\Big)$$

$$+ \Delta_{l+1}^\kappa\Big(g(\cdot)\Big(\sum_{i=l}^{n}\Delta_i^\kappa h\Big)f(\cdot)\Big) + \sum_{j=l+1}^{n-1} \Delta_{j+1}^\kappa\Big(g(\cdot)\Big(\sum_{i=j+1}^{n}\Delta_i^\kappa h\Big)f(\cdot)\Big)$$

$$= \Delta^\kappa_{l+1}\big(g(\cdot)(\Delta^\kappa_l h)f(\cdot)\big) + \sum_{j=m+1}^{n} \Delta^\kappa_j\Big(g(\cdot)\Big(\sum_{i=j}^{n}\Delta^\kappa_i h\Big)f(\cdot)\Big)$$

$$= \Delta^\kappa_{l+1}\big(g(\cdot)(\Delta^\kappa_l h)f(\cdot)\big) + S_R(\kappa). \tag{3.100}$$

Now consider the sequence $\{\Delta^\kappa_{l+1}(g(\cdot)(\Delta^\kappa_l h)f(\cdot))\colon l = 1,\ldots,m-1,m+1,\ldots,n-1\}$. By the Lemma of Sec. 2 of Young (1936) and by inequality (2.21), there exists an l for which

$$\|\Delta^\kappa_{l+1}\big(g(\cdot)(\Delta^\kappa_l h)f(\cdot)\big)\| \le \|\Delta^\kappa_l h\|\big(\|g(x_{l+1})\|\,\|\Delta^\kappa_{l+1} f\| + \|\Delta^\kappa_{l+1} g\|\,\|f(x_{l+1})\|\big)$$
$$\le c_q(n-2)^{-(1/p)-1/q} \mathcal{S}^\kappa_{p,q}(g,h,f), \tag{3.101}$$

where c_q is unnecessary (also below) if $g \equiv 1\!\!1$ or $f \equiv 1\!\!1$. If there is more than one such l, take the least such l. Let $S_1 := -S_L(\kappa) + S_R(\kappa)$ and $S_2 := -S_L(\kappa(l)) + S_R(\kappa(l))$. Then, by (3.99), (3.100) and (3.101), we have

$$\|S_1 - S_2\| \le c_q(n-2)^{-(1/p)-1/q} \mathcal{S}^\kappa_{p,q}(g,h,f).$$

Now continue the process recursively, obtaining $S_3, S_4, \ldots$. We get

$$\|S_1 - S_k\| \le c_q\Big(\sum_{i=1}^{k-1}(n-1-i)^{-(1/p)-1/q}\Big)\mathcal{S}^\kappa_{p,q}(g,h,f)$$
$$\le c_q\zeta(1/p + 1/q)\mathcal{S}^\kappa_{p,q}(g,h,f), \tag{3.102}$$

for $k = 2,\ldots,n-1$. In S_{n-1} all $n-2$ partition points $x_1,\ldots,x_{m-1},x_{m+1},\ldots,x_{n-1}$ have been deleted, so we have

$$S_{n-1} = -S_L(\{x_0,x_m,x_n\}) + S_R(\{x_0,x_m,x_n\}) = S_R(\{x_0,x_m,x_n\})$$

and $\|S_{n-1}\| \le \mathcal{S}^\kappa_{p,q}(g,h,f)$. The lemma follows from this and (3.102). $\qquad\square$

Lemma 3.31. *Let $h \in \mathcal{W}_p([a,b];\mathbb{B})$ and $f \in \mathcal{W}_q([a,b];\mathbb{B})$. For any partition κ of $[a,b]$ and for $p_1 > p > 0$, $q_1 > q > 0$, $1/p_1 + 1/q_1 \ge 1$, we have*

$$\sum_i V_{p_1}(h;[\Delta^\kappa_i])V_{q_1}(f;[\Delta^\kappa_i]) \le \max_i\Big(Osc_{[\Delta^\kappa_i]}(h)^{(p_1-p)/p_1} \wedge Osc_{[\Delta^\kappa_i]}(f)^{(q_1-q)/q_1}\Big)\times$$

$$\times\Big(Osc(h)^{(p_1-p)/p_1} \vee Osc(f)^{(q_1-q)/q_1}\Big)V_p(h)^{p/p_1}V_q(f)^{q/q_1}.$$

Proof. The lemma is a simple consequence of the relation

$$V_{p_1}(f;\tau) \le Osc_\tau(f)^{(p_1-p)/p_1}V_p^{p/p_1}(f;\tau),$$

the identity $uv = (u \wedge v)(u \vee v)$ and Hölder's inequality (3.95). $\qquad\square$

Proof of Theorem 3.27. Let f, h and g be $\mathbb{B}$-valued functions on $[a,b]$ and let $\kappa = \{c = x_0 < x_1 < \cdots < d = x_n\}$ be a partition of an interval $[c,d] \subset [a,b]$. Given any point $x_m \in \kappa$, $m = 1, \ldots, n-1$, we have

$$D(\kappa) := \sum_{i=1}^{n} g(x_i)(\Delta_i^{\kappa} h) f(x_i) - g(x_m)(h(d) - h(c)) f(x_m)$$

$$= \sum_{i=m+1}^{n} \left[g(x_i)(\Delta_i^{\kappa} h) f(x_i) - g(x_m)(\Delta_i^{\kappa} h) f(x_m) \right]$$

$$+ \sum_{i=1}^{m} \left[g(x_i)(\Delta_i^{\kappa} h) f(x_i) - g(x_m)(\Delta_i^{\kappa} h) f(x_m) \right]$$

$$= \sum_{m+1 \leq j \leq i \leq n} \Delta_j^{\kappa} \left(g(\cdot)(\Delta_i^{\kappa} h) f(\cdot) \right) - \sum_{1 \leq i < j \leq m} \Delta_j^{\kappa} \left(g(\cdot)(\Delta_i^{\kappa} h) f(\cdot) \right). \quad (3.103)$$

Let $p_1 > p$ and $q_1 > q$ be such that $1/p_1 + 1/q_1 > 1$. By Lemma 3.30 and by (3.98), it then follows that

$$\|D(\kappa)\| \leq c_{q_1}(1 + \zeta(1/p_1 + 1/q_1)) V_{p_1}(h; [c,d]) \times$$
$$\times \left(V_{q_1}(g; [c,d]) \|f\|_{\infty} + V_{q_1}(f; [c,d]) \|g\|_{\infty} \right). \quad (3.104)$$

If now κ' and κ'' are any two partitions of $[a,b]$ and we have points $y_i' \in [\Delta_i'] := [\Delta_i^{\kappa'}]$ and $y_j'' \in [\Delta_j''] := [\Delta_j^{\kappa''}]$, take a refined partition κ of $[a,b]$ including both partitions and containing all y_i' and y_j''. Denoting by I_k', $k = 1, \ldots, n'$ and by I_l'', $l = 1, \ldots, n''$ those indices i from κ such that $[\Delta_k'] = \cup_{i \in I_k'} [\Delta_i^{\kappa}]$ and $[\Delta_l''] = \cup_{i \in I_l''} [\Delta_i^{\kappa}]$, respectively, by (3.104), it follows that

$$\left\| \sum_{k=1}^{n'} g(y_k')(\Delta_k' h) f(y_k') - \sum_{l=1}^{n''} g(y_l'')(\Delta_l'' h) f(y_l'') \right\|$$

$$\leq \sum_{k=1}^{n'} \left\| g(y_k')(\Delta_k' h) f(y_k') - \sum_{i \in I_k'} g(x_{i-1})(\Delta_i^{\kappa} h) f(x_{i-1}) \right\|$$

$$+ \sum_{l=1}^{n''} \left\| g(y_l'')(\Delta_l'' h) f(y_l'') - \sum_{i \in I_l''} g(x_{i-1})(\Delta_i^{\kappa} h) f(x_{i-1}) \right\|$$

$$\leq c_{q_1}(1 + \zeta(1/p_1 + 1/q_1)) \left[\sum_{k=1}^{n'} V_{p_1}(h; [\Delta_k']) (V_{q_1}(g; [\Delta_k']) \|f\|_{\infty} + V_{q_1}(f; [\Delta_k']) \|g\|_{\infty}) \right.$$

$$\left. + \sum_{l=1}^{n''} V_{p_1}(h; [\Delta_l'']) (V_{q_1}(g; [\Delta_l'']) \|f\|_{\infty} + V_{q_1}(f; [\Delta_l'']) \|g\|_{\infty}) \right]. \quad (3.105)$$

Let $\epsilon > 0$. By Lemmas 2.2 and 2.4, the set E of all x such that

$$\max(\|\Delta^+ u(x)\|, \|\Delta_- u(x)\|) \geq \epsilon$$

for $u = f$, g or h is finite. Using Lemma 2.15 one can find a number

$$0 < \delta_0 < \min\{|x - y|: x, y \in E, \quad x \neq y\}$$

such that the oscillations on any interval $[c, d] \subset [a, b] \setminus E$ of length $< \delta_0$, of any of the three functions, are less than 2ϵ. First assume that the pairs h, g and h, f have no common discontinuities. Then again using Lemma 2.15 one can find a number $0 < \delta \leq \delta_0$ such that, for each interval $[c, d]$ containing at most one point from E, of length $< \delta$, we have

$$\begin{cases} \text{either} & Osc_{[c,d]}(h) < 2\epsilon \quad \text{or} \quad Osc_{[c,d]}(g) < 2\epsilon \\ \text{and} \\ \text{either} & Osc_{[c,d]}(h) < 2\epsilon \quad \text{or} \quad Osc_{[c,d]}(f) < 2\epsilon. \end{cases} \tag{3.106}$$

Hence, by Lemma 3.31, for all partitions κ' and κ'' of $[a, b]$ such that $|\kappa'| \vee |\kappa''| \leq \delta$, the right side of (3.105) is

$$\begin{aligned} &\leq 2c_{q_1}(1 + \zeta(1/p_1 + 1/q_1))[(2\epsilon)^{(p_1-p)/p} \vee (2\epsilon)^{(q_1-q)/q}] V_p(h)^{p/p_1} \times \\ &\quad \times [V_q(g)^{q/q_1}\|f\|_\infty + V_q(f)^{q/q_1}\|g\|_\infty] \times \\ &\quad \times \left(\|h\|_\infty^{(p_1-p)/p_1} \vee \|f\|_\infty^{(q_1-q)/q_1} \vee \|g\|_\infty^{(q_1-q)/q_1}\right), \end{aligned} \tag{3.107}$$

which can be made arbitrarily small by choosing an appropriate ϵ. This proves part (1) of Theorem 3.27.

Second, assume that the pairs h, g and h, f have no common one-sided discontinuities. Then, by existence of left and right limits, for each $x \in E$, one can find $c_x \in [a, x)$ and $d_x \in (x, b]$ such that (3.106), for both $[c_x, x]$ and $[x, d_x]$ instead of $[c, d]$, holds. Denote by κ_0 any partition of $[a, b]$ which contains all points of $E \cup \{c_x, d_x: x \in E\}$ and such that $|\kappa_0| < \delta_0$. Then, for any two refinements κ' and κ'' of κ_0, the right side of (3.105) is bounded as in (3.107). This proves part (2) of Theorem 3.27.

Finally, if f, g and h are unrestricted in the classes $\mathcal{W}_q$, $\mathcal{W}_q$ and $\mathcal{W}_p$, respectively, the functions f_+, g_+ and h_- belong to the same classes and the pairs f_+, h_- and g_+, h_- have no common one-sided discontinuities. Thus, the integral

$$(MPS) \int_a^b g_+^{(b)} dh_-^{(a)} f_+^{(b)}$$

exists by part (2). Moreover, the series

$$\sum_{(a,b)} [g(\Delta_-^+ h)f - g_+(\Delta_-^+ h)f_+] = -\sum_{(a,b)} (\Delta^+ g)(\Delta_-^+ h)f - \sum_{(a,b)} g_+(\Delta_-^+ h)(\Delta^+ f),$$

converges in $\mathbb{B}$ by Hölder's inequality (3.95), since the series

$$\left(\sum_{(a,b)} \|\Delta^+ f\|^q\right)^{1/q}, \quad \left(\sum_{(a,b)} \|\Delta^+ g\|^q\right)^{1/q} \quad \text{and} \quad \left(\sum_{(a,b)} \|\Delta_-^+ h\|^p\right)^{1/p}$$

are majorized by $\mathfrak{S}_q(f)$, $\mathfrak{S}_q(g)$ and $2\mathfrak{S}_p(h)$, respectively (cf. (2.4)). The integral (3.1) thus exists in the (CY) sense in this case.

Turning to the proof of inequality (3.93) take any $x_0 \in [a, b]$. Then, by the definitions in cases (1) and (2) and by Proposition 3.10 in case (3), one can construct a partition κ of $[a, b]$ and a sequence $y_i \in [\Delta_i^\kappa]$ such that $x = x_m$ for some $x_m \in \kappa$ and the Riemann-Stieltjes sum $\sum_i g(y_i)(\Delta_i^\kappa h)f(y_i)$ is close to the integral (3.1). Then (3.93) follows using the representation (3.103), Lemma 3.30 and (3.98). Finally, (3.94) is a consequence of (3.93) and the triangle inequality ($V_{p,\infty}$ was defined in subsection 2.3). □

The following statement complements Lemmas 3.23 and 3.26.

Proposition 3.32. *Let $h \in \mathcal{W}_p([a, b]; \mathbb{B})$ and $f, g \in \mathcal{W}_q([a, b]; \mathbb{B})$, with $p, q > 0$, $1/p + 1/q > 1$. Then the indefinite integrals*

$$\int_a^{(\cdot)} g\, dh\, f \quad and \quad \int_{(\cdot)}^b g\, dh\, f \in \mathcal{W}_p([a, b]; \mathbb{B}),$$

where $\int$ denotes either of the integrals $(CY)\int$, $(LY)\int = (RY)\int$. Moreover, setting as usual $c_p = 1 \vee 2^{(1/p)-1}$, we have

$$V_p(\textstyle\int g\, dh\, f; \tau) \leq c_p c_q (1 + \zeta(1/p + 1/q)) V_p(h; \tau) V_{q,\infty}(g; \tau) V_{q,\infty}(f; \tau), \qquad (3.108)$$

where $\int$ denotes either $(CY)\int_a^{(\cdot)}$ or $(CY)\int_{(\cdot)}^b$, and

$$V_p\Big(\sum [g_+(\Delta h)f - g(\Delta h)f_-]; [a, b]\Big)$$
$$\leq c_p \mathfrak{S}_p(h; (a, b))\big[\|f\|_\infty \mathfrak{S}_q(g; (a, b)) + \mathfrak{S}_q(f; (a, b))\|g\|_\infty\big], \quad (3.109)$$

where $\sum$ denotes either $\sum_{(a, \cdot)}$ or $\sum_{(\cdot, b)}$ and Δh denotes either $\Delta^+ h$ or $\Delta_- h$. If $g \equiv \mathbb{1}$ then $V_{q,\infty}(g) = 1$, c_q in (3.108) and c_p in (3.109) can be omitted, and likewise for f.

Remark. The statement cannot be improved to $\Psi_a := \int_a^{(\cdot)} g\, dh\, f \in \mathcal{W}_r$ for some $r < p$ in general, because, if $f \equiv g \equiv \mathbb{1}$ and $h \in \mathcal{W}_p \setminus \mathcal{W}_r$ for all $r < p$ then $\Psi_a = h - h(a) \notin \mathcal{W}_r$ for all $r < p$.

Proof. It is enough to prove the bounds (3.108) and (3.109). Then the indefinite integrals $(LY)\int$ and $(RY)\int$ have bounded p-variation by Definition 3.11, Theorem 3.12 and Corollary 3.28. To show (3.108) let $\kappa = \{x_i : i = 0, \ldots, n\}$ be a partition of $[a, b]$. Using inequality (2.21), by (3.93), it follows that

$$\left(\sum_{i=1}^n \left\|(CY)\int_{x_{i-1}}^{x_i} g\, dh\, f\right\|^p\right)^{1/p} \leq c_p\left[\left(\sum_{i=1}^n \|g(x_i)(\Delta_i^\kappa h)f(x_i)\|^p\right)^{1/p}\right.$$

$$\left. + \left(\sum_{i=1}^n \left\|(CY)\int_{x_{i-1}}^{x_i} g\, dh\, f - g(x_i)(\Delta_i^\kappa h)f(x_i)\right\|^p\right)^{1/p}\right]$$

$$\leq c_p c_q \big[V_p(h)\|g\|_\infty\|f\|_\infty + (1 + \zeta(1/p + 1/q))V_p(h)(V_q(f)\|g\|_\infty + \|f\|_\infty V_q(g))\big]$$

$$\leq c_p c_q (1 + \zeta(1/p + 1/q)) V_p(h) V_{q,\infty}(g) V_{q,\infty}(f).$$

Thus (3.108) follows for $\int = (CY) \int_a^{(\cdot)}$. Since $(CY) \int_{x_i}^{x_{i-1}} = -(CY) \int_{x_{i-1}}^{x_i}$ for each $i = 1, \ldots, n$ this proves (3.108) also for $\int = (CY) \int_{(\cdot)}^b$. If $g \equiv \mathbb{1}$, then applying Corollary 3.29 instead of (3.93), c_q can be omitted.

To prove (3.109) let $\kappa = \{x_i \colon i = 0, \ldots, n\}$ be again a partition of $[a, b]$. Adding and subtracting $g(\Delta h)f$, by (2.21) and Hölder's inequality (3.95), we get

$$\left(\sum_{i=1}^{n} \| \sum_{[x_{i-1}, x_i) \backslash \{a\}} [g_+(\Delta h)f - g(\Delta h)f_-] \|^p \right)^{1/p}$$

$$\leq c_p \left[\left(\sum_{i=1}^{n} \left(\sum_{[x_{i-1}, x_i) \backslash \{a\}} \|\Delta^+ g\| \|\Delta h\| \|f\| \right)^p \right)^{1/p} \right.$$

$$\left. + \left(\sum_{i=1}^{n} \left(\sum_{[x_{i-1}, x_i) \backslash \{a\}} \|g\| \|\Delta h\| \|\Delta_- f\| \right)^p \right)^{1/p} \right]$$

$$\leq c_p \left[\|f\|_\infty \max_{1 \leq i \leq n} \left(\sum_{[x_{i-1}, x_i) \backslash \{a\}} \|\Delta^+ g\|^q \right)^{1/q} \left(\sum_{(a,b)} \|\Delta h\|^p \right)^{1/p} \right.$$

$$\left. + \|g\|_\infty \max_{1 \leq i \leq n} \left(\sum_{[x_{i-1}, x_i) \backslash \{a\}} \|\Delta_- f\|^q \right)^{1/q} \left(\sum_{(a,b)} \|\Delta h\|^p \right)^{1/p} \right]$$

$$\leq c_p \mathfrak{S}_p(h; (a,b)) [\|f\|_\infty \mathfrak{S}_q(g; (a,b)) + \mathfrak{S}_q(f; (a,b)) \|g\|_\infty].$$

Thus (3.109) follows for $\sum = \sum_{(a,\cdot)}$. Likewise (3.109) follows for $\sum = \sum_{(\cdot,b)}$ and for the case when $f \equiv \mathbb{1}$ or $g \equiv \mathbb{1}$. The proof of Proposition 3.32 is complete. □

3.6. A convergence theorem. We start by recalling some notation. We say that a sequence $\{f_k\} = \{f_k \colon k \geq 1\}$ is a $\mathcal{W}_q([a,b]; \mathbb{B})$-*sequence* if $\sup_k V_{q,\infty}(f_k) < \infty$. Such a sequence *converges locally uniformly* to f at c, if given $\epsilon > 0$, there is a k_0 and a δ, such that, for all $k > k_0$ and all d distant less than δ from c,

$$\|f_k(d) - f(d)\| < \epsilon.$$

We shall also speak of *uniform convergence on the right at c* when the above holds for $d \geq c$, the other conditions being the same, and, similarly, of *uniform convergence on the left at c*.

The following statement is a partial extension of a convergence theorem for (CY) integration due to L. C. Young (1936, p. 269).

Proposition 3.33. *Let $h \in \mathcal{W}_p([a,b]; \mathbb{B})$ and $\{g_k\}$, $\{f_k\}$ be $\mathcal{W}_q([a,b]; \mathbb{B})$-sequences, where $p, q > 0$, $1/p + 1/q > 1$. Suppose $\{g_k\}$ and $\{f_k\}$ converge densely in $[a,b]$ to functions $g, f \in \mathcal{W}_q([a,b]; \mathbb{B})$, respectively, and that both sequences converge locally uniformly to their limits at each discontinuity of h. Then*

$$\lim_{k \to \infty} (CY) \int_a^b g_k \, dh \, f_k = (CY) \int_a^b g \, dh \, f.$$

Proof. Given $\epsilon > 0$, the set E of discontinuities of h with jumps of norm greater than or equal to $\epsilon/2$ is finite. By Lemma 2.15, in any interval whose length does not exceed a certain δ_0 and which contains no points of E, the oscillation of h is less than ϵ. Since $g_k - g$ and $f_k - f$ converge locally uniformly to zero at each point of E one can find a $\delta \in (0, \delta_0)$ and an integer k_0 so that for all $k \geq k_0$, the suprema of $\|g_k - g\|$ and $\|f_k - f\|$ in any interval containing a point of E and having length less than δ, are less than $\epsilon/2$. Therefore, one can determine a partition $\kappa = \{x_i : i = 0, \ldots, n\}$ of $[a, b]$ such that for each interval $[\Delta_i^\kappa]$ either

$$Osc_{[\Delta_i^\kappa]}(h) < \epsilon \quad \text{or} \quad \sup_{[\Delta_i^\kappa]} \|g_k - g\| \vee \sup_{[\Delta_i^\kappa]} \|f_k - f\| < \epsilon/2.$$

For each $i = 1, \ldots, n$, choose $y_i', y_i'' \in [\Delta_i^\kappa]$ such that

$$\lim_{k \to \infty} \|g_k(y_i') - g(y_i')\| = \lim_{k \to \infty} \|f_k(y_i'') - f(y_i'')\| = 0. \tag{3.110}$$

Then using Proposition 3.21, Lemma 3.31 and its proof, and the integral inequality (3.93), we get for any $p_1 > p$ and $q_1 > q$ with $1/p_1 + 1/q_1 > 1$, letting $c := c_{q_1}(1 + \zeta(1/p_1 + 1/q_1))$,

$$\left\| (CY) \int_a^b g_k \, dh \, f_k - (CY) \int_a^b g \, dh \, f \right\|$$

$$\leq \left\| (CY) \int_a^b (g_k - g) \, dh \, f_k \right\| + \left\| (CY) \int_a^b g \, dh \, (f_k - f) \right\|$$

$$\leq \sum_{i=1}^n c V_{p_1}(h; [\Delta_i^\kappa]) \left[V_{q_1}(g_k - g; [\Delta_i^\kappa]) \sup_{[\Delta_i^\kappa]} \|f_k\| + \sup_{[\Delta_i^\kappa]} \|g_k - g\| V_{q_1}(f_k; [\Delta_i^\kappa]) \right]$$

$$+ \sum_{i=1}^n \|(g_k(y_i') - g(y_i'))(\Delta_i^\kappa h) f_k(y_i')\|$$

$$+ \sum_{i=1}^n c V_{p_1}(h; [\Delta_i^\kappa]) \left[V_{q_1}(g; [\Delta_i^\kappa]) \sup_{[\Delta_i^\kappa]} \|f_k - f\| + \sup_{[\Delta_i^\kappa]} \|g\| V_{q_1}(f_k - f; [\Delta_i^\kappa]) \right]$$

$$+ \sum_{i=1}^n \|g(y_i'')(\Delta_i^\kappa h)(f_k(y_i'') - f(y_i''))\| \leq c V_p(h)^{p/p_1} \times$$

$$\times \left[(\epsilon^{(p_1-p)/p_1} \vee \epsilon^{(q_1-q)/q_1})(\|h\|_{(\infty)}^{(p_1-p)/p_1} \vee \|g_k - g\|_{(\infty)}^{(q_1-q)/q_1}) V_q(g_k - g)^{q/q_1} \|f_k\|_\infty \right.$$

$$+ (\epsilon^{(p_1-p)/p_1} \vee \epsilon/2)(\|h\|_{(\infty)}^{(p_1-p)/p_1} \vee \|g_k - g\|_\infty) V_{q_1}(f_k)$$

$$+ (\epsilon^{(p_1-p)/p_1} \vee \epsilon/2)(\|h\|_{(\infty)}^{(p_1-p)/p_1} \vee \|f_k - f\|_\infty) V_{q_1}(g)$$

$$\left. + (\epsilon^{(p_1-p)/p_1} \vee \epsilon^{(q_1-q)/q_1})(\|h\|_{(\infty)}^{(p_1-p)/p_1} \vee \|f_k - f\|_{(\infty)}^{(q_1-q)/q_1}) V_q(f_k - f)^{q/q_1} \|g\|_\infty \right]$$

$$+ (\|f_k\|_\infty + \|g\|_\infty) \max_i (\|g_k(y_i') - g(y_i')\| \vee \|f_k(y_i'') - f(y_i'')\|) \sum_{i=1}^n \|\Delta_i^\kappa h\|.$$

Since $\epsilon > 0$ is arbitrary, by (3.110), the lim sup as $k \to \infty$ of the right side can be made arbitrarily small. This completes the proof of Proposition 3.33. □

4. PRODUCT INTEGRALS

We continue to assume that $\mathbb{B} = (\mathbb{B}, \|\cdot\|)$ is a Banach algebra with an identity $\mathbb{1}$. Recall definition (1.2) of the product integral $\prod_a^b(1+df)$ with respect to a $\mathbb{B}$-valued function f over an interval $[a,b]$. Let

$$\mathcal{P}([a,b];\mathbb{B}) := \{f\colon [a,b] \mapsto \mathbb{B}\colon \prod_a^b(\mathbb{1}+df) \text{ exists and is invertible}\}.$$

This class for $\mathbb{B} = \mathbb{R}$ is treated in Subsection 4.1. The general case is considered starting from Subsection 4.3. The main results are a characterization of $\mathcal{P}([a,b];\mathbb{R})$ (cf. Theorem 4.4) and sufficient conditions for a function to belong to $\mathcal{P}([a,b];\mathbb{B})$ (cf. Theorem 4.23 and Proposition 4.30).

4.1. Necessary and sufficient conditions. In this subsection we restrict ourselves to real-valued functions f. We start with necessary conditions for the existence of the product integral.

Lemma 4.1. *If* $f \in \mathcal{P}([a,b]:\mathbb{R})$ *then* f *is bounded.*

Proof. Suppose not. Take a partition κ of $[a,b]$ and an $M < \infty$ such that for all refinements λ of κ, $1/M \le |\prod_\lambda(1+\Delta^\lambda f)| \le M$. Let $\kappa := \{x_i\colon i = 0,\ldots,n\}$. For some i_0, f is unbounded on the open interval (x_{i_0-1}, x_{i_0}). Fix such an i_0 and consider partitions $\kappa(x) := \kappa \cup \{x\}$ for $x \in (x_{i_0-1}, x_{i_0})$. Let $y := f(x)$, $c := f(x_{i_0-1})$, $d := f(x_{i_0})$. Then by choice of κ, the products $\prod_{i<i_0}(1+\Delta_i^\kappa f)$ and $\prod_{i>i_0}(1+\Delta_i^\kappa f)$ are non-zero and

$$1/M \le 1/|\prod_{\kappa(x)}(1+\Delta^{\kappa(x)}f)|$$
$$= 1/\Big(|\prod_{i<i_0}(1+\Delta_i^\kappa f)||(1+y-c)(1+d-y)||\prod_{i_0<i}(1+\Delta_i^\kappa f)|\Big)$$
$$\to 0 \quad \text{as } |y| \to \infty,$$

a contradiction, proving that f is bounded. The proof of Lemma 4.1 is complete.$\square$

Lemma 4.2. *If* $f \in \mathcal{P}([a,b];\mathbb{R})$ *then* f *is regulated.*

Proof. Suppose not. Then by symmetry we can assume that at some $c \in (a,b]$, f does not have a left limit. So

$$-\infty < \liminf_{x\uparrow c} f(x) =: \alpha < \limsup_{x\uparrow c} f(x) =: \beta < \infty.$$

Take a partition $\kappa = \{x_i\colon i = 0,\ldots,n\}$ of $[a,b]$ and $r > 0$ such that for every refinement λ of κ, $\prod_\lambda(1+\Delta^\lambda f) > r$, or for every refinement λ of κ, $\prod_\lambda(1+\Delta^\lambda f) < -r$. Then for some i and $\delta > 0$, $(c-\delta,c] \subset [x_{i-1},x_i]$. Consider partitions $\lambda := \kappa \cup \{z_j\colon j = 1,\ldots,2m+1\}$ where m is an integer, $x_{i-1} < z_1 < z_2 < \cdots z_{2m+1} < x_i$, and $f(z_j)$ is approximately α for j odd and β for j even. Let $\gamma := \beta - \alpha$. First suppose $\gamma^2 < 2$. For any $\epsilon > 0$ we can choose z_j such that

$$\left|\prod_\lambda (1 + \Delta^\lambda f) / \prod_\kappa (1 + \Delta^\kappa f)\right| = \left|\frac{[1 + f(z_1) - u][1 + v - f(z_{2m+1})]}{1 + v - u}\right| \times$$

$$\times \left|\prod_{j=1}^{2m} [1 + f(z_{j+1}) - f(z_j)]\right| \le \frac{|1 + \alpha - u||1 + v - \alpha|}{|1 + v - u|} |1 - \gamma^2|^m + \epsilon,$$

where $v := f(x_i)$ and $u := f(x_{i-1})$. Letting $m \to \infty$ and $\epsilon \downarrow 0$, the ratio approaches 0, contradicting the choice of κ. Or, if $\gamma^2 \ge 2$, for each fixed m, let $f(z_j) \to \alpha$ for j odd and $f(z_j) \to \beta$ for j even. Then $\prod_\lambda (1 + \Delta^\lambda f)$ will have different signs for m odd and m even, again contradicting the choice of κ. Therefore the left limit for f at c must exist. This proves Lemma 4.2. $\qquad\square$

Theorem 4.3. *If* $f \in \mathcal{P}([a,b]; \mathbb{R})$ *then* $f \in \mathcal{W}_2([a,b]; \mathbb{R})$.

Proof. Since f is regulated, the number of points $x+$ and $x-$ such that $|\Delta^+ f(x)| > 1/2$ or $|\Delta_- f(x)| > 1/2$ respectively is finite, and

$$M := \sup_x \max(|\Delta^+ f(x)|, |\Delta_- f(x)|) < \infty.$$

By Lemma 2.21, there exists a partition $\kappa = \{x_i : i = 0, \ldots, n\} \in Q^\pm([a,b])$ such that, for each $i \notin J_\pm(\kappa)$ (cf. (2.10)), we have $Osc_{[x_{i-1},x_i]}(f) \le 1/2$. If $f \notin \mathcal{W}_2([a,b]; \mathbb{R})$, then $V_2^*(f) = +\infty$ by Lemma 2.3. We have the Taylor series with remainder

$$\log(1 + u) = u - \theta(u)u^2, \tag{4.1}$$

where

$$\theta(u) = -\frac{1}{2} \frac{d^2}{dv^2} \log(1 + v) = \frac{1}{2(1 + v)^2} \tag{4.2}$$

for some v between 0 and u. Thus for $|u| \le 1/2$, we have $2/9 \le \theta(u) \le 2$. Let $\lambda = \{y_j : j = 0, \ldots, m\}$ be a partition of $[a,b]$ which is a refinement of κ. Let $J_\pm(\lambda, \kappa)$ be the set of values of j such that $[y_{j-1}, y_j] = [x_{i-1}, x_i]$ for some $i \in J_\pm(\kappa)$. Then letting $v_j := \Delta_j^\lambda f$, we have

$$\left|\prod_\lambda (1 + \Delta^\lambda f)\right| = \left|\left(\prod_{j \in J_\pm(\lambda,\kappa)} (1 + v_j)\right)\left(\prod_{j \notin J_\pm(\lambda,\kappa)} (1 + v_j)\right)\right|$$

$$\le (1 + M)^N \exp\left(\sum_{j \notin J_\pm(\lambda,\kappa)} \log(1 + v_j)\right),$$

where $N := \operatorname{card} J_\pm(\kappa)$. Now,

$$\sum_{j \notin J_\pm(\lambda,\kappa)} \log(1 + v_j) = \sum_{j \notin J_\pm(\lambda,\kappa)} (v_j - \theta(v_j)v_j^2) \le \sum_{j \notin J_\pm(\lambda,\kappa)} \left(v_j - \frac{2}{9}v_j^2\right),$$

and

$$\sum_{j \notin J_\pm(\lambda,\kappa)} v_j = \sum_{j=1}^m v_j - \sum_{j \in J_\pm(\lambda,\kappa)} v_j \le f(b) - f(a) + NM.$$

Since $\sum_{j \in J_\pm(\lambda,\kappa)} v_j^2 \le NM^2$, and $V_2^*(f) = +\infty$, we can choose λ to make the sum $\sum_{j \notin J_\pm(\lambda,\kappa)} v_j^2$ as large as we like, so $\left| \prod_\lambda (1 + \Delta^\lambda f) \right|$ is arbitrarily small. Since we can also take λ to be a refinement of any partition of $[a,b]$, this contradicts $f \in \mathcal{P}([a,b]; \mathbb{R})$ and proves Theorem 4.3.   □

Next, recalling the notation (2.29) of $\mathcal{W}_p^*$ we formulate necessary and sufficient conditions for the existence of a non-degenerate product integral with respect to a real-valued function.

Theorem 4.4. *We have $f \in \mathcal{P}([a,b]; \mathbb{R})$ if and only if the following two conditions hold:*

(1) $f \in \mathcal{W}_2^*([a,b]; \mathbb{R})$;
(2) $\Delta^+ f(x) \ne -1 \ne \Delta_- f(y)$ for $a \le x < b$, $a < y \le b$.

Remark. It will follow from the proof of Theorem 4.4, specifically from (4.26) and the inequality after it (cf. also Theorem 4.17 below for a more general statement) that

$$\prod_a^b (\mathbb{1} + df) = e^{f(b) - f(a)} \prod_{a \le x \le b} \left[(1 + \Delta_- f)(1 + \Delta^+ f) \right](x) e^{-\Delta_- f(x) - \Delta^+ f(x)} \qquad (4.3)$$

whenever $f \in \mathcal{P}([a,b]; \mathbb{R})$. Recall that $\Delta_- f(a)$ and $\Delta^+ f(b)$ in (4.3) are replaced by 0 as in the definition (2.3).

To prove Theorem 4.4 we need a few more auxiliary results. Let $\mathcal{P}_\pm([a,b]; \mathbb{R})$ be the set of all real-valued functions f on $[a,b]$ such that $\prod_a^b (1 + df)$ exists and is not 0, when points $x+$, $x-$ are allowed as members of partitions.

Theorem 4.5. $\mathcal{P}([a,b]; \mathbb{R}) = \mathcal{P}_\pm([a,b]; \mathbb{R})$, *and for f in these sets, the value of* $\prod_a^b (1 + df)$ *is the same for both.*

Proof. Every product $\prod_\lambda (1 + \Delta^\lambda f)$ for a partition λ containing points $x-$ and/or $x+$ can be approximated arbitrary closely by such a product for a partition λ' not containing points $x-$, $x+$, and if $\lambda \supset \kappa$ for some partition κ not containing points $x-$, $x+$, we can take $\lambda' \supset \kappa$ also. It follows that

$$\mathcal{P}([a,b]; \mathbb{R}) \subset \mathcal{P}_\pm([a,b]; \mathbb{R}). \qquad (4.4)$$

Conversely, suppose $f \in \mathcal{P}_\pm([a,b]; \mathbb{R})$. Let $\kappa = \{x_i : i = 0, \dots, n\}$ be a partition of $[a,b]$ such that $\prod_\lambda (1 + \Delta^\lambda f) \ne 0$ for each refinement λ of κ. By adjoining additional points to κ if necessary, we can assume that whenever a point $x_i = x+$ for some x, then $x_{i-1} = x$ and x_{i+1} is not a point $y+$ or $y-$, while if $x_i = x-$, then $x_{i+1} = x$ and x_{i-1} is not a point $y+$ or $y-$. Suppose for example that for some i and x, $x_{i-1} = x$, $x_i = x+$, $v := x_{i+1} > x$. Then v is not a point $y-$ or $y+$. Let λ_+ be the partition $\{x, x+, v\}$ of $[x, v]$. For $x < u < v$ let $\lambda_u := \lambda(u)$ be the partition $\{x, u, v\}$ of $[x, v]$. For a given partition λ let

$$S(\lambda) := \sup \Big\{ \prod_\mu (1 + \Delta^\mu f) \colon \mu \supset \lambda \Big\},$$

where λ and μ do not contain any points $y-$ or $y+$. Write $S_\pm(\lambda)$ for the supremum when λ and μ may contain points $y-$, $y+$.

Lemma 4.6. $S(\lambda_u) \to S_\pm(\lambda_+)$ as $u \downarrow x$.

Proof. Let λ be any refinement of λ_+, so $\lambda = \{y_j \colon j = 0, \ldots, m\}$, $y_0 = x$, $y_1 = x+$, $\ldots$, $y_m = v$. Consider partitions $\lambda_k = \{z_j^k \colon j = 0, \ldots, m\}$, $k = 1, 2, \ldots$, where $z_j^k = y_j$ for any y_j which is not a point $y-$ or $y+$, $z_j^k \downarrow y_j$ as $k \to \infty$ if $y_j = y+$ for some y, and $z_j^k \uparrow y_j$ as $k \to \infty$ if $y_j = y-$ for some y. Then, clearly,

$$\prod_{\lambda_k} (1 + \Delta^{\lambda_k} f) \to \prod_\lambda (1 + \Delta^\lambda f)$$

as $k \to \infty$ and $\prod_{\lambda_k} (1 + \Delta^{\lambda_k} f) \leq S(\lambda(y_1^k))$ for all $k \geq 1$. It follows that $S_\pm(\lambda_+) \leq \liminf_{u \downarrow x} S(\lambda_u)$. Conversely, let μ_u be a partition of $[x, v]$ containing u but not containing $x+$, and let $\mu_{u,+} := \mu_u \cup \{x+\}$. Let w be the smallest member of μ_u other than x. Then noting that $\Delta^+ f(x) \neq -1$,

$$\frac{\prod_{\mu_u} (1 + \Delta^{\mu_u} f)}{\prod_{\mu_{u,+}} (1 + \Delta^{\mu_u,+} f)} = \frac{1 + f(w) - f(x)}{[1 + f(w) - f(x+)][1 + f(x+) - f(x)]} \to 1 \quad \text{as } u \downarrow x$$

since then $w \to x$. So $S_\pm(\lambda_+) \geq \limsup_{u \downarrow x} S(\lambda_u)$, proving the lemma. $\square$

Let $\epsilon > 0$. Applying Lemma 4.6, and the symmetric statement for $v-$ in place of $x+$, we get that there is a partition κ', containing the same number of points as κ, containing each point of κ which is not a point $y-$ or $y+$, and not containing any points $y-$ or $y+$, such that

$$\sup_{\lambda \supset \kappa'} \prod_\lambda (1 + \Delta^\lambda f) \leq \sup_{\mu \supset \kappa} \prod_\mu (1 + \Delta^\mu f) + \epsilon, \qquad (4.5)$$

where λ does not contain any points $y-$, $y+$. By other symmetric forms of Lemma 4.6, we get a partition κ'' not containing any points $y-$ or $y+$ and such that

$$\inf_{\lambda \supset \kappa''} \prod_\lambda (1 + \Delta^\lambda f) \geq \inf_{\mu \supset \kappa} \prod_\mu (1 + \Delta^\mu f) - \epsilon, \qquad (4.6)$$

where λ does not contain any points $y-$, $y+$. Moreover, since to obtain (4.5) and (4.6) we only need to replace each $x+$ or $x-$ in κ by a point $y > x$ or $y < x$ respectively, close enough to x, we can take $\kappa' = \kappa''$. Thus (4.5) and (4.6) imply $f \in \mathcal{P}([a, b]; \mathbb{R})$, which together with (4.4) completes the proof of Theorem 4.5. $\square$

Theorem 4.7. *Let $1 < p < \infty$ and $f \in W_p([a,b]; \mathbb{R})$. Then*

$$\mathfrak{S}_p(f) = \sup\Big\{ \inf_{\lambda \supset \kappa} \Big(\sum_\lambda |\Delta^\lambda f|^p\Big)^{1/p} : \kappa \in Q^\pm([a,b])\Big\}.$$

Proof. Since the inequality

$$\mathfrak{S}_p(f) \le \sup\Big\{ \inf_{\lambda \supset \kappa} \Big(\sum_\lambda |\Delta^\lambda f|^p\Big)^{1/p} : \kappa \in Q^\pm([a,b])\Big\}. \qquad (4.7)$$

always holds it suffices to prove the converse inequality. To this end suppose $\kappa \in Q^\pm([a,b])$ is such that

$$U(\kappa) := \inf_{\lambda \supset \kappa} \sum_\lambda |\Delta^\lambda f|^p > \mathfrak{S}_p(f)^p. \qquad (4.8)$$

Let $V(\kappa) := U(\kappa) - \mathfrak{S}_p(f)^p > 0$. Note that since $p > 1$, $2^p + 1 < 3^p$. Let

$$0 < \epsilon < V(\kappa)3^{-p}\min(1/2, (3^p - 2^p - 1)/8). \qquad (4.9)$$

If we replace κ by a refinement κ' of κ, then $U(\kappa') \ge U(\kappa)$ and $V(\kappa') \ge V(\kappa)$, so (4.9) will hold for κ' in place of κ. Therefore we can and do assume that for all partitions $\lambda \supset \kappa$,

$$\sum_\lambda |\Delta^\lambda f|^p < V_p^*(f)^p + \epsilon \qquad (4.10)$$

and

$$\sum_{j \in J_\pm(\lambda)} |\Delta_j^\lambda f|^p \ge \sum_{i \in J_\pm(\kappa)} |\Delta_i^\kappa f|^p \ge \mathfrak{S}_p(f)^p - \epsilon. \qquad (4.11)$$

Choose a partition $\lambda \supset \kappa$ such that

$$\sum_\lambda |\Delta^\lambda f|^p < U(\kappa) + \epsilon. \qquad (4.12)$$

By definition of $U(\kappa)$, for any $\rho \supset \kappa$,

$$\sum_\rho |\Delta^\rho f|^p \ge U(\kappa). \qquad (4.13)$$

However, we will construct a partition $\lambda' \supset \lambda \supset \kappa$ in such a way that

$$\sum_{\lambda'} |\Delta^{\lambda'} f|^p < U(\kappa). \qquad (4.14)$$

This will contradict (4.8) and prove the inequality

$$\mathfrak{S}_p(f) \ge \sup\Big\{ \inf_{\lambda \supset \kappa} \Big(\sum_\lambda |\Delta^\lambda f|^p\Big)^{1/p} : \kappa \in Q^\pm([a,b])\Big\}. \qquad (4.15)$$

To begin the construction of λ' take any $j \notin J_{\pm}(\lambda)$ (defined in (2.10)), where the partition $\lambda = \{x_j: j = 0, \ldots, n\}$. So, $[x_{j-1}, x_j]$ is not an interval $[x-, x]$ or $[x, x+]$. By symmetry (replacing f by $-f$ if necessary) we can assume that $f(x_{j-1}) \leq f(x_j)$. If $f(x_{j-1}) = f(x_j)$, we will not insert a new point in (x_{j-1}, x_j). If $f(x_{j-1}) < f(x_j)$ we have two cases. Let $[s, t] := [f(x_{j-1}), f(x_j)]$ and let $[q, r]$ be the middle third of $[s, t]$, i.e.

$$q := s + \frac{1}{3}(t - s) \quad \text{and} \quad r := s + \frac{2}{3}(t - s).$$

Case 1. For some $y \in (x_{j-1}, x_j)$, $f(y) \in [q, r]$. We choose such a y and make it the one point in $\lambda' \cap (x_{j-1}, x_j)$. Then we have

$$|f(y) - s|^p + |t - f(y)|^p \leq \left[\left(\frac{2}{3}\right)^p + \left(\frac{1}{3}\right)^p\right](t - s)^p. \tag{4.16}$$

Case 2. For any $y \in (x_{j-1}, x_j)$, $f(y) \notin [q, r]$. Set

$$\xi := \sup\{y \in [x_{j-1}, x_j]: f(y) < q\},$$

where the supremum is defined in the natural way in a set containing points $x-$ and $x+$. If $\xi = x+$ then $f(x+) < q$ (hence $x+ < x_j$) and $f(z) > r > q$ for all $x+ < z \leq x_j$, a contradiction, showing that ξ cannot be a point $x+$.

If $\xi = y-$ then $f(y-) \leq q$ while $f(y) > r$, so $\Delta_- f(y) > r - q$. Then let $\lambda' \cap [x_{j-1}, x_j] := \{x_{j-1}, y-, y, x_j\}$, where possibly $y- = x_{j-1}$ or $y = x_j$, but not both, so $\lambda' \cap (x_{j-1}, x_j)$ contains at least one point. We have

$$|\Delta_- f(y)|^p > |t - s|^p / 3^p. \tag{4.17}$$

The other possibility is that ξ is an ordinary point. Then $\xi < x_j$ (possibly $x_j = \xi+$) and $f(\xi) < q$, while $f(\xi+) \geq r$. Let $\lambda' \cap [x_{j-1}, x_j] := \{x_{j-1}, \xi, \xi+, x_j\}$, where possibly $\xi = x_{j-1}$ or $\xi+ = x_j$, but not both. Then

$$|\Delta^+ f(\xi)|^p > |t - s|^p / 3^p. \tag{4.18}$$

This completes the construction of $\lambda' := \{z_i: i = 0, \ldots, k\}$.

To show (4.14) let $I(\lambda)$ be the set of all $j = 1, \ldots, n$, $j \notin J_{\pm}(\lambda)$ for which Case 1 holds, and $II(\lambda)$ the remaining $j \notin J_{\pm}(\lambda)$ for which Case 2 holds. By (4.17) and (4.18), we have

$$\sum_{i \in J_{\pm}(\lambda')} |\Delta_i^{\lambda'} f|^p \geq \sum_{j \in J_{\pm}(\lambda)} |\Delta_j^{\lambda} f|^p + \frac{1}{3^p} \sum_{j \in II(\lambda)} |\Delta_j^{\lambda} f|^p.$$

Thus by (4.11),

$$\sum_{j \in II(\lambda)} |\Delta_j^{\lambda} f|^p \leq 3^p \epsilon \tag{4.19}$$

and

$$\sum_{j \in II(\lambda)} \sum_{i \in A_j \cap J_{\pm}(\lambda')} |\Delta_i^{\lambda'} f|^p \leq \epsilon, \tag{4.20}$$

where $A_j := \{i = 1, \ldots, k: [z_{i-1}, z_i] \subset [x_{j-1}, x_j]\}$. Each term $\Delta_j^\lambda f = t - s$ for $j \in II(\lambda)$ is a sum $u + v + w$ where $u = \Delta_i^{\lambda'} f$ for some $i \in J_\pm(\lambda')$, $|w| \le |u|$ and $|v| \le |u|$ in all cases. Thus

$$\sum_{j \in II(\lambda)} \sum_{i \in A_j \setminus J_\pm(\lambda')} |\Delta_i^{\lambda'} f|^p \le 2\epsilon. \tag{4.21}$$

On the other hand, by (4.16),

$$\sum_{j \in I(\lambda)} \sum_{i \in A_j} |\Delta_i^{\lambda'} f|^p \le \frac{2^p + 1}{3^p} \sum_{j \in I(\lambda)} |\Delta_j^\lambda f|^p. \tag{4.22}$$

Therefore, by (4.20), (4.21) and (4.22),

$$\sum_{\lambda'} |\Delta^{\lambda'} f|^p \le \sum_{i \in J_\pm(\lambda)} |\Delta_i^\lambda f|^p + \frac{2^p + 1}{3^p} \sum_{j \in I(\lambda)} |\Delta_j^\lambda f|^p + 3\epsilon. \tag{4.23}$$

Since $\sum_{i \in J_\pm(\lambda)} |\Delta_i^\lambda f|^p \le \mathfrak{S}_p(f)^p$, we have by (4.13) and (4.19),

$$\sum_{j \in I(\lambda)} |\Delta_j^\lambda f|^p = \sum_\lambda |\Delta^\lambda f|^p - \sum_{j \in II(\lambda)} |\Delta_j^\lambda f|^p - \sum_{j \in J_\pm(\lambda)} |\Delta_j^\lambda f|^p$$

$$\ge V(\kappa) - 3^p \epsilon.$$

Thus by (4.23),

$$\sum_{\lambda'} |\Delta^{\lambda'} f|^p \le \sum_\lambda |\Delta^\lambda f|^p - \left(1 - \frac{2^p + 1}{3^p}\right) \sum_{j \in I(\lambda)} |\Delta_j^\lambda f|^p + 3\epsilon$$

$$\le \sum_\lambda |\Delta^\lambda f|^p - \left(1 - \frac{2^p + 1}{3^p}\right)(V(\kappa) - 3^p \epsilon) + 3\epsilon$$

by (4.12) and (4.9) $< U(\kappa) + 4\epsilon - \left(1 - \frac{2^p + 1}{3^p}\right)\frac{V(\kappa)}{2}$

by (4.9) again $< U(\kappa) + 4\epsilon - 4\epsilon = U(\kappa)$

which implies (4.14) and, hence, (4.15). This in conjunction with (4.7) proves Theorem 4.7. $\qquad \square$

Corollary 4.8. *Let $1 < p < \infty$ and $f \in W_p([a,b]; \mathbb{R})$. Then $f \in W_p^*([a,b]; \mathbb{R})$ if and only if the limit*

$$\lim_{(\kappa)} \sum_\kappa |\Delta^\kappa f|^p \left(= \mathfrak{S}_p(f)^p \right) \tag{4.24}$$

exists under refinements of partitions in $Q^\pm([a,b])$.

Proof. Let $f \in W_p^*([a,b]; \mathbb{R})$. Then the existence of the limit (4.24) follows from Lemma 2.22 and Theorem 4.7 directly. Conversely, suppose that the limit (4.24) exists but

$$\mathfrak{S}_p(f) < V_p^*(f) \tag{4.25}$$

(we recall that (2.12) always holds). Let $\kappa \in Q^\pm([a,b])$. Recursively, there are partitions $\kappa \subset \lambda_1 \subset \lambda_2 \subset \cdots$ such that $\sum_{\lambda_j} |\Delta^{\lambda_j} f|^p > V_p^*(f)^p - 1/j$ for j odd, by Lemma 2.22, and $\sum_{\lambda_j} |\Delta^{\lambda_j} f|^p < \mathfrak{S}_p(f)^p + 1/j$ for j even, by Theorem 4.7. Hence, by (4.25), the limit (4.24) cannot exist. This contradiction implies that $f \in \mathcal{W}_p^*([a,b]; \mathbb{R})$. The proof of Corollary 4.8 is complete. $\qquad\square$

Now we are ready to prove the main result of this subsection.

Proof of Theorem 4.4. By Theorem 4.5, it suffices to prove that $f \in \mathcal{P}_\pm([a,b]; \mathbb{R})$ if and only if f satisfies (1) and (2). For any partition λ of $[a,b]$, let

$$Q(\lambda) := Q(\lambda, f) := \sum_\lambda (\Delta^\lambda f)^2$$

and

$$Q_\pm(\lambda) := Q_\pm(\lambda, f) := \sum_{j \in J_\pm(\lambda)} (\Delta_j^\lambda f)^2.$$

To prove the "if" part suppose f satisfies (1) and (2). Given $\epsilon > 0$, take a partition κ of $[a,b]$ such that $\{x, x+\} \subset \kappa$ whenever $|\Delta^+ f(x)| > 1/2$, $a \le x < b$, and $\{x-, x\} \subset \kappa$ whenever $|\Delta_- f(x)| > 1/2$, $a < x \le b$. Also, choose κ by Lemma 2.21 such that for all $i \notin J_\pm(\kappa)$, $\kappa = \{x_i: i = 0, \ldots, m\}$, $Osc_{[x_{i-1}, x_i]}(f) \le 1/2$. Further by Lemma 2.22, one can choose κ so that $Q(\lambda) < V_2^*(f)^2 + \epsilon$ for all $\lambda \supset \kappa$, and $Q_\pm(\kappa) > \mathfrak{S}_2(f)^2 - \epsilon$. Thus for any $\lambda \supset \kappa$, $Q_\pm(\lambda) > \mathfrak{S}_2(f)^2 - \epsilon$. Let $J_\pm(\kappa, \lambda) := \{j \in J_\pm(\lambda): \{y_{j-1}, y_j\} \subset \kappa\}$, where $\lambda = \{y_j: j = 0, \ldots, k\}$. Then since $f \in \mathcal{W}_2^*([a,b]; \mathbb{R})$, for all $\lambda \supset \kappa$,

$$\sum_{j \notin J_\pm(\kappa, \lambda)} (\Delta_j^\lambda f)^2 = Q(\lambda) - Q_\pm(\kappa) < 2\epsilon.$$

Also, for all $\lambda \supset \kappa$, $\sum_{j \notin J_\pm(\kappa, \lambda)} \Delta_j^\lambda f \equiv f(b) - f(a) - \sum_{i \in J_\pm(\kappa)} \Delta_i^\kappa f$, not depending on λ. For any $\lambda \supset \kappa$ and $j = 0, \ldots, k$, $1 + \Delta_j^\lambda f \ne 0$, since $\Delta_j^\lambda f \ne -1$ for $j \in J_\pm(\kappa, \lambda)$ by assumption (2), and $|\Delta_j^\kappa f| \le 1/2$ for other j by choice of κ. Thus letting $d_j := \Delta_j^\kappa f$, by the Taylor series with remainder (4.1),

$$\prod_\lambda (1 + \Delta^\lambda f) = \prod_{j \in J_\pm(\kappa, \lambda)} (1 + \Delta_j^\lambda f) \exp\Big\{ \sum_{j \notin J_\pm(\kappa, \lambda)} \log(1 + \Delta_j^\lambda f) \Big\}$$

$$= e^{f(b) - f(a)} \prod_{i \in J_\pm(\kappa)} (1 + \Delta_i^\kappa f) \times$$

$$\times \exp\Big\{ - \sum_{i \in J_\pm(\kappa)} \Delta_i^\kappa f - \sum_{j \notin J_\pm(\kappa, \lambda)} \theta(d_j) d_j^2 \Big\}. \qquad (4.26)$$

We have $0 \le \sum_{j \notin J_\pm(\kappa, \lambda)} \theta(d_j) d_j^2 < 4\epsilon$. Thus, for $\lambda \supset \kappa$, $\prod_\lambda(1 + \Delta^\lambda f)$ is finite and is determined within a factor of e^ϵ. Letting $\epsilon \downarrow 0$, we get that $\prod_a^b(1 + df)$ exists and is finite and non-zero, so $f \in \mathcal{P}_\pm([a,b]; \mathbb{R})$ as desired.

Conversely, suppose $f \in \mathcal{P}_\pm([a,b]; \mathbb{R})$. Suppose f has a jump of -1. By symmetry, we can assume $\Delta_- f(x) = -1$ for some $x \in (a,b)$. Let κ be a partition of $[a,b]$ containing $x-$ and x. Then for any $\lambda \supset \kappa$, $\prod_\lambda(1 + \Delta^\lambda f) = 0$, a contradiction.

So (2) holds. To prove (1), by Theorem 4.3, we have $f \in \mathcal{W}_2([a,b];\mathbb{R})$. Suppose $V_2^*(f) > \mathfrak{S}_2(f)$. Then one can choose $\delta \in (0,1)$ so that

$$(1-\delta)^2 V_2^*(f)^2 > (1+\delta)^2 \mathfrak{S}_2(f)^2. \tag{4.27}$$

Choose a partition $\kappa = \{x_i : i = 0,\ldots,n\} \in Q^{\pm}([a,b])$ by Lemma 2.21 so that for all $i \notin J_{\pm}(\kappa)$, $Osc_{[x_{i-1},x_i]}(f) \le \delta$. Thus for any $\lambda \supset \kappa$, $\max\{|d_j| \colon j \notin J_{\pm}(\kappa,\lambda)\} \le \delta$, where $d_j := \Delta_j^\lambda f$. For $|u| \le \delta$ and $\theta(u)$ as in (4.2), we have

$$\frac{1}{2(1+\delta)^2} \le \theta(u) \le \frac{1}{2(1-\delta)^2}.$$

Thus in equation (4.26), for $Q_0(\lambda) := \sum_{j \notin J_{\pm}(\kappa,\lambda)} d_j^2$,

$$\frac{Q_0(\lambda)}{2(1+\delta)^2} \le \sum_{j \notin J_{\pm}(\kappa,\lambda)} \theta(d_j) d_j^2 \le \frac{Q_0(\lambda)}{2(1-\delta)^2}.$$

Now by the proof of Corollary 4.8, for different partitions $\lambda', \lambda'' \supset \kappa$, we can let $Q_0(\lambda')$ approach $V_2^*(f)^2 - \sum_{i \in J_{\pm}(\kappa)}(\Delta_i^\kappa f)^2 =: T$ and $Q_0(\lambda'')$ approach $\mathfrak{S}_2(f)^2 - \sum_{i \in J_{\pm}(\kappa)}(\Delta_i^\kappa f)^2 =: U$. Then $T > U \ge 0$. It follows from (4.27) that $T/(1+\delta)^2 > U/(1-\delta)^2$. Thus, as partitions become more refined, the sums $\sum_{j \notin J_{\pm}(\kappa,\lambda)} \theta(d_j) d_j^2$ will not converge. So, by (4.26), the products $\prod_\lambda (1 + \Delta^\lambda f)$ will not converge, a contradiction. So (1) and, hence, Theorem 4.4 is proved. □

Now we provide two examples of applications of Theorem 4.4. First, by a special case of Theorem 2 of Fristedt and Taylor (1973, see also p. 264), for any stable Lévy process $X = \{X(t) \colon t \in [0,T]\}$, $T < \infty$, of index $\alpha \in (0,2)$ with right-continuous sample paths, which has no drift if $\alpha < 1$ and has a Lévy measure symmetric about 0 if $\alpha = 1$, $\lim_{|\kappa| \to 0} \left(\sum_\kappa |\Delta^\kappa X|^p\right)^{1/p} = \mathfrak{S}_p(X)$ almost surely and the right side is finite if and only if $p > \alpha$. Therefore, by Lemmas 2.3 and 2.14 and Theorem 4.4, for almost all sample paths of X, there exist product integrals

$$\mathcal{P}_0(X)(t) := \prod_0^t (1 + dX) = \exp\{X(t) - X(0)\} \prod_{[0,t]} (1 + \Delta X)e^{-\Delta X} \quad \text{for } t \in [0,T].$$

By Corollary 5.23 below, almost all sample paths of $\{\mathcal{P}_0(X)(t) \colon t \in [0,T]\}$ satisfy a certain linear integral equation with respect to X.

Here is the second example. Let $B_H = \{B_H(t) \colon t \in [0,T]\}$ be a fractional Brownian motion with index H, $0 < H < 1$, i.e. a mean zero Gaussian process with the covariance function

$$Cov(B_H(t), B_H(s)) = \frac{1}{2}\{|t|^{2H} + |s|^{2H} - |t - s|^{2H}\}$$

for $s, t \in [0,T]$. The result of Fernique (1964) implies that B_H has almost all sample paths continuous. Therefore Theorem 3 of Kawada and Kôno (1973) implies that $\lim_{\delta \downarrow 0} V_p^{(\delta)}(B_H; [0,T]) < \infty$ (cf. (2.3)) and, hence, $V_p^*(B_H; [0,T]) < \infty$ almost surely for any $p > 1/H$. Then by Lemma 2.3, almost all sample paths of B_H have finite p-variation for $p > 1/H$. Hence Lemma 2.14 and Theorem 4.4 again imply that, for almost all sample paths of B_H, the product integrals

$$\mathcal{P}_0(B_H)(t) := \prod_0^t (1 + dB_H) = \exp\{B_H(t) - B_H(0)\} \quad \text{for } t \in [0, T]$$

exist whenever $H \in (1/2, 1)$.

If $H = 1/2$ and $B_H(0) = 0$ almost surely in the last example then $B := B_{1/2}$ is a (standard) Brownian motion. By Taylor (1972), we have that

$$\eta := \sup \Big\{ \sum_{i=1}^n \psi\big(|B(t_i) - B(t_{i-1})|\big) : 0 = t_0 < t_1 < \cdots < t_n = T \Big\}$$

is finite with probability 1 if $\psi(u) := u^2/(2LL(u))$ for $L(u) := 1 \vee |\log u|$ but η will be infinite with probability 1 if ψ is replaced by any function ϕ such that $\phi(u)/\psi(u) \to +\infty$ as $u \to 0$. Hence, since the 2-variation of B is infinite with probability 1, Theorem 4.4 is not applicable to sample paths of a Brownian motion. The three results in the rest of this subsection apply to functions which have, or in Proposition 4.10 may have, infinite 2-variation.

Proposition 4.9. *If $f \in \mathcal{R}([a,b]; \mathbb{R})$ then for any sequence $\{\kappa(m): m \geq 1\}$ of partitions of $[a,b]$ such that $\sum_{\kappa(m)}(\Delta^{\kappa(m)}f)^2 \to \infty$ as $m \to \infty$ we have*

$$\lim_{m \to \infty} \prod_{\kappa(m)} (1 + \Delta^{\kappa(m)}f) = 0.$$

Proof. Take $M < \infty$ such that $|f(x)| \leq M$ for $a \leq x \leq b$. There is an $n < \infty$ such that for any partition κ of $[a,b]$, $|\Delta_i^\kappa f| > 1/2$ for at most n values of i.

It can be checked by derivatives that $\log(1 + u) \leq u - u^2/5$ for $|u| \leq 1/2$. For a given partition $\kappa = \{x_i : i = 0, \ldots, k\}$ of $[a,b]$ let $I := \{i = 1, \ldots, k : |\Delta_i^\kappa f| > 1/2\}$ and $J := \{1, \ldots, k\} \setminus I$. Then

$$\Big| \prod_\kappa (1 + \Delta^\kappa f) \Big| = \Big| \prod_{i \in I} (1 + \Delta_i^\kappa f) \Big| \cdot \Big| \prod_{i \in J} (1 + \Delta_i^\kappa f) \Big|$$

$$\leq (1 + 2M)^n \exp\Big\{ \sum_{i \in J} \Delta_i^\kappa f - \sum_{i \in J} (\Delta_i^\kappa f)^2/5 \Big\}$$

$$\leq (1 + 2M)^n \exp\Big\{ 2M - \sum_{i \in I} \Delta_i^\kappa f + \sum_{i \in I} (\Delta_i^\kappa f)^2/5 - \sum_\kappa (\Delta^\kappa f)^2/5 \Big\}$$

$$\leq (1 + 2M)^n \exp\Big\{ 2M(n+1) + 4nM^2/5 - \sum_\kappa (\Delta^\kappa f)^2/5 \Big\}.$$

The conclusion follows.   □

If one does not require the uniform convergence over all partitions as in (1.2) then such a limit may not be degenerate for real-valued functions with infinite 2-variation. More precisely, the following statement holds:

Proposition 4.10. *Suppose* $f: [0, t] \mapsto \mathbb{R}$ *and* $\{\kappa(m): m \geq 1\}$ *is a sequence of partitions of* $[0, t]$ *such that*

$$\lim_m \max_i |\Delta_i^{\kappa(m)} f| = 0 \tag{4.28}$$

and

$$\lim_m \sum_i (\Delta_i^{\kappa(m)} f)^2 = C. \tag{4.29}$$

Then

$$\lim_m \prod_{\kappa(m)} (1 + \Delta^{\kappa(m)} f) = \exp\{f(t) - f(0) - \frac{C}{2}\}. \tag{4.30}$$

Proof. By (4.28), there exists an integer m_0 such that $\max_i |\Delta_i^{\kappa(m)} f| < 1/2$ for all $m \geq m_0$. By Taylor's theorem with remainder

$$\log(1 + u) = u - u^2/2 + 3\theta u^3 \qquad \text{for } |u| \leq 1/2,$$

where $|\theta| = |\theta(u)| \leq 1$. Thus

$$\left| \log \left(\prod_{\kappa(m)} (1 + \Delta^{\kappa(m)} f) \right) - \left[f(t) - f(0) - \frac{C}{2} \right] \right|$$

$$\leq \frac{1}{2} \left| \sum_i (\Delta_i^{\kappa(m)} f)^2 - C \right| + 3 \max_i |\Delta_i^{\kappa(m)} f| \sum_i (\Delta_i^{\kappa(m)} f)^2$$

for all $m \geq m_0$. (4.30) now follows from (4.28) and (4.29).   □

As noted above almost all sample paths of a Brownian motion provide examples of functions with infinite 2-variation but still satisfying the conditions of Proposition 4.10. More precisely, using Theorem 4.5 of Dudley (1973) and the Fernique-Landau-Marcus-Shepp inequality for Gaussian processes (cf. Marcus and Shepp, 1971, or Fernique, 1971, Théorème 8), by Proposition 4.10, we get

Corollary 4.11. *Let* $B = \{B(t): t \geq 0\}$ *be a Brownian motion. If* $\{\kappa(m): m \geq 1\}$ *is a sequence of partitions of* $[0, t]$ *such that* $|\kappa(m)| = o(1/\log m)$, *then, with probability 1,*

$$\lim_m \prod_{\kappa(m)} (1 + \Delta^{\kappa(m)} B) = \exp\{B(t) - t/2\}.$$

4.2. Extensions to some Banach algebras. Now, some examples will be given to show that for many Banach algebras $\mathbb{B}$, $f \in \mathcal{P}([0, 1]; \mathbb{B})$ does not imply that f or $\|f\|$ is measurable or that $\|f\|$ is bounded or even integrable.

Definition 4.12. A subset S of an algebra $\mathcal{A}$ over $\mathbb{R}$ will be called *nil-quadratic* if $AB = 0$ for all $A, B \in S$.

Example. If $A^2 = 0$, then the set of all multiples cA, $c \in \mathbb{R}$, is nil-quadratic.

Lemma 4.13. *Let A be an algebra with an identity $\mathbb{I}$ and S a nil-quadratic subset of A. Then for any $A_1, ..., A_n \in S$,*
(a) $(\mathbb{I} + A_n - A_{n-1}) \cdots (\mathbb{I} + A_2 - A_1)(\mathbb{I} + A_1) = \mathbb{I} + A_n;$
(b) $(\mathbb{I} - A_n)(\mathbb{I} + A_n - A_{n-1}) \cdots (\mathbb{I} + A_2 - A_1)(\mathbb{I} + A_1) = \mathbb{I}.$

Proof. For (a), we use induction on n. It holds for $n = 1$. Suppose it holds for n. Then $(\mathbb{I} + A_{n+1} - A_n)(\mathbb{I} + A_n) = \mathbb{I} + A_{n+1}$, so (a) holds for all $n = 1, 2, \cdots$. Then (b) follows from (a).   □

Proposition 4.14. *Let A be an algebra with identity $\mathbb{I}$ and S a nil-quadratic subset of A. Let $-\infty < a < b < \infty$. Let f be a function from $[a, b]$ into A such that $f(a) = f(b) = 0$ and $f(x) \in S$ for $a < x < b$. Then for any partition κ of $[a, b]$, $\Pi_\kappa(\mathbb{I} + \Delta^\kappa f) = \mathbb{I}$. Thus for any topology τ on A, including the discrete topology τ, if the product integral $\prod_a^b(\mathbb{I} + df)$ is defined as the limit with respect to τ, if it exists, of products $\Pi_\kappa(\mathbb{I} + \Delta^\kappa f)$ for partitions κ of $[a, b]$, either as $\mathrm{mesh}(\kappa) \to 0$ or as partitions are refined, then $\prod_a^b(\mathbb{I} + df)$ exists and equals $\mathbb{I}$.*

Proof. The first conclusion follows from (b) of Lemma 4.13. The rest is then clear. □

It is easily seen that an algebra A over $\mathbb{R}$ includes an infinite nil-quadratic set S if and only if it contains an $A \neq 0$ with $A^2 = 0$ (let S be the set of all multiples cA, $c \in \mathbb{R}$).

Proposition 4.15. *Let $\mathbb{B}$ be any Banach algebra with an identity $\mathbb{I}$ such that for some $0 \neq A \in \mathbb{B}$, $A^2 = 0$. One such $\mathbb{B}$ is the algebra M_2 of all 2×2 real matrices. Then for an arbitrary function h from $(0, 1)$ into $[0, \infty)$, there is a function f from $[0, 1]$ into $\mathbb{B}$ such that the product integral $\prod_a^b(\mathbb{I} + df)$ exists, even for the discrete topology on $\mathbb{B}$ and as the mesh of partitions $\to 0$, with $\|f\| = h$ on $(0, 1)$. Thus, f and $\|f\|$ need not be bounded or Lebesgue measurable, or if $\|f\|$ is measurable, one can have $\int_0^1 g(\|f(x)\|)dx = +\infty$ for any unbounded function g on $[0, \infty)$, such as $g(y) = LL...Ly$ where $Lu := \max(1, \log(u))$, $L0 := 1$.*

Proof. Let $A := \left(\begin{smallmatrix} 0 & 1 \\ 0 & 0 \end{smallmatrix}\right)$. Then $A^2 = 0$, so M_2 has the stated property.

Let $\mathbb{B}$ be a Banach algebra with identity containing some $A \neq 0$ with $A^2 = 0$. Then let S be an infinite nil-quadratic subset of $\mathbb{B}$. If f is any function from $[0, 1]$ into $\mathbb{B}$ with $f(0) = f(1) = 0$ and $f(x) \in S$ for $0 < x < 1$, the product integral $\prod_0^1(\mathbb{I} + df) = \mathbb{I}$ in the strong sense described, by the previous Proposition 4.14.

We can assume that for each $A \in S$ and $c \in \mathbb{R}$, $cA \in S$. So we can make $\|f\| = h$ on $(0, 1)$. Let A, B be any two elements of S, where we can assume $\|A\| \neq \|B\|$. Let J be a non-Lebesgue measurable subset of $(0, 1)$. Let $f(x) = A$ for $x \in J$ and $f(x) = B$ for $x \in (0, 1) \setminus J$. Here f and $\|f\|$ are non-measurable.

Or, let g be an unbounded real function on $[0, \infty)$. Choose $y_n \in [0, \infty)$, $n \geq 1$, such that $|g(y_n)| > 4^n$ for all n. By symmetry, we can assume that $g(y_n) > 0$ for all n. There are $C_n \in S$ such that $\|C_n\| = y_n$ for all n. Let $f(x) = C_n$ for $2^{-n} \leq x < 2^{1-n}$, $n = 1, 2, \cdots$. Then $\int_0^1 g(\|f(x)\|)dx = +\infty$.   □

Next we show that $f \in \mathcal{P}([a, b]; \mathbb{B})$ implies boundedness of f for certain commutative Banach algebras $\mathbb{B}$. By results of I. M. Gelfand, for each commutative

Banach algebra $\mathbb{B}$ with unit, real or complex, there is a continuous homomorphism of $\mathbb{B}$ into the Banach algebra $C(K)$ of complex-valued continuous functions on K for some compact K. In particular (cf. Proposition B.6.12 in Doran and Belfi, 1986), if $\mathbb{B}$ is a commutative complex Banach algebra with unit then this homomorphism is a homeomorphic isomorphism onto the image subalgebra of $C(K)$, then called the Gelfand representation, if and only if there exists a positive constant C such that

$$\|u^2\| \geq C\|u\|^2 \qquad \forall u \in \mathbb{B}. \tag{4.31}$$

In the light of what has been proved in this subsection this condition seems to be natural for the next statement.

Proposition 4.16. *Let $\mathbb{B} = (\mathbb{B}, \|\cdot\|)$ be a commutative (real) Banach algebra with an identity $\mathbb{1}$ such that (4.31) holds for some positive constant C. If $f \in \mathcal{P}([a,b];\mathbb{B})$ then f is bounded.*

Proof. Suppose not. Let $P := \prod_a^b (1 + df)$ and $\Pi_\lambda := \Pi_\lambda (\mathbb{1} + \Delta^\lambda f)$ for any partition λ of $[a,b]$. Since the set of invertible elements in $\mathbb{B}$ is open one can choose a partition κ_1 of $[a,b]$ such that, for all refinements λ of κ_1, Π_λ is invertible and

$$\|\Pi_\lambda - P\| \leq 1/[2\|P^{-1}\|]. \tag{4.32}$$

Then by Lemma 2.5 in Bonsall and Duncan (1973), we have

$$\|\Pi_\lambda^{-1} - P^{-1}\| \leq 2\|P^{-1}\|^2 \|\Pi_\lambda - P\| \qquad \forall \lambda \supset \kappa_1. \tag{4.33}$$

Let $\epsilon > 0$. There exists a partition κ_2 of $[a,b]$ such that, for all refinements λ of κ_2,

$$\|\Pi_\lambda - P\| \leq \epsilon/[2\|P^{-1}\|(1 + \|P\|\,\|P^{-1}\|)].$$

Hence, by (4.32) and (4.33), for any two refinements λ_1 and λ_2 of $\kappa_1 \cup \kappa_2$, we get

$$\|\Pi_{\lambda_1}\Pi_{\lambda_2}^{-1} - \mathbb{1}\| = \|\Pi_{\lambda_1}\Pi_{\lambda_2}^{-1} - PP^{-1}\| \tag{4.34}$$
$$\leq \|\Pi_{\lambda_1}\|\,\|\Pi_{\lambda_2}^{-1} - P^{-1}\| + \|\Pi_{\lambda_1} - P\|\,\|P^{-1}\| \leq \epsilon.$$

Let $\kappa := \kappa_1 \cup \kappa_2 =: \{x_i : i = 0, \ldots, n\}$. For some i_0, f is unbounded on the open interval (x_{i_0-1}, x_{i_0}). As in the proof of Lemma 4.1, fix such an i_0 and consider partitions $\kappa(x) := \kappa \cup \{x\}$ for $x \in (x_{i_0-1}, x_{i_0})$. Let $c := f(x_{i_0-1})$ and $d := f(x_{i_0})$. Since each factor in an invertible product is also invertible, by (4.34), we get

$$\epsilon \geq \|\Pi_{\kappa(x)}\Pi_\kappa^{-1} - \mathbb{1}\| = \|(\mathbb{1} + f(x) - c)(\mathbb{1} + d - f(x))(\mathbb{1} + d - c)^{-1} - \mathbb{1}\|$$
$$= \|(f(x) - c)(d - f(x))(\mathbb{1} + d - c)^{-1}\|. \tag{4.35}$$

To show that the right side of (4.35) cannot be small when $\|f(x)\| \to \infty$ we use the Gelfand representation. Let Φ_B be the set of all non-zero homomorphisms of $\mathbb{B}$ into $\mathbb{C}$. Endowed with an analogue of the weak-star topology on $\mathbb{B}^*$, Φ_B is called the carrier space for $\mathbb{B}$ (cf. Definition 18.11 in Bonsall and Duncan, 1973). Let $C(\Phi_B)$ be the Banach algebra of complex-valued continuous functions on Φ_B with the sup-norm $\|\cdot\|_\infty$. Then the Gelfand representation of $\mathbb{B}$ is the mapping $u \mapsto \hat{u}$ of $\mathbb{B}$ into

$C(\Phi_B)$ defined by $\widehat{u}(\phi) := \phi(u)$ for each $u \in \mathbb{B}$ and $\phi \in \Phi_B$. By Theorem 18.13 in Bonsall and Duncan (1973), the carrier space Φ_B is a compact Hausdorff space, the mapping $u \mapsto \widehat{u}$ is a homomorphism of $\mathbb{B}$ into $C(\Phi_B)$ and, for all $u \in \mathbb{B}$,

$$\|\widehat{u}\|_\infty = \lim_{n\to\infty} \|u^n\|^{1/n}. \tag{4.36}$$

By (4.31), we get that $\|u^{2^n}\|^{1/2^n} \geq C^{1-2^{-n}}\|u\|$ for each $n \geq 1$. Then by (4.36), it follows that $\|\widehat{u}\|_\infty \geq C\|u\|$, for all $u \in \mathbb{B}$. The converse inequality $\|\widehat{u}\|_\infty \leq \|u\|$ for all $u \in \mathbb{B}$ follows just from (4.36) and the inequality $\|uv\| \leq \|u\|\,\|v\|$ for all $u, v \in \mathbb{B}$. Therefore the set $\{\|(f(x))^\wedge\|_\infty \colon x \in (x_{i_0-1}, x_{i_0})\}$ is unbounded. The function $\left((\mathbb{1} + d - c)^{-1}\right)^\wedge$ is bounded away from 0. Thus the right side of (4.35) is unbounded too. This contradiction proves Proposition 4.16. □

Let $\mathbb{B} = (\mathbb{B}, \|\cdot\|)$ be a commutative Banach algebra with an identity $\mathbb{1}$. We finish this subsection by extending the sufficiency part of Theorem 4.4 to $\mathbb{B}$-valued functions f. For this we will need the exponential and logarithmic functions defined, as usual, by the equations

$$\exp(u) := \mathbb{1} + \sum_{n=1}^{\infty} \frac{u^n}{n!}, \quad \text{for } u \in \mathbb{B},$$

and

$$\log(u) := -\sum_{n=1}^{\infty} \frac{(\mathbb{1} - u)^n}{n}, \quad \text{for } \|\mathbb{1} - u\| < 1,$$

respectively. They are continuous in their domains of definition and

$$\exp(\log(u)) = u, \quad \text{for } \|\mathbb{1} - u\| < 1.$$

Since $\mathbb{B}$ is commutative (in the rest of this subsection) we have also

$$\exp(u)\exp(v) = \exp(u + v), \quad \text{for all } u, v \in \mathbb{B}.$$

Theorem 4.17. *Let $\mathbb{B} = (\mathbb{B}, \|\cdot\|)$ be a commutative Banach algebra with identity. If $f \in \mathcal{W}_2^*([a,b], \mathbb{B})$ then the product integral with respect to f over $[a,b]$ exists and*

$$\prod_a^b (\mathbb{1} + df) = \lim_{\delta \downarrow 0} \prod_{x \in I_\pm(\delta)} [(\mathbb{1} + \Delta_- f)(\mathbb{1} + \Delta^+ f)](x)\exp\left(-\Delta_- f(x) - \Delta^+ f(x)\right)$$

$$= \exp[f(b) - f(a)] \prod_{[a,b]} (\mathbb{1} + \Delta f)\exp(-\Delta f), \tag{4.37}$$

where $I_\pm(\delta) := \{x \in [a,b) \colon |\Delta^+ f(x)| > \delta\} \cup \{x \in (a,b] \colon |\Delta_- f(x)| > \delta\}$, $\Delta_- f(a) := 0$ and $\Delta^+ f(b) := 0$.

Proof. To show that the product on the right side of (4.37) exists first note that, for each finite set $\nu \subset \{x \colon \max\left(\|\Delta^+ f(x)\|, \|\Delta_- f(x)\|\right) \leq 1/2\}$,

$$\prod_\nu [(\mathbb{1} + \Delta_- f)(\mathbb{1} + \Delta^+ f)\exp(-\Delta_- f - \Delta^+ f)] =$$

$$\exp\left(\sum_\nu [\log(\mathbb{1} + \Delta_- f) + \log(\mathbb{1} + \Delta^+ f) - \Delta_- f - \Delta^+ f]\right).$$

Second, for all $\|u\| \le 1/2$,

$$\|\log(\mathbb{1} + u) - u\| \le 2\|u\|^2. \tag{4.38}$$

Therefore, the product $\prod_{[a,b]}$ exists in $\mathbb{B}$ because the set of discontinuities of f is countable and $\mathfrak{S}_2(f) < \infty$.

Denote the whole expression on the right side of (4.37) by A. To prove that the limit on the left side of (4.37) exists and equals A, let $0 < \epsilon < 3e^{\|f(b)-f(a)\|}$ and let

$$c(\epsilon) := \frac{\epsilon e^{-\|f(b)-f(a)\|-2v_2(f)}}{18[1 + (1 + Osc(f))^{2M} e^{2MOsc(f)+2\mathfrak{S}_2(f)^2}]}, \tag{4.39}$$

where

$$M := \text{card}\{x \in [a,b] : \|\Delta_- f(x)\| > 1/2 \text{ or } \|\Delta^+ f(x)\| > 1/2\} < \infty.$$

First, by Lemma 2.22 there exists a partition λ of $[a,b]$ such that, for all refinements $\kappa \supset \lambda$,

$$\sum_{\kappa} \|\Delta^\kappa f\|^2 \le V_2^*(f)^2 + c(\epsilon). \tag{4.40}$$

Second, there exists a finite set $\mu \subset [a,b]$ such that

$$\max\{\|\Delta_- f(x)\|, \|\Delta^+ f(x)\|\} \le 1/4, \qquad \forall x \notin \mu, \tag{4.41}$$

$$B_\mu := \sum_{x \in \mu} [\|\Delta_- f(x)\|^2 + \|\Delta^+ f(x)\|^2] \ge \mathfrak{S}_2(f)^2 - c(\epsilon), \tag{4.42}$$

and

$$\|A - \exp(f(b) - f(a))A_\mu\| < \epsilon/3, \tag{4.43}$$

where

$$A_\mu := \prod_{x \in \mu} [(\mathbb{1} + \Delta_- f)(\mathbb{1} + \Delta^+ f)](x)\exp(-\Delta_- f(x) - \Delta^+ f(x)).$$

Third, by Lemma 2.15 and (4.41), one can find a $\delta_1 > 0$ such that

$$\|f(z) - f(y)\| \le 1/2 \tag{4.44}$$

whenever $[y,z] \subset [a,b] \setminus \mu$ and $z - y < \delta_1$. Let $\mu := \{\xi_1, \ldots, \xi_N\}$. Finally, since f is regulated there exists a $\delta_2 > 0$ such that, letting $d_j^- := f(\xi_j) - f(y_j)$ for $y_j \in [\xi_j - \delta_2, \xi_j)$ and $d_j^+ := f(z_j) - f(\xi_j)$ for $z_j \in (\xi_j, \xi_j + \delta_2]$, for each $j = 1, \ldots, N$, $d_j^- := 0$ if $\xi_j = a$, $d_j^+ := 0$ if $\xi_j = b$, we have for any such y_j and ξ_j

$$\left| B_\mu - \sum_{j=1}^N [\|d_j^-\|^2 + \|d_j^+\|^2] \right| < c(\epsilon) \tag{4.45}$$

and

$$\left\| A_\mu - \prod_{j=1}^N [(1 + d_j^-)(1 + d_j^+)e^{-d_j^- - d_j^+}] \right\| < \epsilon/[3e^{\|f(b)-f(a)\|}]. \tag{4.46}$$

Choose a partition $\kappa(\epsilon) \supset \lambda \cup \mu$ such that $|\kappa(\epsilon)| < \delta_1 \wedge \delta_2 \wedge \delta_3$, where

$$0 < \delta_3 := \min\{|\xi_i - \xi_j| : i, j = 1, \ldots, N, \quad i \neq j\}.$$

We will prove that, for each partition $\kappa \supset \kappa_\epsilon$,

$$Z(\kappa) := \left\| A - \prod_\kappa (\mathbb{1} + \Delta^\kappa f) \right\| < \epsilon. \tag{4.47}$$

To this end let $\kappa := \{x_i : i = 0, \ldots, n\}$ and let $I_\mu(\kappa) := \{i = 0, \ldots, n : x_i \in \mu$ or $x_{i-1} \in \mu\}$. Then we have

$$\sum_{i \notin I_\mu(\kappa)} \|\Delta_i^\kappa f\|^2 = \sum_\kappa \|\Delta^\kappa f\|^2 - \sum_{i \in I_\mu(\kappa)} \|\Delta_i^\kappa f\|^2$$

by (4.40) and (4.45) $\leq V_2^*(f)^2 + c(\epsilon) - B_\mu + c(\epsilon)$

by (4.42) and $f \in \mathcal{W}_2^*$ $\leq 3c(\epsilon).$

Thus, using the inequalities $\|\mathbb{1} - \exp(u)\| \leq \|u\|e^{\|u\|}$, (4.38) and (4.44) we get

$$D_\mu := \left\| \exp(f(b) - f(a)) - \prod_{i \notin I_\mu(\kappa)} (\mathbb{1} + \Delta_i^\kappa f) \prod_{i \in I_\mu(\kappa)} \exp(\Delta_i^\kappa f) \right\|$$

$$\leq e^{\|f(b) - f(a)\|} \left\| \mathbb{1} - \exp\left(\sum_{i \notin I_\mu(\kappa)} [\log(\mathbb{1} + \Delta_i^\kappa f) - \Delta_i^\kappa f]\right) \right\|$$

$$\leq 2e^{\|f(b) - f(a)\| + 2v_2(f)} \sum_{i \notin I_\mu(\kappa)} \|\Delta_i^\kappa f\|^2$$

$$\leq 6c(\epsilon)e^{\|f(b) - f(a)\| + 2v_2(f)}. \tag{4.48}$$

By (4.46) and (4.39), it follows that

$$\left\| \prod_{i \in I_\mu(\kappa)} (\mathbb{1} + \Delta_i^\kappa f)\exp(-\Delta_i^\kappa f) \right\| \leq \epsilon / \left[3e^{\|f(b) - f(a)\|} \right] + \|A_\mu\|$$

$$\leq 1 + (1 + Osc(f))^{2M} e^{2M Osc(f) + 2\Theta_2(f)^2} \leq \epsilon / \left[18c(\epsilon)e^{\|f(b) - f(a)\| + 2v_2(f)} \right].$$

Hence using (4.43), (4.46) again and (4.48) we arrive at

$$Z(\kappa) \leq \epsilon/3 + \Big\| \exp(f(b) - f(a)) A_\mu$$

$$- \Big(\prod_{i \notin I_\mu(\kappa)} (\mathbb{1} + \Delta_i^\kappa f) \prod_{i \in I_\mu(\kappa)} \exp(\Delta_i^\kappa f) \Big) \Big(\prod_{i \in I_\mu(\kappa)} (\mathbb{1} + \Delta_i^\kappa f)\exp(-\Delta_i^\kappa f) \Big) \Big\|$$

$$\leq \epsilon/3 + e^{\|f(b) - f(a)\|} \Big\| A_\mu - \prod_{i \in I_\mu(\kappa)} (\mathbb{1} + \Delta_i^\kappa f)\exp(-\Delta_i^\kappa f) \Big\|$$

$$+ \Big\| \prod_{i \in I_\mu(\kappa)} (\mathbb{1} + \Delta_i^\kappa f)\exp(-\Delta_i^\kappa f) \Big\| D_\mu < \epsilon/3 + \epsilon/3 + \epsilon/3 = \epsilon.$$

This implies (4.47). Since $\epsilon > 0$ is arbitrary the proof of Theorem 4.17 is complete.
$\square$

4.3. Preliminary bounds. Let f be a $\mathbb{B}$-valued function defined on an interval τ and let $\kappa = \{x_0, x_1, \ldots, x_n\}$ be a partition of τ. Since $\mathbb{B}$ may not be commutative we will use the product sign to denote

$$\prod_{\kappa} (\mathbb{I} + \Delta^{\kappa} f) := \prod_{i=1}^{n} (\mathbb{I} + \Delta_i^{\kappa} f)$$

$$:= (\mathbb{I} + f(x_n) - f(x_{n-1})) \cdots (\mathbb{I} + f(x_1) - f(x_0)),$$

with the prescribed order. Some authors use the reverse order. Define also the *product for f based on κ* by $P(\kappa, \Delta, f) := \prod_{i=i_1+1}^{i_2} (\mathbb{I} + \Delta_i^{\kappa} f)$, where $\Delta = [x_{i_1}, x_{i_2}]$ for some $0 \le i_1 < i_2 \le n$.

To begin with, we state two algebraic identities to be used in the rest of the paper. Let $a_n, \ldots, a_1$ and $b_n, \ldots, b_1$ be arbitrary elements of $\mathbb{B}$. Put $b_0 := a_{n+1} := \mathbb{I}$. Then

$$\prod_{j=1}^{n} a_j - \prod_{j=1}^{n} b_j = \sum_{i=1}^{n+1} \left(\prod_{j=i+1}^{n} a_j \right) (a_i - b_i) \left(\prod_{j=0}^{i-1} b_j \right) \tag{4.49}$$

and

$$\prod_{j=1}^{n} (\mathbb{I} + a_j) = \mathbb{I} + \sum_{k=1}^{n} \sum_{1 \le j_k < \cdots < j_1 \le n} a_{j_1} \cdots a_{j_k}. \tag{4.50}$$

Equation (4.49) is a telescoping sum. Equation (4.50) holds by induction. We generalize these equations to product integrals in Subsections 5.2 and 4.5, respectively.

The following bounds will be useful in the rest of the paper. Their proofs follow the pattern of the proof of Lemma 5.1 of Freedman (1983a). We give them here for easier reference and to correct some small errors.

Lemma 4.18. *Let $f \in W_p([a, b]; \mathbb{B})$, $0 < p < 2$, and let*

$$\rho := 2c_p(1 + \zeta(2/p))V_p(f; [a, b]) < 1,$$

where $c_p = 1 \vee 2^{(1/p)-1}$. Then, for each partition $\kappa \in Q([c, d])$ of any subinterval $[c, d]$ of $[a, b]$, where c, d may be points $x+$, $x-$, we have

$$\|P(\kappa, [c, d], f)\| \le 1/(1 - \rho), \tag{4.51}$$

$$\|P(\kappa, [c, d], f) - \mathbb{I}\| \le V_p(f; [c, d])/(1 - \rho), \tag{4.52}$$

$$\|P(\kappa, [c, d], f) - \mathbb{I} - [f(d) - f(c)]\| \le 2c_p(1 + \zeta(2/p))V_p(f; [c, d])^2/(1 - \rho)$$

$$\le \rho V_p(f; [c, d])/(1 - \rho). \tag{4.53}$$

The same inequalities hold also when $\kappa \in Q^{\pm}([c, d])$.

Proof. The last statement will follow from the others using regularity of f and Lemmas 2.5, 2.15 and 2.16. So it will be enough to prove the inequalities without points $x\pm$. Let $\kappa = \{x_0, \ldots, x_n\}$ be a partition of $[c, d]$. Define $H_0(l) := \mathbb{I}$ for each $l = 0, 1, \ldots, n$ and

$$H_k(l) := \sum_{j=k}^{l} (\Delta_j^{\kappa} f) H_{k-1}(j - 1)$$

for each $k = 1, \ldots, n$ and all $l = k, \ldots, n$. Then, by (4.50),

$$P(\kappa, [c, d], f) = \mathbb{1} + \sum_{k=1}^{n} H_k(n). \tag{4.54}$$

Define also, for $k = 0, 1, \ldots, n$, $\alpha_k := \max_{k \leq l \leq n} \|H_k(l)\|$ and

$$\beta_k := \max \left\{ \left(\sum_{r=1}^{m} \|H_k(l(r)) - H_k(l(r-1))\|^p \right)^{1/p} : k = l(0) < \cdots < l(m) = n \right\}.$$

We note that $\alpha_0 = 1$, $\beta_0 = 0$,

$$\alpha_1 = \max_{1 \leq l \leq n} \|f(x_l) - f(c)\| \leq V_p(f; [c, d]) \tag{4.55}$$

and

$$\beta_1 = \max \left\{ \left(\sum_{r=1}^{m} \|f(x_{l(r)}) - f(x_{l(r-1)})\|^p \right)^{1/p} : 1 = l(0) < \cdots < l(m) = n \right\}$$
$$\leq V_p(f; [c, d]). \tag{4.56}$$

Next, for each $k = 2, \ldots, n$ and $l = k, \ldots, n$, we have

$$H_k(l) = - \left\{ \sum_{k \leq j \leq i \leq l} (\Delta_j^\kappa f)[H_{k-1}(i) - H_{k-1}(i-1)] \right\} + [f(x_l) - f(x_{k-1})]H_{k-1}(l).$$

Let

$$h(x) := \begin{cases} H_{k-1}(i), & \text{for } x_{i-1} < x \leq x_i, \ i = k, \ldots, l-1 \\ H_{k-1}(l), & \text{for } x > x_{l-1} \\ H_{k-1}(k-1), & \text{for } x \leq x_{k-1}. \end{cases}$$

Using Lemma 3.30, with $g \equiv \mathbb{1}$, noting that $\Delta_i^\kappa h = H_{k-1}(i) - H_{k-1}(i-1)$ for $i = k, \ldots, l$ and is 0 for the other indices i, and (m, n) in Lemma 3.30 $= (k-1, l)$, we get

$$\alpha_k \leq (1 + \zeta(2/p))V_p(f; [c, d])\beta_{k-1} + V_p(f; [c, d])\alpha_{k-1} \tag{4.57}$$

for all $k = 2, \ldots, n$. Moreover, for each $k = 2, \ldots, n$ and $k \leq l(r-1) < l(r) \leq n$,

$$H_k(l(r)) - H_k(l(r-1)) = - \left\{ \sum_{l(r-1) < j \leq i \leq l(r)} (\Delta_j^\kappa f)[H_{k-1}(i) - H_{k-1}(i-1)] \right\}$$
$$+ [f(x_{l(r)}) - f(x_{l(r-1)})]H_{k-1}(l(r)).$$

Applying first inequality (2.21) and then Lemma 3.30 as above gives

$$\beta_k \leq c_p(1 + \zeta(2/p))V_p(f; [c, d])\beta_{k-1} + c_p V_p(f; [c, d])\alpha_{k-1} \tag{4.58}$$

for all $k = 2, \ldots, n$. Let

$$Q := c_p(1 + \zeta(2/p))V_p(f; [c, d]) \begin{pmatrix} 1 & 1 \\ 1 & 1 \end{pmatrix}.$$

By (4.55) - (4.58), it follows that, for all $k = 1, \ldots, n$,

$$\begin{pmatrix} \alpha_k \\ \beta_k \end{pmatrix} \le Q^{k-1} \begin{pmatrix} \alpha_1 \\ \beta_1 \end{pmatrix} \le V_p(f; [c,d]) Q^{k-1} \begin{pmatrix} 1 \\ 1 \end{pmatrix}. \tag{4.59}$$

Since for $k \ge 2$

$$\begin{pmatrix} 1 & 1 \\ 1 & 1 \end{pmatrix}^{k-1} = 2^{k-2} \begin{pmatrix} 1 & 1 \\ 1 & 1 \end{pmatrix},$$

(4.59) yields

$$\alpha_k \le \left[2c_p(1 + \zeta(2/p)) \right]^{k-1} V_p(f; [c,d])^k$$

for all $k = 1, \ldots, n$. Hence, by (4.54),

$$\left\| P(\kappa, [c,d], f) \right\| \le 1 + \sum_{k=1}^{n} \alpha_k \le 1/(1 - \rho),$$

$$\left\| P(\kappa, [c,d], f) - \mathbb{1} \right\| \le V_p(f; [c,d]) \sum_{k=1}^{n} \rho^{k-1} \le V_p(f; [c,d])/(1 - \rho)$$

and, similarly,

$$\left\| P(\kappa, [c,d], f) - \mathbb{1} - [f(d) - f(c)] \right\| \le 2c_p(1 + \zeta(2/p)) V_p(f; [c,d])^2/(1 - \rho).$$

The proof of Lemma 4.18 is complete. □

Next using our Lemma 2.20 above, we extend Lemma 5.1 of Freedman (1983a) to not necessarily continuous functions.

Lemma 4.19. *Let* $f \in \mathcal{W}_p([a,b]; \mathbb{B})$, $0 < p < 2$. *Then, for each partition* $\kappa \in Q^\pm(\Delta)$ *of a subinterval* Δ *of* $[a,b]$, *we have*

$$\left\| P(\kappa, \Delta, f) \right\| \le \left[2(1 + V_p(f; [a,b]))^2 \right]^N,$$

where $N \le 4 + 3v_p(f; [a,b])/c^p$ *and* $c = \left[4(1 \vee 2^{(1/p)-1})(1 + \zeta(2/p)) \right]^{-1}$.

Proof. First let $\kappa = \{x_0, \ldots, x_n\}$ be a partition of $[a,b]$ and let $\kappa^* = \{y_0, \ldots, y_N\}$ be another partition of $[a,b]$ given by Lemma 2.20 with the number c^p. First, for all $i = 1, \ldots, n$, we have $\|\Delta_i^\kappa f\| \le V_p(f)$. Second, there are at most $2N$ values of i where $[x_{i-1}, x_i]$ contains some y_j. For all other values of i, $[x_{i-1}, x_i]$ is included in the interior of some $[y_{j-1}, y_j]$, which is then not an interval $[x-, x]$ or $[x, x+]$. Let $I(j) := \{i = 1, \ldots, n \colon [x_{i-1}, x_i] \subset [y_{j-1}, y_j]\}$. Then, by (4.51) and (2.11),

$$\left\| \prod_{i \in I(j)} (\mathbb{1} + \Delta_i^\kappa f) \right\| \le 2.$$

There are at most N such $I(j)$. So

$$\left\| P(\kappa, [a,b], f) \right\| \le \left[2(1 + V_p(f; [a,b]))^2 \right]^N$$

as desired. On a subinterval Δ, V_p and v_p are no larger, so the same bound applies.
□

4.4. The product integral and p-variation. Here we define a product integral and show that it exists for an arbitrary $\mathbb{B}$-valued function with finite p-variation with $0 < p < 2$. To begin with we extend the definition (1.2) of the product integral replacing the interval $[a, b]$ by an arbitrary interval τ.

Definition 4.20. Let $\mathbb{B} = (\mathbb{B}, \| \cdot \|)$ be a Banach algebra and let f be a $\mathbb{B}$-valued function on an interval τ which may be open or closed at either end.

(1) *The product integral with respect to f over the interval τ,* $\prod_\tau (1 + df)$, *is defined as the $R \in \mathbb{B}$, if it exists, such that for every $\epsilon > 0$ there is a partition λ of τ such that, for each refinement κ of λ, $\|P(\kappa, \tau, f) - R\| \le \epsilon$. The product integral will be called degenerate if $R = 0$.*

(2) *If $c, d \in \tau$ then, setting $a- := a$ and $b+ := b$, product integrals will be defined over intervals $[c\pm, d]$, $[c, d\pm]$ or $[c\pm, d\pm]$ as in (1), except for the following: if the left endpoint is $c\pm$, then $f(c\pm)$ is defined and partitions $\kappa = \{x_0, \ldots, x_n\}$ with $x_0 = c\pm$ are used in defining products $P(\kappa, [c\pm, d], f)$ for f. Likewise if the endpoint is $d\pm$, then $f(d\pm)$ is defined and $x_n = d\pm$ for each partition $\kappa = \{x_0, \ldots, x_n\}$. (Points $x+$, $x-$ are not allowed as members of partitions except as endpoints.)*

The following relations between product integrals over different intervals hold as for Moore-Pollard-Stieltjes integrals (cf. Lemma 3.1). Recalling the notation $f^{(b-)}$ and $f^{(a+)}$ as defined in (2.17) and (2.18) we have:

Lemma 4.21. *Let f be a $\mathbb{B}$-valued function on $[a, b]$.*

(1) *If f has a left limit at b then*

$$\prod_{[a,b)} (1 + df) = \prod_a^{b-} (1 + df) = \prod_a^b (1 + df^{(b-)}),$$

where each product integral exists if and only if both the others exist.

(2) *If f has a right limit at a then*

$$\prod_{(a,b]} (1 + df) = \prod_{a+}^b (1 + df) = \prod_a^b (1 + df^{(a+)}),$$

where each product integral exists if and only if both the others exist.

Proof. We will prove part (1) only and note that a proof of (2) is based on symmetric arguments. First assume $R_1 := \prod_{[a,b)} (1 + df)$ exists. Then there exists a constant $K(f)$ and a partition κ_0 of $[a, b)$ such that, for each refinement κ' of κ_0,

$$\|P(\kappa', [a, b), f)\| \le K(f).$$

Given $\epsilon > 0$, take a partition $\kappa = \{x_0, x_1, \ldots, x_{n-1}\} \supset \kappa_0$ of $[a, b)$ such that $\|P(\kappa', [a, b), f) - R_1\| \le \epsilon/2$ for each refinement κ' of κ. We can assume x_{n-1} is close enough to b so that $Osc_{[x_{n-1}, b)}(f) < \epsilon/[2K(f)]$. Let ζ be the partition $\{x_0, \ldots, x_n\}$ of $[a, b-]$ with $x_n = b-$. For any refinement $\zeta' = \{y_0, \ldots, y_m\}$ of ζ, we have $y_m = b-$, while $\kappa' = \{y_0, \ldots, y_{m-1}\}$ is a partition of $[a, b)$ and a refinement of κ. Thus

$$\|P(\zeta', [a, b-], f) - R_1\| \le \|P(\kappa', [a, b), f) - R_1\|$$

$$+\|P(\kappa',[a,b),f)\|\,\|f(b-)-f(y_{m-1})\| \le \epsilon/2 + \epsilon/2 = \epsilon.$$

Therefore $R_2 := \prod_a^{b-}(\mathbb{1}+df)$ exists and equals R_1.

Next, suppose R_2 exists. Given $\epsilon > 0$, take a partition $\kappa = \{x_0,\ldots,x_n\}$ of $[a,b-]$ with $x_n = b-$ such that $\|P(\kappa',[a,b-],f) - R_2\| \le \epsilon$ for each refinement κ' of κ. Let $\kappa_b = \{y_0,\ldots,y_n\}$, where $y_i = x_i$ for $i = 0,1,\ldots,n-1$ and $y_n = b$. Then κ_b is a partition of $[a,b]$ and, for each refinement $\kappa_b' = \{z_0,\ldots,z_m\}$ of κ_b, the partition $\kappa' = \{z_0,\ldots,z_{m-1},b-\}$ is a refinement of κ and $P(\kappa_b',[a,b],f^{(b-)}) = P(\kappa',[a,b-],f)$, so $R_3 := \prod_a^b(\mathbb{1}+df^{(b-)})$ exists and equals R_2.

Now assume R_3 exists. Take a partition $\kappa_0 = \{x_0,\ldots,x_n\}$ of $[a,b]$ such that $\|P(\kappa',[a,b],f^{(b-)})\| \le C(f)/2$ for some finite constant $C(f)$ and all refinements κ' of κ_0, and such that $\|f(b-)-f(y)\| \le 1/2$ for all $y \in [x_{n-1},b)$. Then $\zeta_0 = \{x_0,\ldots,x_{n-1}\}$ is a partition of $[a,b)$ and, for any refinement $\zeta' = \{y_0,\ldots,y_{m-1}\}$ of ζ_0, the partition $\kappa' = \{y_0,\ldots,y_{m-1},b\}$ is a refinement of κ_0, $\mathbb{1}+f(b-)-f(y_{m-1})$ is invertible and $\|(\mathbb{1}+f(b-)-f(y_{m-1}))^{-1}\| \le 2$. Thus

$$\|P(\zeta',[a,b),f)\| \le \|P(\kappa',[a,b],f)\|\,\|(\mathbb{1}+f(b-)-f(y_{m-1}))^{-1}\| \le C(f)$$

for all refinements ζ' of ζ_0. Given $\epsilon > 0$, take a partition $\kappa = \{x_0,\ldots,x_n\}$ of $[a,b]$ such that $\kappa \supset \kappa_0$ and, for each $\kappa' \supset \kappa$, $\|P(\kappa',[a,b],f^{(b-)}) - R_3\| \le \epsilon/2$, where we can assume that $Osc_{[x_{n-1},b)}(f) \le \epsilon/[2C(f)]$. Then $\zeta = \{x_0,\ldots,x_{n-1}\}$ is a partition of $[a,b)$ such that, for each refinement $\zeta' = \{y_0,\ldots,y_{m-1}\}$ of ζ, the partition $\kappa' = \{y_0,\ldots,y_{m-1},b\}$ is a refinement of κ and

$$\|P(\zeta',[a,b),f) - R_3\| \le \epsilon/2 + \|f(b-)-f(y_{m-1})\|\,\|P(\zeta',[a,b),f)\|$$
$$\le \epsilon/2 + \epsilon/2 = \epsilon.$$

Therefore R_1 exists and equals R_3. Lemma 4.21 is proved. □

We defined the product integral $\prod_\tau(\mathbb{1}+df)$ to be the limit of the products $P(\kappa,\tau,f)$ if it exists, where the limit is taken under refinements of partitions κ. By definition, these partitions κ do not contain points $x+$ or $x-$ except when they may appear as endpoints. Given a function $f \in \mathcal{R}(\tau;\mathbb{B})$, we say that $\prod_\tau(\mathbb{1}+df)_\pm$ exists and equals $R \in \mathbb{B}$ if the partitions κ and λ in Definition 4.20 may contain points $x+$ or $x-$.

Lemma 4.22. *Let* $f \in \mathcal{R}(\tau;\mathbb{B})$ *where* $\tau = [a,b]$, $[a,b-]$, $[a+,b]$ *or* $[a+,b-]$. *If* $\prod_\tau(\mathbb{1}+df)$ *exists then so does* $\prod_\tau(\mathbb{1}+df)_\pm$ *and both are equal. The converse implication holds if, in addition, there is a finite constant* $C(f)$ *such that*

$$\|P(\kappa,\Delta,f)\| \le C(f) \tag{4.60}$$

for each partition $\kappa \in Q^\pm(\Delta)$ *and for any subinterval* Δ *of* τ.

Proof. We prove the statement when $\tau = [a,b]$ only and note that the proof for other intervals τ is similar. If $\prod_a^b(1+df)$ exists then the argument used to prove (4.3) implies that points $x+$ and $x-$ can be included as members of partitions in Definition 4.20(1).

Conversely, suppose that $\prod_a^b(\mathbb{1} + df)_\pm$ exists. Given $\epsilon > 0$, let $\kappa^* = \{x_i\colon i = 0, \dots, n\} \in Q^\pm([a,b])$ be such that, for all refinements $\lambda^* \in Q^\pm([a,b])$ of κ^*,

$$\left\| \prod_a^b(\mathbb{1} + df)_\pm - P(\lambda^*, [a,b], f) \right\| \le \epsilon/2. \qquad (4.61)$$

By adjoining additional points to κ^* if necessary, we can assume that whenever a point $x_i = x+$ for some x, then $x_{i-1} = x$ and x_{i+1} is not a point $y+$ or $y-$, while if $x_i = x-$, then $x_{i+1} = x$ and x_{i-1} is not a point $y+$ or $y-$. Let $J_- := \{i = 1, \dots, n\colon [x_{i-1}, x_i] = [x-, x]$ for some $x\}$ and let $J_+ := \{i = 0, \dots, n-1\colon [x_i, x_{i+1}] = [x, x+]$ for some $x\}$. We note that $1 \notin J_-$, $n-1 \notin J_+$ and $J_- \cap J_+$ may not be empty. Since f is regulated one can choose $y_i^l \in (x_{i-2}, x_i)$ for $i \in J_-$ and $y_i^r \in (x_i, x_{i+2})$ for $i \in J_+$ such that

$$\sum_{i \in J_-} \|f(w_i) - f(x_i-)\| + \sum_{i \in J_+} \|f(x_i+) - f(z_i)\| \le \epsilon/[4C(f)^2] \qquad (4.62)$$

holds for all $w_i \in [y_i^l, x_i)$, $i \in J_-$, and all $z_i \in (x_i, y_i^r]$, $i \in J_+$. Let κ be the partition of $[a,b]$ equal to κ^* except that, for each $i \in J_-$, $x_{i-1} (= x_i-)$ is replaced by y_i^l and, for each $i \in J_+$, $x_{i+1} (= x_i+)$ is replaced by y_i^r. Let $\lambda = \{y_j\colon j = 0, \dots, m\}$ be a refinement of κ and let $\{j(i)\colon i \in J_- \cup J_+\}$ be the subset of $\{j = 0, \dots, m\}$ such that $y_{j(i)} = x_i$ for some $i \in J_- \cup J_+$. Compose a partition λ^* from the points of λ except that, for each $i \in J_-$, $y_{j(i)-1}$ is replaced by x_i- and, for each $i \in J_+$, $y_{j(i)+1}$ is replaced by x_i+. Then, using the telescoping sum representation (4.49) we get

$$P(\lambda^*, [a,b], f) - P(\lambda, [a,b], f)$$

$$= \sum_{i \in J_-} \big\{ P(\lambda^*, [x_i-, b], f)[-f(y_{j(i)-1}) + f(x_i-)]P(\lambda, [a, y_{j(i)-2}], f)$$

$$+ P(\lambda^*, [x,b], f)[f(y_{j(i)-1}) - f(x_i-)]P(\lambda, [a, y_{j(i)-1}], f) \big\}$$

$$+ \sum_{i \in J_+} \big\{ P(\lambda^*, [y_{j(i)+2}, b], f)[-f(x_i+) + f(y_{j(i)+1})]P(\lambda, [a, y_{j(i)+1}], f)$$

$$+ P(\lambda^*, [x_i+, b], f)[f(x_i+) - f(y_{j(i)+1})]P(\lambda, [a, x_i], f) \big\}.$$

Since f is regulated each product $P(\kappa^*, \Delta, f)$ baised on partitions κ^* containing points $x+$ and $x-$ can be approximated arbitrary closely by products $P(\kappa, \Delta, f)$, where $\kappa \in Q([a,b])$. Hence, (4.60) holds also for products with κ replaced by κ^*. Then, invoking (4.62) we get

$$\|P(\lambda^*, [a,b], f) - P(\lambda, [a,b], f)\|$$

$$\le 2C(f)^2 \Big\{ \sum_{i \in J_-} \|f(y_{j(i)-1}) - f(x_i-)\| + \sum_{i \in J_+} \|f(y_{j(i)+1}) - f(x_i+)\| \Big\} \le \epsilon/2.$$

This together with (4.61) implies that $\| \prod_a^b(\mathbb{1} + df)_\pm - P(\lambda, [a,b], f)\| \le \epsilon$ for any refinement λ of κ. The proof of Lemma 4.22 is complete. $\qquad \square$

Next is the main statement of this subsection.

Theorem 4.23. *Let $\mathbb{B} = (\mathbb{B}, \|\cdot\|)$ be a Banach algebra and let $f \in \mathcal{W}_p(\tau, \mathbb{B})$, $0 < p < 2$, $\tau = [a,b]$, $[a,b-]$, $[a+,b]$ or $[a+,b-]$. Then the product integral with respect to f over τ exists.*

Remark. Using the Banach fixed point theorem Freedman (1983a, Theorem 5.1) proved the existence of the product integral with respect to each (uniformly) *continuous* operator-valued function with bounded p-variation, $0 < p < 2$, as a solution of an evolution integral equation. For improvements see Lyons (1994).

Proof. We will prove the theorem when $\tau = [a,b]$ only and note that the proof for other τ is the same. Therefore, according to Lemma 4.22, it will be enough to prove the existence of $\prod_a^b (\mathbb{1} + df)_\pm$. To this aim it suffices to show that given an $\epsilon > 0$ there is a partition $\kappa^* \in \mathcal{Q}^\pm([a,b])$ such that, for each refinement $\lambda^* \in \mathcal{Q}^\pm([a,b])$ of κ^*,

$$\|P(\lambda^*, [a,b], f) - P(\kappa^*, [a,b], f)\| \le \epsilon. \qquad (4.63)$$

By Lemma 4.19, there is a $C(f) < \infty$ such that (4.60) holds. Let $\epsilon > 0$ and let $\kappa^* = \{x_0, x_1, \ldots, x_n\}$ be the partition of $[a,b]$ constructed in Lemma 2.20 using the number

$$c = \left(\epsilon / \left[(4c_p(1 + \zeta(2/p))C(f)^2 v_p(f; [a,b])]\right)\right)^{p/(2-p)} \wedge \left(1/4c_p(1 + \zeta(2/p))\right)^p,$$

where $c_p = 1 \vee 2^{(1/p)-1}$. Hence, for each $i \notin J_\pm(\kappa^*)$, we have

$$V_p(f; [\Delta_i^{\kappa^*}])^{2-p} \le \epsilon / \left[4c_p(1 + \zeta(2/p))C(f)^2 v_p(f; [a,b])\right] \qquad (4.64)$$

and

$$2c_p(1 + \zeta(2/p))V_p(f; [\Delta_i^{\kappa^*}]) \le 1/2. \qquad (4.65)$$

Let λ^* be a refinement of κ^*. Using the telescoping sum representation (4.49) we get

$$P(\lambda^*, [a,b], f) - P(\kappa^*, [a,b], f) = \prod_{i=1}^n P(\lambda^*, [\Delta_i^{\kappa^*}], f) - \prod_{i=1}^n (\mathbb{1} + \Delta_i^{\kappa^*} f)$$

$$= \sum_{i \notin J_\pm(\kappa^*)} P(\lambda^*, [x_i, b], f) \left[P(\lambda^*, [\Delta_i^{\kappa^*}], f) - \mathbb{1} - \Delta_i^{\kappa^*} f\right] P(\kappa^*, [a, x_{i-1}], f).$$

By (4.53) of Lemma 4.18, we have

$$\left\|P(\lambda^*, [\Delta_i^{\kappa^*}], f) - \mathbb{1} - \Delta_i^{\kappa^*} f\right\| \le 4c_p(1 + \zeta(2/p))V_p(f; [\Delta_i^{\kappa^*}])^2,$$

whenever $V_p(f; [\Delta_i^{\kappa^*}]) \le (4c_p(1 + \zeta(2/p)))^{-1}$. Therefore, by (4.65), we get

$$\|P(\lambda^*, [a,b], f) - P(\kappa^*, [a,b], f)\| \le 4c_p(1 + \zeta(2/p))C(f)^2 \sum_{i \notin J_\pm(\kappa^*)} V_p(f; [\Delta_i^{\kappa^*}])^2$$

$$\le 4c_p(1 + \zeta(2/p))C(f)^2 v_p(f; [a,b]) \max_{i \notin J_\pm(\kappa^*)} V_p(f; [\Delta_i^{\kappa^*}])^{2-p} \le \epsilon,$$

where the last inequality follows from (4.64). Since λ^* is an arbitrary refinement of κ^*, (4.63), and, hence Theorem 4.23, is proved. $\qquad \square$

The following statement is a simple consequence of Theorem 4.23 and Definition 4.20.

Corollary 4.24. *Let* $f \in \mathcal{W}_p(\tau; \mathbb{B})$, $0 < p < 2$. *Then, for each* $a \leq y \leq z \leq b$, *the product integral with respect to* f *over the interval* $[y, z]$ *exists and*

$$\prod_a^b (\mathbb{1} + df) = \prod_y^b (\mathbb{1} + df) \prod_a^y (\mathbb{1} + df), \quad \forall\, a \leq y \leq b,\ a, b \in \tau.$$

We finish this subsection by proving that the product integral with respect to a function f exists in a stronger sense if, in addition, f has only one-sided jumps. The following auxiliary statement is similar to Lemma 7 of Dobrushin (1953).

Lemma 4.25. *Suppose* $f \in \mathcal{W}_p(\tau; \mathbb{B})$, $0 < p < 2$, *and, at each point of* τ, f *is either right-continuous or left-continuous. Then for each point* $\xi \in \tau$ *and any* $\epsilon > 0$, *there exists* $\delta > 0$ *such that, for each subinterval* $[x, y] \subset \tau$ *of length* $< \delta$ *containing the point* ξ *and for any partition* κ *of* $[x, y]$,

$$\left\| [\mathbb{1} + f(y) - f(x)] - P(\kappa, [x, y], f) \right\| < \epsilon.$$

Proof. We prove the statement when $\tau = (a, b)$ and f is right-continuous at some $\xi \in (a, b)$. The proof for other cases is similar and therefore is omitted. Let $\epsilon > 0$. By Lemma 4.19, there exists a finite constant $C(f)$ such that, for any partition λ of (a subinterval of) (a, b), we have

$$\left\| P(\lambda, \tau, f) \right\| \leq C(f). \tag{4.66}$$

Since f is right-continuous at ξ, by Lemmas 2.18 and 2.19, there exists $\delta > 0$ such that

$$\left.\begin{array}{c} V_p(f; [x, \xi)) \vee V_p(f; [\xi, y]) < \{\epsilon/[6C(f)]\} \wedge \{1/[4c_p(1 + \zeta(2/p))]\} \\ \left\| [f(y) - f(x)] - [f(y_1) - f(x_1)] \right\| < \epsilon/[3C(f)] \end{array}\right\} \tag{4.67}$$

whenever $a < x \leq x_1 < \xi \leq y_1 \leq y < b$ and $y - x < \delta$. Let $\kappa = \{x_i : i = 0, \ldots, n\}$ be a partition of any such subinterval $[x, y] \subset (a, b)$ and let i_0 be the integer such that $x_{i_0 - 1} < \xi \leq x_{i_0}$. Then by (4.49), (4.52), (4.66) and (4.67), it follows that

$$\left\| [\mathbb{1} + f(y) - f(x)] - P(\kappa, [x, y], f) \right\|$$

$$= \left\| \mathbb{1}[\mathbb{1} + f(y) - f(x)]\mathbb{1} - \Big(\prod_{i < i_0}(\mathbb{1} + \Delta_i^\kappa f)\Big)(\mathbb{1} + \Delta_{i_0}^\kappa f)\Big(\prod_{i > i_0}(\mathbb{1} + \Delta_i^\kappa f)\Big) \right\|$$

$$\leq C(f)\Big\{ \Big\| \mathbb{1} - \prod_{i < i_0}(\mathbb{1} + \Delta_i^\kappa f) \Big\| + \left\| f(y) - f(x) - \Delta_{i_0}^\kappa f \right\| + \Big\| \mathbb{1} - \prod_{i > i_0}(\mathbb{1} + \Delta_i^\kappa f) \Big\| \Big\}$$

$$\leq C(f)\Big\{ 2V_p\big(f; [x, x_{i_0 - 1}]\big) + \left\| f(y) - f(x) - \Delta_{i_0}^\kappa f \right\| + 2V_p\big(f; [x_{i_0}, y]\big) \Big\}$$

$$\leq \epsilon/3 + \epsilon/3 + \epsilon/3 = \epsilon.$$

The proof of Lemma 4.25 is complete. $\qquad\qquad\qquad\qquad\qquad\qquad\qquad\qquad\quad \square$

Theorem 4.26. *If, in addition to the assumptions of Theorem 4.23, f is either right-continuous or left-continuous at each interior point of τ then*

$$\prod_\tau (\mathbb{1} + df) = \lim_{|\kappa|\downarrow 0} P(\kappa, \tau, f).$$

Remark. The following example of Hildebrandt (1959, p. 368) shows that the one-sided continuity assumption in Theorem 4.26 cannot be weakened in general. Let

$$f(x) = \begin{cases} 0, & \text{for } 0 \le x < 1/2 \\ u, & \text{for } x = 1/2 \\ v, & \text{for } 1/2 < x \le 1. \end{cases}$$

If a partition κ of $[0, 1]$ does not contain the point $1/2$ then $P(\kappa, [0, 1], f) = 1 + v$, while if κ contains the point $1/2$, then $P(\kappa, [0, 1], f) = (1 + u)(1 + v - u)$, so the limit exists as the mesh of κ tends to zero if and only if $u(v - u) = 0$, i.e., if and only if f is right- or left-continuous.

Proof. It suffices to show that given an $\epsilon > 0$ there is a $\delta > 0$ such that, for each partition κ of τ with mesh $|\kappa| < \delta$ and any refinement λ of κ,

$$\|P(\kappa, \tau, f) - P(\lambda, \tau, f)\| < \epsilon. \tag{4.68}$$

Let $\epsilon > 0$ and let $p_1 \in (p, 2)$. By Lemma 4.19, there exists a finite constant $C(f)$ such that, for all partitions κ of τ, (4.66) holds. By Lemma 2.2, there is a finite set $\{\xi_j : j = 1, \ldots, N\} \subset \tau$ such that

$$\|\Delta^+ f(x)\| \vee \|\Delta_- f(x)\| < \{\epsilon / [8c_{p_1}(1 + \zeta(2/p_1))C(f)^2 v_p(f; \tau)^{2/p_1}]\}^{p_1/(2(p_1 - p))}/2$$

if $x \ne \xi_j$ for each $j = 1, \ldots, N$. Then using Lemmas 2.15, 2.18 and 2.19 one can find $\delta_1 > 0$ such that

$$Osc_\Delta(f)^{2(p_1 - p)/p_1} < \epsilon / [8c_{p_1}(1 + \zeta(2/p_1))C(f)^2 v_p(f; \tau)^{2/p_1}] \tag{4.69}$$

and

$$2c_{p_1}(1 + \zeta(2/p_1))V_{p_1}(f; \Delta) \le 1/2 \tag{4.70}$$

for each subinterval Δ of length $< \delta_1$ which either does not contain any ξ_j for $j = 1, \ldots, N$, or ξ_j may be the left endpoint of Δ if $f(\xi_j+) = f(\xi_j)$, or ξ_j may be the right endpoint of Δ if $f(\xi_j-) = f(\xi_j)$. Moreover, using Lemma 4.25 one can find $\delta_2 > 0$ such that

$$\|[\mathbb{1} + f(y_j) - f(x_j)] - P(\kappa_j, [x_j, y_j], f)\| < \epsilon / [2NC(f)^2] \tag{4.71}$$

for each $j = 1, \ldots, N$, each subinterval $[x_j, y_j] \subset \tau$ of length $< \delta_2$ containing ξ_j and any partition κ_j of $[x_j, y_j]$. Let $\delta := \delta_1 \wedge \delta_2$. To show (4.68) let $\kappa = \{x_i : i = 0, \ldots, n\}$ be a partition of τ with mesh $|\kappa| = \max(x_i - x_{i-1}) < \delta$ and let λ be a refinement of κ. Let

$$I := \{i = 1, \ldots, n : \text{ for all } j = 1, \ldots, N, \text{ either } \xi_j \notin [x_{i-1}, x_i],$$

or $x_{i-1} = \xi_j$ and $f(\xi_j+) = f(\xi_j)$, or $x_i = \xi_j$ and $f(\xi_j-) = f(\xi_j)\}$.

Then by (4.53) and (4.70), we get

$$\sum_{i \in I} \left\| (\mathbb{1} + \Delta_i^\kappa f) - P(\kappa_i', [\Delta_i^\kappa], f) \right\| \leq 4c_{p_1}(1 + \zeta(2/p_1)) \sum_{i \in I} V_{p_1}^2(f; [\Delta_i^\kappa])$$

$$\leq 4c_{p_1}(1 + \zeta(2/p_1)) v_p(f; \tau)^{2/p_1} \max_{i \in I} Osc_{[\Delta_i^\kappa]}(f)^{2(p_1-p)/p_1} \leq \epsilon/[2C(f)^2], \quad (4.72)$$

where (4.69) was used in the last inequality. Hence, using the telescoping sum representation (4.49) in conjunction with (4.71) and (4.72) it follows that

$$\left\| P(\kappa, \tau, f) - P(\lambda, \tau, f) \right\| \leq C(f)^2 \Big(\sum_{i \in I} + \sum_{i \notin I} \Big) \left\| (\mathbb{1} + \Delta_i^\kappa f) - P(\lambda, [\Delta_i^\kappa], f) \right\|$$

$$\leq \epsilon/2 + \epsilon/2 = \epsilon.$$

Therefore (4.68) holds and the proof of Theorem 4.26 is complete .   □

4.5. Local series representation. Here equation (4.50) is generalized to product integrals with respect to functions $f \in \mathcal{R}([a, b]; \mathbb{B})$ such that, for some $0 < p < 2$,

$$\gamma := 2c_p^2(2 + \zeta(2/p))V_p(f; [a, b]) < 1, \quad (4.73)$$

where $c_p = 1 \vee 2^{(1/p)-1}$. The generalization leads to a series of iterated integrals which is usually called a Peano series. Peano (1886-7) used such a series to solve linear differential equations (cf. also p. 58 in the same reference for a remark concerning priority in establishing this method).

Let $a \leq c \leq y \leq d \leq b$ and let $H_0^L(f; c)(y) := H_0^R(f; d)(y) := \mathbb{1}$. Define recursively, for each $k \geq 1$ and all $c \leq y \leq d$,

$$H_k^L(f; c)(y) := (LY) \int_c^y df\, H_{k-1}^L(f; c) = (CY) \int_c^y df(x)\, H_{k-1}^L(f; c)_-^{(c)}(x)$$

$$+ \sum_{c < x < y} [\Delta^+ f \Delta_-(H_{k-1}^L(f; c))](x) \quad (4.74)$$

and

$$H_k^R(f; d)(y) := (RY) \int_y^d H_{k-1}^R(f; d)\, df = (CY) \int_y^d H_{k-1}^R(f; d)_+^{(d)}(x)\, df(x)$$

$$- \sum_{y < x < d} [\Delta^+(H_{k-1}^R(f; d))\Delta_- f](x). \quad (4.75)$$

Next, using the integrals $H_k^L(f; c)$ and $H_k^R(f; d)$ we define two Peano series G^L and G^R, respectively. Theorem 4.29 below shows that both series coincide and gives a product integral representation whenever the function f has bounded p-variation, $0 < p < 2$ and (4.73) holds.

Proposition 4.27. *Let $f \in \mathcal{R}([a, b]; \mathbb{B})$ be such that (4.73) holds for some $p \in (0, 2)$ and let $a \leq c \leq d \leq b$. Then the series*

$$G^L(f; c, d) := \sum_{k \geq 0} H_k^L(f; c)(d) \quad and \quad G^R(f; c, d) := \sum_{k \geq 0} H_k^R(f; d)(c) \qquad (4.76)$$

converge absolutely in $\mathbb{B}$. Moreover, the bounds

$$\|G(f; c, d)\| \leq 1/(1 - \gamma_{cd}), \qquad (4.77)$$
$$\|G(f; c, d) - \mathbb{1}\| \leq V_p(f; [c, d])/(1 - \gamma_{cd}) \qquad (4.78)$$
$$\|G(f; c, d) - \mathbb{1} - [f(d) - f(c)]\| \leq (2 + \zeta(2/p))V_p(f; (c, d])V_p(f; [c, d))/(1 - \gamma_{cd})$$
$$\leq (2 + \zeta(2/p))V_p(f; [c, d])^2/(1 - \gamma_{cd}) \qquad (4.79)$$

hold, where G denotes either G^L or G^R and

$$\gamma_{cd} := 2c_p^2(2 + \zeta(2/p))V_p(f; [c, d]) \leq \gamma.$$

Proof. To show absolute convergence of the series (4.76) we bound the norms of each $H_k^L(f; c)(d)$ and $H_k^R(f; d)(c)$ for $k \geq 1$ defined by (4.74) and (4.75), respectively. Clearly, $\int_u^v dg(x) = g(v) - g(u)$ for any $u \leq v$ and function g for the (MPS) or Riemann-Stieltjes integral. If g is regulated the same holds for the (CY) integral. So

$$\|H_1^L(f; c)(d)\| = \|H_1^R(f; d)(c)\| = \|f(d) - f(c)\| \leq V_p(f; [c, d]). \qquad (4.80)$$

Since $H_k^L(f; c)(c) = 0$ for each $k \geq 1$, we have for $k \geq 2$ by Lemma 3.26 that $\Delta^+(H_k^L(f; c))(c) = 0$ and for $c \leq y \leq d$,

$$H_k^L(f; c)(y) = (LY) \int_c^y df_{(c+)} H_{k-1}^L(f; c).$$

Hence by Lemmas 2.5 and 3.26, relations (2.12) and (2.14), Corollary 3.29 with $x_0 = c$ and Hölder's inequality (3.95), we have

$$\|H_2^L(f; c)(d)\| = \left\|(CY) \int_c^d df^{(c+)} \left[f_-^{(c)} - f(c)\right] + \sum_{(c,d)} (\Delta^+ f)(\Delta_- f)\right\|$$
$$\leq (1 + \zeta(2/p))V_p(f; (c, d])V_p(f; [c, d)) + \mathfrak{S}_p(f; (c, d))^2$$
$$\leq (2 + \zeta(2/p))V_p(f; (c, d])V_p(f; [c, d)). \qquad (4.81)$$

Likewise we have

$$\|H_2^R(f; d)(d)\| = \left\|(CY) \int_c^d \left[f(d) - f_+^{(d)}\right] df^{(d-)} + \sum_{(c,d)} (\Delta^+ f)(\Delta_- f)\right\|$$
$$\leq (2 + \zeta(2/p))V_p(f; (c, d])V_p(f; [c, d)). \qquad (4.82)$$

For $k \geq 3$ we have to bound also the p-variation of the functions $H_{k-1}^L(f;c)$ and $H_{k-1}^R(f;d)$ over $[c,d)$ and $(c,d]$, respectively. Invoking, in addition, Proposition 3.32 and using the bound $\|g\|_\infty \leq V_p(g)$ if $g(c) = 0$, by (2.21), we get recursively

$$
\begin{aligned}
V_p(H_{k-1}^L(f;c);[c,d)) &\leq c_p \Big[V_p\Big((CY)\int_c^{(\cdot)} df\, H_{k-2}^L(f;c)_-^{(c)};[c,d)\Big) \\
&\quad + V_p\Big(\sum_{(c,\cdot)}[\Delta^+ f \Delta_-(H_{k-2}^L(f;c))];[c,d)\Big)\Big] \\
&\leq c_p\big[2c_p(1+\zeta(2/p))V_p(f;[c,d))V_p(H_{k-2}^L(f;c);[c,d)) \\
&\quad + \mathfrak{S}_p(f;(c,d))\mathfrak{S}_p(H_{k-2}^L(f;c);(c,d))\big] \\
&\leq \big[2c_p^2(2+\zeta(2/p))\big]V_p(f;[c,d))V_p(H_{k-2}^L(f;c);[c,d)) \\
&\leq \cdots \\
&\leq \big[2c_p^2(2+\zeta(2/p))\big]^{k-2}V_p(f;[c,d))^{k-1}. \qquad (4.83)
\end{aligned}
$$

Likewise it follows that for $k \geq 3$

$$
V_p(H_{k-1}^R(f;d);(c,d]) \leq \big[2c_p^2(2+\zeta(2/p))\big]^{k-2}V_p(f;(c,d])^{k-1}.
$$

Finally, as in (4.81), we get

$$
\begin{aligned}
\|H_k^L(f;d)(d)\| &\leq \Big\|(CY)\int_c^d df_{(c+)}\, H_{k-1}^L(f;c)_-^{(c)} + \sum_{(c,d)}[\Delta^+ f \Delta_-(H_{k-1}^L(f;c))]\Big\| \\
&\leq (1+\zeta(2/p))V_p(f;(c,d])V_p(H_{k-1}^L(f;c);[c,d)) \\
&\quad + \mathfrak{S}_p(f;(c,d))\mathfrak{S}_p(H_{k-1}^L(f;c);(c,d)) \\
&\leq (2+\zeta(2/p))V_p(f;(c,d])V_p(H_{k-1}^L(f;c);[c,d)) \\
\text{by (4.83)} \quad &\leq \big(2c_p^2\big)^{k-2}(2+\zeta(2/p))^{k-1}V_p(f;(c,d])V_p(f;[c,d))^{k-1} \qquad (4.84)
\end{aligned}
$$

and, likewise,

$$
\|H_k^R(f;d)(c)\| \leq \big(2c_p^2\big)^{k-2}(2+\zeta(2/p))^{k-1}V_p(f;(c,d])V_p(f;[c,d))^{k-1}. \qquad (4.85)
$$

Now using (4.80), (4.81) and (4.84) when $G = G^L$ and (4.80), (4.82) and (4.85) when $G = G^R$ we get the desired bounds

$$
\|G(f;c,d)\| \leq \sum_{k=0}^\infty \gamma_{cd}^k = 1/(1-\gamma_{cd}),
$$

$$
\begin{aligned}
\|G(f;c,d) - \mathbb{1}\| &\leq V_p(f;[c,d])\sum_{k=1}^\infty \big[2c_p^2(2+\zeta(2/p))V_p(f;[c,d])\big]^{k-1} \\
&= V_p(f;[c,d])/(1-\gamma_{cd}),
\end{aligned}
$$

and

$$
\begin{aligned}
\|G(f;c,d) - \mathbb{1} - [f(d)-f(c)]\| &\leq (2+\zeta(2/p))V_p(f;(c,d])V_p(f;[c,d))\times \\
&\quad \times \sum_{k=2}^\infty \big[2c_p^2(2+\zeta(2/p))V_p(f;[c,d])\big]^{k-2} \\
&= (2+\zeta(2/p))V_p(f;(c,d])V_p(f;[c,d))/(1-\gamma_{cd}).
\end{aligned}
$$

The proof of Proposition 4.27 is complete. □

The next statement shows that both Peano series are multiplicative interval functions.

Lemma 4.28. *Let $f \in \mathcal{R}([a,b]; \mathbb{B})$ be such that (4.73) holds for some $p \in (0,2)$ and let G denote either G^L or G^R defined by (4.76). Then, for any $c \in [a,b]$, which may be also a point $x\pm$, $G(f; a, b) = G(f; c, b)G(f; a, c)$. Moreover, for any $c \in (a, b]$, $G(f; c-, c) = \mathbb{I} + (\Delta_- f)(c)$ and, for any $c \in [a, b)$, $G(f; c, c+) = \mathbb{I} + (\Delta^+ f)(c)$.*

Proof. Since $G(f; c, c) = \mathbb{I}$ for any c, we can assume $a < c < b$. The first part of the statement is a consequence of the additivity property (3.68) of the (LY) and the (RY) integrals (cf. Proposition 3.25). Indeed, for each $k \geq 1$, breaking the interval $[a, b]$ at a point c in each integral consecutively we get

$$H_k^L(f; a)(b) = (LY) \int_a^c df(x) \, H_{k-1}^L(f; a)(x) + (LY) \int_c^b df(x) \, H_{k-1}^L(f; a)(x)$$

$$= H_k^L(f; a)(c) + H_1^L(f; c)(b)H_{k-1}^L(f; a)(c)$$

$$+ (LY) \int_c^b df(x) \, (LY) \int_c^x df(y) \, H_{k-2}^L(f; a)(y)$$

$$= \cdots = \sum_{i=0}^k H_i^L(f; c)(b)H_{k-i}^L(f; a)(c).$$

Likewise, for each $k \geq 1$, we get

$$H_k^R(f; b)(a) = (RY) \int_c^b dH_{k-1}^R(f; b)(x) \, df(x) + (RY) \int_a^c H_{k-1}^R(f; b)(x) \, df(x)$$

$$= \cdots = \sum_{i=0}^k H_{k-i}^R(f; b)(c)H_i^R(f; c)(a).$$

Then changing the order of summation we arrive at the first desired relation

$$G^L(f; a, b) = \mathbb{I} + \sum_{k=1}^\infty \sum_{i=0}^k H_i^L(f; c)(b)H_{k-i}^L(f; a)(c)$$

$$= \mathbb{I} + \sum_{k=1}^\infty H_k^L(f; c)(b) + \sum_{k=1}^\infty H_k^L(f; a)(c) + \left(\sum_{k=1}^\infty H_k^L(f; c)(b)\right)\left(\sum_{k=1}^\infty H_k^L(f; a)(c)\right)$$

$$= \left(\mathbb{I} + \sum_{k=1}^\infty H_k^L(f; c)(b)\right)\left(\mathbb{I} + \sum_{k=1}^\infty H_k^L(f; a)(c)\right) = G^L(f; c, b) \, G^L(f; a, c).$$

Likewise, the relation $G^R(f; a, b) = G^R(f; c, b)G^R(f; a, c)$ follows. Turning to the second part of the statement, by Lemma 3.23, for each $k \geq 2$, we have

$$H_k^L(f; c-)(c) = (CY) \int_{c-}^c df(x) \, H_{k-1}^L(f; c-)_-^{(c-)}(x)$$

$$= (\Delta_- f)(c)H_{k-1}^L(f; c-)(c-) = 0$$

and

$$H_k^R(f;c)(c-) = (CY)\int_{c-}^c H_{k-1}^R(f;c)_+^{(c)}\,df(x)$$
$$= H_{k-1}^R(f;c)(c)(\Delta_- f)(c) = 0$$

whenever $c \in (a,b]$. Likewise

$$H_k^L(f;c)(c+) = (CY)\int_c^{c+} df(x)\,H_{k-1}^L(f;c)_-^{(c)}(x) = (\Delta^+ f)(c)H_{k-1}^L(f;c)(c) = 0$$

and

$$H_k^R(f;c+)(c) = (CY)\int_c^{c+} H_{k-1}^R(f;c+)_+^{(c+)}(x)\,df(x)$$
$$= H_{k-1}^R(f;c+)(c+)(\Delta^+ f)(c) = 0$$

whenever $c \in [a,b)$. This implies the last two relations of the statement and completes the proof of Lemma 4.28. □

Here is the main result of this subsection. It gives a local Peano series representation of the product integral.

Theorem 4.29. *Let $f \in \mathcal{R}([a,b];\mathbb{B})$ be such that (4.73) holds for some $p \in (0,2)$. Then*

$$\prod_a^b(\mathbb{1} + df) = G^L(f;a,b) = G^R(f;a,b).$$

Proof. By Lemma 4.19, there exists a finite constant $C(f)$ such that (4.60) holds. Hence, by Theorem 4.23 and Lemma 4.22, given $\epsilon > 0$ there exists a partition κ of the interval $[a,b]$ such that, for all refinements $\lambda \in Q^\pm([a,b])$ of κ,

$$\left\| \prod_a^b(\mathbb{1} + df) - P(\lambda,[a,b],f) \right\| \le \epsilon/2. \qquad (4.86)$$

Next, by Lemma 2.20, one can construct another partition κ^* of $[a,b]$ using the number

$$c = \left\{ \epsilon(1-\gamma)^2 / \left[2C(f)(2 + \zeta(2/p))v_p(f;[a,b]) \right] \right\}^{p/(2-p)}.$$

Let $\kappa^*(\epsilon) := \kappa^* \cup \kappa_0 = \{x_0, x_1, \ldots, x_n\}$ and let $\Delta_i^* := \Delta_i^{\kappa^*(\epsilon)}$. Then, for each $i \notin J_\pm(\kappa^*(\epsilon))$, we have

$$V_p(f;[\Delta_i^*])^{2-p} \le \epsilon(1-\gamma)^2 / \left[2C(f)(2 + \zeta(2/p))v_p(f;[a,b]) \right]. \qquad (4.87)$$

By Lemma 4.28, for each $i \in J_\pm(\kappa^*(\epsilon))$, we have $G(f;x_{i-1},x_i) = \mathbb{1} + \Delta_i^* f$, where G denotes either G^L or G^R. Hence, by Lemma 4.28 again and by the telescoping sum representation (4.49), it follows that

$$G(f;a,b) - P(\kappa^*(\epsilon),[a,b],f)$$
$$= \sum_{i \notin J_\pm(\kappa^*(\epsilon))} G(f;x_i,b)\left[G(f;x_{i-1},x_i) - \mathbb{1} - \Delta_i^* f \right] P(\kappa^*(\epsilon),[a,x_{i-1}],f). \qquad (4.88)$$

Moreover, by Proposition 4.27, for each $i \notin J_\pm(\kappa^*(\epsilon))$, we have the bounds

$$\|G(f; x_{i-1}, x_i) - \mathbb{1} - \Delta_i^* f\| \le \left[(2 + \zeta(2/p))V_p(f; [\Delta_i^*])^2\right]/(1 - \gamma)$$

and $\|G(f; x_i, b)\| \le 1/(1 - \gamma)$. Therefore, by (4.88) and (4.60), we get

$$\|G(f; a, b) - P(\kappa^*(\epsilon), [a, b], f)\|$$
$$\le \left[C(f)(2 + \zeta(2/p))v_p(f; [a, b])/(1 - \gamma)^2\right] \max_{i \in I} V_p(f; [\Delta_i^*])^{2-p} \le \epsilon/2, \quad (4.89)$$

where (4.87) was used in the last inequality. Using (4.86) for $\lambda = \kappa^*(\epsilon)$ together with (4.89) we arrive at $\left\| \prod_a^b (1 + df) - G(f; a, b) \right\| \le \epsilon$. Since $\epsilon > 0$ is arbitrary this completes the proof of Theorem 4.29. $\qquad\square$

We finish this section with a condition for a product integral to be non-degenerate (non-zero). We recall that v is the inverse of u if and only if $uv = \mathbb{1} = vu$. The inverse of u will be denoted by u^{-1}.

Proposition 4.30. *Let $f \in W_p(\tau, \mathbb{B})$, $0 < p < 2$, and let τ be a finite interval. Suppose that for every $x \in \tau$, $\mathbb{1} + (\Delta_- f)(x)$ and $\mathbb{1} + (\Delta^+ f)(x)$ are invertible, if x is not the left or right endpoint of τ respectively. Then the product integral $\prod_\tau (\mathbb{1} + df)$ is non-degenerate and in fact invertible with inverse $(\prod_\tau (\mathbb{1} + df))^{-1}$ such that*

$$\left\| \left(\prod_\tau (\mathbb{1} + df) \right)^{-1} \right\| \le [2D(f)]^N, \quad (4.90)$$

where $N \le 4 + 3v_p(f; \tau)/c$, with $c = 1/\left[4c_p^2(2 + \zeta(2/p))\right]^p$, where $c_p = 1 \vee 2^{(1/p)-1}$, and

$$D(f) := \max \left\{ \|(\mathbb{1} + (\Delta f)(x))^{-1}\| : \|(\Delta f)(x)\| > 1/2, \; x \in \tau, \; \Delta = \Delta^+ \text{ or } \Delta_- \right\}.$$

Remark. For any $u \in \mathbb{B}$ with $\|u\| < 1$, $\mathbb{1} + u$ is invertible, and there will only be finitely many jumps $(\Delta_- f)(x)$ or $(\Delta^+ f)(x)$ of norm $> 1/2$.

Proof. We will give a proof for a closed interval $\tau = [a, b]$. The proof can easily be adapted to other intervals. Let $\kappa^* = \{x_0, x_1, \dots, x_n\}$ be the partition of $[a, b]$ given by Lemma 2.20 using the number c. Then, for each $\tau_i := [\Delta_i^{\kappa^*}]$ with $i \notin J_\pm(\kappa^*)$, $V_p(f; \tau_i) \le 1/\left[4c_p^2(2 + \zeta(2/p))\right]$. Thus we can apply Proposition 4.27 with $\rho_i = 2c_p^2(2 + \zeta(2/p))V_p(f; \tau_i) \le 1/2$. Then by (4.78) and Theorem 4.29

$$\left\| \prod_{\tau_i} (\mathbb{1} + df) - \mathbb{1} \right\| \le 2V_p(f; \tau_i) \le 1/2.$$

Thus each $\prod_{\tau_i} (\mathbb{1} + df)$ is invertible and for the inverse $(\prod_{\tau_i} (\mathbb{1} + df))^{-1}$ we have

$$\left\| \left(\prod_{\tau_i} (\mathbb{1} + df) \right)^{-1} \right\| \le 2. \quad (4.91)$$

For intervals $[x-, x]$ or $[x, x+]$, the product integrals $\prod_{[x-, x]} (\mathbb{1} + df) = \mathbb{1} + (\Delta_- f)(x)$ or $\prod_{[x, x+]} (\mathbb{1} + df) = \mathbb{1} + (\Delta^+ f)(x)$ are invertible by assumption and (4.91) holds for each $\tau_i = [x-, x]$ or $\tau_i = [x, x+]$ such that $\|(\Delta_- f)(x)\| \le 1/2$ or $\|(\Delta^+ f)(x)\| \le 1/2$, respectively. Thus by Corollary 4.24, $\prod_a^b (\mathbb{1} + df)$ is a product of invertible elements of $\mathbb{B}$ and therefore invertible. The bound for N in (4.90) is given by Lemma 2.20. $\qquad\square$

5. Indefinite Product Integrals

Let $f \in \mathcal{W}_p([a,b], \mathbb{B})$. Any such function also has bounded p-variation over each subinterval of $[a,b]$. Therefore Theorem 4.23 implies the existence of the indefinite product integrals

$$\mathcal{P}_a(f)(y) := \prod_a^y (\mathbb{1} + df), \quad y \in [a,b],$$

and

$$\mathcal{P}^b(f)(y) := \prod_y^b (\mathbb{1} + df), \quad y \in [a,b],$$

whenever $0 < p < 2$. Moreover, by Corollary 4.24, for each $y \in [a,b]$,

$$\prod_a^b (\mathbb{1} + df) = \prod_y^b (\mathbb{1} + df) \prod_a^y (\mathbb{1} + df) = \mathcal{P}^b(f)(y) \mathcal{P}_a(f)(y). \tag{5.1}$$

We consider $\mathcal{P}_a$ and $\mathcal{P}^b$ as (nonlinear) operators defined on $\mathcal{W}_p$, $0 < p < 2$, and call them *product integral operators*. By Lemma 2.3 and Proposition 5.3 below, product integral operators have values also in $\mathcal{W}_p$, $0 < p < 2$. The main result concerning differentiability of product integral operators is given in Subsection 5.3.

5.1. Regulated indefinite product integrals. To describe left- and right-continuous modifications of the indefinite product integrals $\mathcal{P}_a$ and $\mathcal{P}^b$ we recall Definition 4.20(2) of a product integral over intervals $[a, y\pm]$ and $[y\pm, b]$. The following statement shows that the first equality in (5.1) holds also in the case when y is an $x+$ or an $x-$.

Lemma 5.1. *Let $f \in \mathcal{W}_p([a,b]; \mathbb{B})$, $0 < p < 2$. Then*

$$\prod_a^{y+} (\mathbb{1} + df) = [\mathbb{1} + (\Delta^+ f)(y)] \prod_a^y (\mathbb{1} + df) \tag{5.2}$$

and

$$\prod_y^b (\mathbb{1} + df) = \prod_{y+}^b (\mathbb{1} + df)[\mathbb{1} + (\Delta^+ f)(y)] \tag{5.3}$$

for all $y \in [a,b)$. Also

$$\prod_a^{y-} (\mathbb{1} + df) = [\mathbb{1} + (\Delta_- f)(y)] \prod_a^{y-} (\mathbb{1} + df) \tag{5.4}$$

and

$$\prod_{y-}^b (\mathbb{1} + df) = \prod_y^b (\mathbb{1} + df)[\mathbb{1} + (\Delta_- f)(y)] \tag{5.5}$$

for all $y \in (a,b]$.

Proof. To show (5.2) we can consider partitions $\{x_0, x_1, \ldots, x_n\}$ of $[a, y+]$ with $x_{n-1} = y$, so that $\{x_0, x_1, \ldots, x_{n-1}\}$ is a partition of $[a,y]$. This proves (5.2). The analogous argument implies also (5.5).

To show (5.3) let $y \in [a, b)$. If $(\Delta^+ f)(y) = 0$ then (5.3) holds because the value of f at the endpoints y and $y+$ is the same. Suppose $(\Delta^+ f)(y) \neq 0$. By Lemma 4.19, there exists a finite constant $C(f)$ such that, for any partition κ of any subinterval Δ of $[a, b]$,

$$\|P(\kappa, \Delta, f)\| \leq C(f). \qquad (5.6)$$

Let $\epsilon > 0$. By Theorem 4.23, there exists a partition $\kappa = \{y < x_1 < \cdots < b = x_n\}$ of $[y, b]$ such that

$$\left\| \prod_y^b (\mathbb{1} + df) - P(\kappa, [x_1, b], f)[\mathbb{1} + f(x_1) - f(y)] \right\| \leq \epsilon/3$$

and such that for the partition $\kappa^+ = \{y+ < x_1 < \cdots < b = x_n\}$ of $[y+, b]$

$$\left\| \prod_{y+}^b (\mathbb{1} + df) - P(\kappa^+, [x_1, b], f)(\mathbb{1} + f(x_1) - f(y+)) \right\| \leq \epsilon/[3\|\mathbb{1} + (\Delta^+ f)(y)\|].$$

We can assume x_1 is close enough to y so that

$$\|f(x_1) - f(y+)\| \leq \epsilon/[3C(f)\|(\Delta^+ f)(y)\|].$$

Then we have

$$\left\| \prod_y^b (\mathbb{1} + df) - \prod_{y+}^b (\mathbb{1} + df)[\mathbb{1} + (\Delta^+ f)(y)] \right\|$$

$$\leq 2\epsilon/3 + \left\| P(\kappa, [x_1, b], f) \right\| \|[f(x_1) - f(y+)](\Delta^+ f)(y)\| \leq \epsilon.$$

Since $\epsilon > 0$ is arbitrary this proves (5.3). The proof of (5.4) is similar and, hence, is omitted. The proof of Lemma 5.1 is complete. $\qquad \square$

The following statement assures that the above extensions of the definition of the product integral are consistent with left- and right-continuous modifications of $\mathcal{P}_a$ and $\mathcal{P}^b$.

Lemma 5.2. *Let $f \in \mathcal{W}_p([a, b]; \mathbb{B})$, $0 < p < 2$. Then the indefinite product integrals $\mathcal{P}_a(f)$ and $\mathcal{P}^b(f)$ are regulated. Moreover,*

$$\mathcal{P}_a(f)(y+) = \prod_a^{y+} (\mathbb{1} + df) \quad and \quad \mathcal{P}^b(f)(y+) = \prod_{y+}^b (\mathbb{1} + df) \qquad (5.7)$$

for all $y \in [a, b)$ and

$$\mathcal{P}_a(f)(y-) = \prod_a^{y-} (\mathbb{1} + df) \quad and \quad \mathcal{P}^b(f)(y-) = \prod_{y-}^b (\mathbb{1} + df) \qquad (5.8)$$

for all $y \in (a, b]$.

Proof. We start by proving the first equality in (5.7) for an arbitrary but fixed $y \in [a, b)$. Let $C(f)$ be the constant in (5.6) and let $\epsilon > 0$. By Lemmas 2.5 and 2.18, there exists $\delta \in (0, b - y)$ such that

$$V_p(f; [y+, y + \delta]) \leq [4c_p(1 + \zeta(2/p))]^{-1} \wedge \{\epsilon/[6\|\mathbb{1} + (\Delta^+ f)(y)\| \|\mathcal{P}_a(f)\|_\infty]\}, \quad (5.9)$$

where $c_p = 1 \vee 2^{(1/p)-1}$. Let $z \in (y, y+\delta]$. By Theorem 4.23, there exists a partition $\kappa = \{y = x_0 < x_1 < \cdots < z = x_n\}$ of $[y, z]$ such that

$$\left\| \prod_y^z (\mathbb{1} + df) - P(\kappa, [y, z], f) \right\| \le \epsilon / \left[3 \|\mathcal{P}_a(f)\|_\infty \right]. \tag{5.10}$$

We can assume x_1 is close enough to y so that

$$\|f(x_1) - f(y+)\| \le \epsilon / \left[3C(f) \|\mathcal{P}_a(f)\| \right]. \tag{5.11}$$

By (4.52) and (5.9), we have

$$\|P(\kappa, [x_1, z], f) - \mathbb{1}\| \le 2V_p(f; [x_1, z])$$
$$\le \epsilon / \left[3 \|\mathbb{1} + (\Delta^+ f)(y)\| \, \|\mathcal{P}_a(f)\|_\infty \right]. \tag{5.12}$$

Then using (5.1) and (5.2), by (5.6), (5.10), (5.11) and (5.12), we get

$$\left\| \mathcal{P}_a(f)(z) - \prod_a^{y+} (\mathbb{1} + df) \right\| \le \|\mathcal{P}_a(f)\|_\infty \left\{ \left\| \prod_y^z (\mathbb{1} + df) - P(\kappa, [y, z], f) \right\| \right.$$
$$+ \left\| P(\kappa, [x_1, z], f)[f(x_1) - f(y+)] \right\|$$
$$+ \left. \left\| P(\kappa, [x_1, z], f)[\mathbb{1} + (\Delta^+ f)(y)] - [\mathbb{1} + (\Delta^+ f)(y)] \right\| \right\}$$
$$\le \epsilon/3 + \epsilon/3 + \epsilon/3 = \epsilon.$$

Since $\epsilon > 0$ is arbitrary this proves the first equality in (5.7). Symmetric arguments show that the second equality in (5.8) holds. The proofs of the other two equalities are simpler and we omit them. The proof of Lemma 5.2 is complete. $\qquad\square$

Recall the definition (2.7) of V_p^*. It follows from the next statement in conjunction with Lemma 2.3 that the product integral operators $\mathcal{P}_a$ and $\mathcal{P}^b$ take the space $\mathcal{W}_p$ into itself, $0 < p < 2$.

Proposition 5.3. *Let $f \in W_p([a, b]; \mathbb{B})$, $0 < p < 2$. Then*

$$V_p^*(\mathcal{P}_a(f)) \le c_p V_p^*(f) \|\mathcal{P}_a(f)\|_\infty < \infty, \tag{5.13}$$

where $c_p = 1 \vee 2^{(1/p)-1}$. The same statement with $\mathcal{P}^b(f)$ instead of $\mathcal{P}_a(f)$ holds also.

Proof. We will only provide a proof for $\mathcal{P}_a(f)$. The statement for $\mathcal{P}^b(f)$ is proven in the same way. By Lemma 4.19 and Definition 4.20, one can conclude that $\|\mathcal{P}_a(f)\|_\infty < \infty$. Let $\epsilon > 0$ and let $\kappa^* = \{x_i : i = 0, \ldots, n\} \in Q^\pm([a, b])$ be the partition constructed in Lemma 2.20 using the number

$$c = \left\{ 1 \wedge \epsilon^p / \left[\|\mathcal{P}_a(f)\|_\infty V_p(f; [a, b]) \right]^p \right\} / \left[4c_p^2 (2 + \zeta(2/p)) \right]^p. \tag{5.14}$$

Let ξ_j, $j = 1, \ldots, N-1$, denote those $x_{i_j} \in \kappa^* \setminus \{a, b\}$ for which $x_{i_j - 1} = x_{i_j} -$ or $x_{i_j + 1} = x_{i_j} +$. We put $\xi_0 := x_0 = a$ and $\xi_N := x_n = b$. It may be assumed that κ^* contains all points $\xi_{j-1}+$ and ξ_j- for $j = 1,\dots,N$ (even if there are no corresponding one-sided jumps). We can also assume that κ^* is fine enough so that

$$\Big(\sum_{i=1}^{n} v_p(f; [x_{i-1}, x_i])\Big)^{1/p} < V_p^*(f; [a,b]) + \epsilon.$$

Since $\mathcal{P}_a(f)$ is regulated (Lemma 5.2) one can choose points $y_{j-1}^r \in (\xi_{j-1}, x_{i_{j-1}+2})$ and $y_j^l \in (x_{i_j-2}, \xi_j)$ for $j = 1,\dots,N$ such that

$$A := \sum_{j=1}^{N}\Big[\big\|\mathcal{P}_a(f)(\xi_j-) - \mathcal{P}_a(f)(w_j^l)\big\|^p + \big\|\mathcal{P}_a(f)(w_j^r) - \mathcal{P}_a(f)(\xi_{j-1}+)\big\|^p\Big]$$

$$\leq (\epsilon/[2c_p^2])^p \tag{5.15}$$

holds for all $w_j^r \in (\xi_{j-1}, y_{j-1}^r]$ and all $w_j^l \in [y_j^l, \xi_j)$ for $j = 1,\dots,N$. Denote the partition

$$\{a = \xi_0, y_0^r, x_2, \dots, x_{i_j-2}, y_j^r, x_{i_j} = \xi_j, y_j^r, x_{i_j+2}, \dots, x_{n-2}, y_N^l, b = \xi_N\}$$

by $\kappa(\epsilon)$ and take any refinement $\kappa := \{z_l: l = 0,\dots,m\}$ of $\kappa(\epsilon)$. For $j = 1,\dots,N$, let z_{j-1}^r and z_j^l denote the points of the partition κ which are on the right nearest to ξ_{j-1} and on the left nearest to ξ_j, respectively. By the construction of κ^* (cf. Lemma 2.20 and (5.14)), we have

$$V_p(f; [\Delta_l^\kappa]) \leq \epsilon/\big[4c_p^2(2 + \zeta(2/p))\|\mathcal{P}_a(f)\|_\infty V_p(f; [a,b])\big] \tag{5.16}$$

and

$$2c_p^2(2 + \zeta(2/p))V_p(f; [\Delta_l^\kappa]) \leq 1/2 \tag{5.17}$$

whenever the interval $[\Delta_l^\kappa]$ contains no points ξ_j for $j = 0, 1,\dots,N$, say $l \in M$ in this case. By Corollary 4.24 and Theorem 4.29, with $G := G^L \equiv G^R$, for these intervals we can write

$$\Delta_l^\kappa(\mathcal{P}_a(f)) = (\Delta_l^\kappa f)\mathcal{P}_a(f)(z_{l-1}) + (G(f; z_{l-1}, z_l) - \mathbb{1} - \Delta_l^\kappa f)\mathcal{P}_a(f)(z_{l-1}), \quad \forall l \in M.$$

By (4.79) and (5.17), for all $l \in M$, we have

$$\big\|G(f; z_{l-1}, z_l) - \mathbb{1} - \Delta_l^\kappa f\big\| \leq 2(2 + \zeta(2/p))V_p(f; [\Delta_l^\kappa])^2. \tag{5.18}$$

For the other intervals, which are either $[\xi_{j-1}, z_{j-1}^r]$ or $[z_j^l, \xi_j]$ for $j = 1,\dots,N$, by (5.2) and (5.4), we can write either

$$\Delta_l^\kappa \mathcal{P}_a(f) = (\Delta^+ f)(\xi_{j-1})\mathcal{P}_a(f)(\xi_{j-1}) + [\mathcal{P}_a(f)(z_{j-1}^r) - \mathcal{P}_a(f)(\xi_{j-1}+)],$$

or

$$\Delta_l^\kappa \mathcal{P}_a(f) = (\Delta_- f)(\xi_j)\mathcal{P}_a(f)(\xi_j-) + [\mathcal{P}_a(f)(\xi_j-) - \mathcal{P}_a(f)(z_j^l)],$$

respectively, for some $j = 1,\dots,N$. Therefore, using inequalities (2.21) and (2.9), by (5.15), (5.16) and (5.18), we conclude that

$$\Big(\sum_{l=1}^{m}\|\Delta_l^\kappa \mathcal{P}_a(f)\|^p\Big)^{1/p} \leq c_p\big[(V_p^*(f; [a,b]) + \epsilon)\|\mathcal{P}_a(f)\|_\infty + c_p A^{1/p}$$

$$+ c_p\Big(\sum_{l \in M}\|G(f; z_{l-1}, z_l) - \mathbb{1} - \Delta_l^\kappa f\|^p\Big)^{1/p}\|\mathcal{P}_a(f)\|_\infty\big]$$

$$\leq c_p V_p^*(f; [a,b])\|\mathcal{P}_a(f)\|_\infty + \epsilon(1 + c_p\|\mathcal{P}_a(f)\|_\infty).$$

Since $\epsilon > 0$ is arbitrary this proves Proposition 5.3. $\qquad\square$

5.2. The Duhamel formula. Here the algebraic identity (4.49) is extended to product integrals. Following Dollard and Friedman (1979, p. 33) we call such an extension a Duhamel formula (cf. Theorem 5.5 below). A simpler variant of this formula known under the same name is the following: for any matrices (or elements of $\mathbb{B}$) A, B and each $y \geq 0$,

$$\exp(yA) - \exp(yB) = \int_0^y \exp((y - x)A)\, dx (A - B) \exp(xB).$$

The proof of our extension of the Duhamel formula is based on a convergence theorem for (CY) integrals as given by Proposition 3.33. We start with a construction of a sequence of step functions giving approximations to product integrals involved in the formula.

Lemma 5.4. *Let $g, f \in W_p([a, b]; \mathbb{B})$, $0 < p < 2$. Then there exists an increasing sequence of partitions $\{\zeta_k \colon k \geq 1\}$ of $[a, b]$ such that the set $D := \cup_k \zeta_k$ is dense in $[a, b]$ and contains $a+$, $b-$ and $\xi-$, ξ, $\xi+$ for each jump ξ of g or f with $a < \xi < b$. Moreover, for each $y \in D$,*

$$\lim_{k \to \infty} P(\zeta_k, [x_{j(y)}^k, b], g) = \prod_y^b (\mathbb{1} + dg), \tag{5.19}$$

$$\lim_{k \to \infty} P(\zeta_k, [a, x_{j(y)}^k], f) = \prod_a^y (\mathbb{1} + df) \tag{5.20}$$

and

$$\lim_{k \to \infty} P(\zeta_k, [a, x_{j(y)-1}^k], f) = \prod_a^{y - \vee a} (\mathbb{1} + df), \tag{5.21}$$

where $(\xi+)- := \xi$ and $(\xi-)- := \xi-$ for any $\xi \in [a, b]$ and for each partition ζ_k, $j(y) := j_k(y) := \min\{j \geq 0 \colon y \leq x_j^k,\ x_j^k \in \zeta_k\}$, and $y = x_{j(y)}^k$ for all sufficiently large k.

Proof. Let $\{\xi_k \colon k \geq 2\}$ and $\{\eta_k \colon k \geq 2\}$ denote the jumps of g and f, respectively, other than a, b. Let $\xi_1 := b$, $\eta_1 := a$. For each $k \geq 1$, put

$$r_k := \left\{ a + l 2^{-k} \colon l = 1, \ldots, [2^k(b - a)] \right\}.$$

We will proceed recursively. To construct ζ_1, by Theorem 4.23, there exists a partition $\zeta(a, g)$ of $[a, b]$ such that $\left\| P(\zeta, [a, b], g) - \prod_a^b (1 + dg) \right\| \leq 1$ for all refinements ζ of $\zeta(a, g)$, and a partition $\zeta(b, f)$ of $[a, b]$ such that $\left\| P(\zeta, [a, b], f) - \prod_a^b (1 + df) \right\| \leq 1$ for all refinements ζ of $\zeta(b, f)$. We may and do assume that each $\zeta(a, g)$ and $\zeta(b, f)$ contains $(a + b)/2$. Let

$$\zeta_1 := \{x_1^1, \ldots, x_{n(1)}^1\} := \{a+, b-\} \cup \zeta(a, g) \cup \zeta(b, f).$$

Assume that $k \geq 2$ and we already have partitions $\zeta_1, \ldots, \zeta_{k-1}$ and $\zeta_{k-1} = \{x_0^{k-1}, \ldots, x_{n(k-1)}^{k-1}\}$. To construct ζ_k using Theorem 4.23, for each $i = 1, \ldots, n(k-1)$, one can find a partition $\zeta_k(x_i^{k-1}, g)$ of $[x_i^{k-1}, b]$ such that

$$\left\| P(\zeta, [x_i^{k-1}, b], g) - \prod_{x_i^{k-1}}^b (\mathbb{1} + dg) \right\| \leq 1/k$$

for all refinements ζ of $\zeta_k(x_i^{k-1}, g)$ and a partition $\zeta_k(x_i^{k-1}, f)$ of $[a, x_i^{k-1}]$ such that

$$\left\| P(\zeta, [a, x_i^{k-1}], f) - \prod_a^{x_i^{k-1}} (\mathbb{1} + df) \right\| \leq 1/k$$

for all refinements ζ of $\zeta_k(x_i^{k-1}, f)$. We may and do assume that the new partitions $\zeta_k(x_i^{k-1}, g)$ and $\zeta_k(x_i^{k-1}, f)$ include also $r_k \cap [x_i^{k-1}, b]$ or $r_k \cap [a, x_i^{k-1}]$, respectively. Let

$$\zeta_k := \{\eta_k-, \eta_k, \eta_k+, \xi_k-, \xi_k, \xi_k+\} \cup \cup_{i=1}^{n(k-1)} \left(\zeta_k(x_i^{k-1}, g) \cup \zeta_k(x_i^{k-1}, f) \right).$$

We arrive recursively at a sequence of partitions $\{\zeta_k : k \geq 1\}$. By the construction the set $D := \cup_k \zeta_k$ is dense in $[a, b]$ and contains $a+$, $b-$ and $\xi-, \xi, \xi+$ for all jumps ξ of g or f, $a < \xi < b$. To check (5.19) take any $\epsilon > 0$ and $y \in D$. Choose k_0 so large that $y \in \zeta_{k_0}$ and $1/k_0 < \epsilon$. Note that $x_{j(y)} = y$ for all $k \geq k_0$. Then, for each $k \geq k_0$, we have

$$\left\| P(\zeta_k, [x_{j(y)}^k, b], g) - \prod_y^b (\mathbb{1} + dg) \right\| \leq 1/k < \epsilon.$$

Similarly (5.20) follows.

We show that (5.21) follows from (5.20). By Lemma 4.19, there exists a finite constant $C(f)$ such that $\|P(\kappa, \Delta, f)\| \leq C(f)$ for any partition κ of any subinterval Δ of $[a, b]$. Thus, for each $y \in D$ and for all k such that $y \in \zeta_k = \{x_i^k : i = 0, \ldots, n(k)\}$, we have

$$\left\| P(\zeta_k, [a, x_{j(y)}^k], f) - P(\zeta_k, [a, x_{j(y)-1}^k], f) \right\| \leq C(f) \|f(y) - f(x_{j(y)-1}^k)\|. \tag{5.22}$$

Therefore (5.21) follows from (5.20) and (5.22) if $y = \xi_i-$ for some i because f is regulated. (5.21) for $y = \xi := \xi_i$ follows from Lemma 5.1 and from (5.20) for $y = \xi-$ because $x_{j(\xi)-1} = \xi-$ for large enough k. If $y \in D$ is not a point $\xi+$ and $(\Delta_- f)(y) = 0$ then, by Lemma 4.21, $\prod_a^{y-\vee a}(1 + df) = \prod_a^y(1 + df)$. Therefore (5.21) again follows from (5.20) and (5.22) because f is left-continuous at y in this case. If $y = \xi+$ then (5.21) follows from (5.20) for $y = \xi$. The proof of Lemma 5.4 is complete. $\qquad \square$

Now we are ready to prove the Duhamel formula, the main result of this subsection.

Theorem 5.5. Let $g, f \in \mathcal{W}_p([a, b]; \mathbb{B})$, $0 < p < 2$. Then

$$\prod_a^b (\mathbb{1} + dg) - \prod_a^b (\mathbb{1} + df) = (LY) \int_a^b \mathcal{P}^b(g) \, d(g - f) \, \mathcal{P}_a(f) \tag{5.23}$$

$$= (RY) \int_a^b \mathcal{P}^b(g) \, d(g - f) \, \mathcal{P}_a(f). \tag{5.24}$$

Proof. By Theorem 4.23, the left side of (5.23) is well-defined. By Lemma 2.3, Proposition 5.3 and (2.15), the indefinite integrals $\mathcal{P}^b(g)$ and $\mathcal{P}_a(f)_-$ belong to

$\mathcal{W}_p([a,b];\mathbb{B})$. Therefore by Corollary 3.28, the (LY) and (RY) integrals in (5.23) and (5.24) exist. Hence, by Theorem 3.12, the second equality holds. To prove the first equality we will show that it is a limit of corresponding equalities for step functions approximating the product integrals.

Given a partition $\kappa = \{x_i : i = 0, \ldots, n\}$ of $[a,b]$ define a function G_κ on $[a,b]$ by

$$G_\kappa(y) := P(\kappa, [x_{\tau(y)}, b], g) = \begin{cases} \prod_{j=i+1}^{n}(\mathbb{1} + \Delta_j^\kappa g), & \text{for } y \in (x_{i-1}, x_i], \ 2 \le i \le n, \\ \prod_{j=2}^{n}(\mathbb{1} + \Delta_j^\kappa g), & \text{for } y \in [a, x_1] \end{cases}$$

and a function F_κ on $[a,b]$ by

$$F_\kappa(y) := P(\kappa, [a, x_{\tau(y)-1}], f) = \begin{cases} \prod_{j=1}^{i-1}(\mathbb{1} + \Delta_j^\kappa f), & \text{for } y \in (x_{i-1}, x_i], \ 2 \le i \le n, \\ \mathbb{1}, & \text{for } y \in [a, x_1], \end{cases}$$

where $\tau(y) := \tau_\kappa(y) := \min\{1 \le i \le n : y \le x_i\}$. Note that for $j(\cdot)$ as in Lemma 5.4, $\tau(\cdot) \equiv j(\cdot)^{(a+)}$. Here and in the proof below products over an empty set of indices are set equal to $\mathbb{1}$. Let $h := g - f$. Since h_+ is right-continuous and G_κ, F_κ are left-continuous (constants) on $[x_{i-1}+, x_i]$, by definition of (Y_2) integral, for each $i = 1, \ldots, n$, we get

$$(CY)\int_{x_{i-1}}^{x_i} G_\kappa\, dh\, F_\kappa = (MPS)\int_{x_{i-1}+}^{x_i}(G_\kappa)_-^{(x_{i-1})}\, dh_+^{(x_i)}(F_\kappa)_-^{(x_{i-1})} + [G_\kappa\Delta^+hF_\kappa](x_{i-1})$$
$$= P(\kappa, [x_i, b], g)(\Delta_i^\kappa h)P(\kappa, [a, x_{i-1}], f)$$
$$- [(G_\kappa)_+\Delta^+h(F_\kappa)_+ - G_\kappa\Delta^+hF_\kappa](x_{i-1}).$$

Therefore by Proposition 3.22 for (CY) integrals and the algebraic identity (4.49), we have

$$(CY)\int_a^b G_\kappa\, dh\, F_\kappa = \sum_{i=1}^{n}(CY)\int_{x_{i-1}}^{x_i} G_\kappa\, dh\, F_\kappa = P(\kappa, [x_1, b], g)(\Delta_1^\kappa h)$$
$$+ \sum_{i=2}^{n-1} P(\kappa, [x_i, b], g)(\Delta_i^\kappa h)P(\kappa, [a, x_{i-1}], f)$$
$$+ (\Delta_n^\kappa h)P(\kappa, [a, x_{n-1}], f)$$
$$- \sum_{i=1}^{n}[(G_\kappa)_+\Delta^+h(F_\kappa)_+ - G_\kappa\Delta^+hF_\kappa](x_{i-1})$$
$$= P(\kappa, [a, b], g) - P(\kappa, [a, b], f)$$
$$- \sum_{i=2}^{n}[(G_\kappa)_+\Delta^+h(F_\kappa)_+ - G_\kappa\Delta^+hF_\kappa](x_{i-1}). \tag{5.25}$$

Take the sequence $\zeta_k = \{x_i^k : i = 0, \ldots, n(k)\}$, $k \ge 1$, of partitions of $[a,b]$ constructed in Lemma 5.4. By (5.19) with $y = a$ and (5.20) with $y = b$, we get

$$\lim_{k\to\infty}[P(\zeta_k, [a,b], g) - P(\zeta_k, [a,b], f)] = \prod_a^b(\mathbb{1} + dg) - \prod_a^b(\mathbb{1} + df). \tag{5.26}$$

For each $k \geq 1$, let $F_k := F_{\zeta_k}$ and $G_k := G_{\zeta_k}$. To show (5.23), by (5.25) and Definition 3.11, it suffices to prove that

$$\lim_{k \to \infty} (CY) \int_a^b G_k \, dh \, F_k = (CY) \int_a^b \mathcal{P}^b(g)^{(a+)} \, dh \, \mathcal{P}_a(f)_-^{(a)} \qquad (5.27)$$

and

$$\lim_{k \to \infty} \sum_{i=2}^{n(k)} [(G_k)_+ \Delta^+ h (F_k)_+ - G_k \Delta^+ h F_k](x_{i-1}^k)$$

$$= \sum_{(a,b)} [\mathcal{P}^b(g)_+ \Delta^+ h \mathcal{P}_a(f) - \mathcal{P}^b(g) \Delta^+ h \mathcal{P}_a(f)_-]. \qquad (5.28)$$

First we will show that the hypotheses of the convergence theorem of Proposition 3.33 hold, and then that (5.27) holds. To begin with, by Lemmas 4.21 and 5.4, we have that

$$\lim_{k \to \infty} G_k = \mathcal{P}^b(g)^{(a+)} \quad \text{and} \quad \lim_{k \to \infty} F_k = \mathcal{P}_a(f)_-^{(a)} \quad \text{on } D, \qquad (5.29)$$

and so densely in $[a, b]$. Next we show that both sequences converge locally uniformly to their limits at each discontinuity point of $h = g - f$. By Lemma 4.19, there exists a finite constant $C(g)$ such that, for any partition κ of any subinterval Δ of $[a, b]$,

$$\|P(\kappa, \Delta, g)\| \leq C(g). \qquad (5.30)$$

Let y be a jump of g or f. Then $y \in D := \cup_k \zeta_k$. We start by showing that if $y > a$,

$$G_k \to \mathcal{P}^b(g) \quad \text{as } k \to \infty, \text{ uniformly on the left at } y. \qquad (5.31)$$

Let $\epsilon > 0$. First using Lemma 5.4 choose an integer k_0 so large that, for all $k \geq k_0$,

$$y-, y \in \zeta_k = \{x_i^k : i = 0, \ldots, n(k)\} \qquad (5.32)$$

and

$$\|G_k(y-) - \mathcal{P}^b(g)(y-)\| \vee \|G_k(y) - \mathcal{P}^b(g)(y)\| \leq \epsilon/3. \qquad (5.33)$$

Second, by Lemmas 2.5 and 2.19, there exists a $\delta \in (0, y - a)$ such that, for all $z \in [y - \delta, y)$,

$$V_p(g; [z, y-]) \leq [4c_p(1 + \zeta(2/p))]^{-1} \wedge (\epsilon/[6C(g)]), \qquad (5.34)$$

where $c_p = 1 \vee 2^{(1/p)-1}$, and, by Lemma 5.2,

$$\|\mathcal{P}^b(g)(z) - \mathcal{P}^b(g)(y-)\| \leq \epsilon/3. \qquad (5.35)$$

By (5.30), (4.52) and (5.34), we have

$$\|G_k(y-) - G_k(z)\| \leq C(g)\|\mathbb{I} - P(\zeta_k, [x_{j(z)}^k, x_{j(y-)}^k], g)\|$$

$$\leq 2C(g)V_p(g; [z, y-]) \leq \epsilon/3,$$

whenever $k \geq k_0$ and $z \in [y - \delta, y)$. Thus using also (5.33) and (5.35) we arrive at

$$\|G_k(z) - \mathcal{P}^b(g)(z)\| \leq \|G_k(z) - G_k(y-)\| + \|G_k(y-) - \mathcal{P}^b(g)(y-)\|$$
$$+ \|\mathcal{P}^b(g)(y-) - \mathcal{P}^b(g)(z)\| \leq \epsilon/3 + \epsilon/3 + \epsilon/3 = \epsilon$$

for all $k \geq k_0$ and all $z \in [y - \delta, y)$. Since the same bound holds when $z = y$ by (5.33) this proves (5.31).

To show that

$$G_k \to \mathcal{P}^b(g)^{(a+)} \quad \text{as } k \to \infty, \text{ uniformly on the right at } y, \qquad (5.36)$$

whenever $y < b$, we proceed as above except that now we choose an interval $(y, x^{k_0}_{j(y)+2}]$ instead of $(y, y + \delta)$ because $z \leq j(z)$ for all $z \in [a, b]$. Moreover, we let $(\xi-)+ := \xi$ for $a < \xi \leq b$ and $(\xi+)+ := \xi+$ for $a \leq \xi < b$. Note that for $\tau(\cdot)$ as in the definition of G_κ and F_κ, $\tau(\cdot) \equiv j(\cdot)^{(a+)}$. Then (5.36) together with (5.31) implies

$$G_k \to \mathcal{P}^b(g)^{(a+)} \quad \text{as } k \to \infty, \text{ locally uniformly at each jump of } g \text{ or } f. \qquad (5.37)$$

Turning to the sequence $\{F_k\}$, first we have to show that if y is a jump of f or g and $y > a$ then

$$F_k \to \mathcal{P}_a(f)_- \quad \text{as } k \to \infty, \text{ uniformly on the left at } y. \qquad (5.38)$$

To this end we argue as in the proof of (5.31) except that now we choose $y \in \zeta_k$ and $\|F_k(y) - \mathcal{P}_a(f)(y-)\| \leq \epsilon/3$, for a given $\epsilon > 0$ and for all large enough k, instead of (5.32) and (5.33).

Now we show that if y is a jump of f or g and $y < b$,

$$F_k \to \mathcal{P}_a(f)^{(a)}_- \quad \text{as } k \to \infty, \text{ uniformly on the right at } y. \qquad (5.39)$$

Let $C(f)$ be the same constant as in (5.30) except that the function g is replaced by f. Let $\epsilon > 0$. Using Lemma 5.4 again choose an integer k_0 so large that, for all $k \geq k_0$, we have $y, y+ \in \zeta_k = \{x^k_i : i = 0, \ldots, n(k)\}$,

$$\|F_k(y) - \mathcal{P}_a(f)^{(a)}_-(y)\| \vee (\|F_k(y+) - \mathcal{P}_a(f)(y)\|[1 + \|(\Delta^+ f)(y)\|]) \leq \epsilon/3, \quad (5.40)$$

and by Lemmas 2.5 and 2.18, for $u := x^{k_0}_{j(y)+2}$,

$$V_p(f; [y+, u]) \leq [4c_p(1 + \zeta(2/p))]^{-1} \wedge (\epsilon/[6C(f)]), \qquad (5.41)$$

and by Lemma 5.2, for any $z \in (y, u]$,

$$\|\mathcal{P}_a(f)(y+) - \mathcal{P}_a(f)(z-)\| \leq \epsilon/3. \qquad (5.42)$$

By (4.52) and (5.41), we have

$$\|F_k(z) - [\mathbb{1} + (\Delta^+ f)(y)]F_k(y+)\| \leq C(f)\|\mathbb{1} - P(\zeta_k, [x^k_{j(y+)}, x^k_{j(z)-1}], f)\|$$
$$\leq 2C(f)V_p(f; [y+, x^k_{j(z)}]) \leq \epsilon/3$$

whenever $k \geq k_0$ and $z \in (y, u]$ since $x^k_{j(z)} \leq u$. Therefore using also (5.2), (5.40) and (5.42) we arrive at

$$\|F_k(z) - \mathcal{P}_a(f)(z-)\| \leq \|F_k(z) - [\mathbb{1} + (\Delta^+ f)(y)]F_k(y+)\|$$

$$+\|F_k(y+) - \mathcal{P}_a(f)(y)\|[1 + \|(\Delta^+ f)(y)\|] + \|\mathcal{P}_a(f)(y+) - \mathcal{P}_a(f)(z-)\|$$

$$\leq \epsilon/3 + \epsilon/3 + \epsilon/3 = \epsilon$$

for all $k \geq k_0$ and all $z \in (y, u]$. Since the same bound holds for $z = y$ by (5.40) this proves (5.39) and together with (5.38) implies that

$$F_k \to \mathcal{P}_a(f)^{(a)}_- \quad \text{as } k \to \infty, \text{ locally uniformly at each jump of } g \text{ or } f. \tag{5.43}$$

By Lemmas 5.2 and 2.3, Proposition 5.3 and (2.15), the indefinite product integrals $\mathcal{P}^b(g)^{(a+)}$ and $\mathcal{P}_a(f)^{(a)}_-$ belong to $\mathcal{W}_p([a, b]; \mathbb{B})$. To apply Proposition 3.33 we still have to show that $\{F_k\}$ and $\{G_k\}$ are $\mathcal{W}_p$-sequences. We will do that for the sequence $\{F_k\}$ and note that similar arguments apply to $\{G_k\}$. Let $\kappa = \{y_l: l = 0, \ldots, m\}$ be the partition of $[a, b]$ given by Lemma 2.20 using the number

$$c = 1/\left[8c_p^2(1 + \zeta(2/p))\right]^p, \tag{5.44}$$

where $c_p = 1 \vee 2^{(1/p)-1}$. Take any $F_k = F_{\zeta_k}$ with $\zeta_k = \{x^k_i: i = 0, 1, \ldots, n(k)\}$. We note that

$$v_p(F_k; [a, b]) = \max\left\{ \sum_{j=1}^r \|\Delta^\eta_j F_k\|^p: \eta := \{x_{i(0)}, x_{i(1)}, \ldots, x_{i(r)}\} \subset \zeta_k \right\}. \tag{5.45}$$

For any such η, let $J := \{i(j): j = 1, \ldots, r\}$ and

$$J_1 := \left\{i(j) \in J: [\Delta^\eta_j] \supset [\Delta^\kappa_l] \text{ for some } l = 1, \ldots, m\right\}.$$

Let $J_2 := J \setminus J_1$. So, if $j \in J_2$ then the interval $[\Delta^\eta_j]$ contains at most one point of κ. Hence, by (2.9), (5.44) and Lemma 2.20,

$$v_p(f; [\Delta^\eta_j]) \leq 1/\left[4c_p(1 + \zeta(2/p))\right]^p, \quad \forall j \in J_2.$$

Therefore one can apply Lemma 4.18 with

$$\rho = 2c_p(1 + \zeta(2/p))V_p(f; [\Delta^\eta_j]) \leq 1/2.$$

Then by (4.52)

$$D_j(f) := \left\|P(\zeta_k, [x^k_{i(j-1)}, x^k_{i(j)}], f) - \mathbb{1}\right\| \leq 2V_p(f; [\Delta^\eta_j]). \tag{5.46}$$

Since $\text{card}(J_1) \leq m$, by (5.46) and (5.6), it follows that

$$\sum_{j=1}^r \|\Delta^\eta_j F_k\|^p \leq \sum_{j \in J_1} \left(\|F_k(x^k_{i(j)})\| + \|F_k(x^k_{i(j-1)})\|\right)^p + C(f)^p \sum_{j \in J_2} D_j(f)^p$$

$$\leq m\left[2C(f)\right]^p + \left[2C(f)\right]^p v_p(f; [a, b]).$$

Since J is an arbitrary subset of $\{1,\ldots,n(k)\}$, by (5.45), we get

$$v_p(F_k;[a,b]) \le [2C(f)]^p (m + v_p(f;[a,b]))$$

for each $k \ge 1$. Therefore $\{F_k\}$ is a $\mathcal{W}_p$-sequence.

Now, by (5.37), (5.29) and (5.43), we can apply Proposition 3.33 to conclude (5.27). To complete the proof of (5.23) we now need to show (5.28). By the definitions of the functions G_k and F_k and Lemma 5.4, for each $y \in D \cap (a,b)$, we have

$$\lim_{k\to\infty} [(G_k)_+ \Delta^+ h(F_k)_+ - G_k \Delta^+ h F_k](y)$$
$$= [\mathcal{P}^b(g)_+ \Delta^+ h \mathcal{P}_a(f) - \mathcal{P}^b(g)\Delta^+ h \mathcal{P}_a(f)_-](y).$$

Therefore it suffices to show that given $\epsilon > 0$ there exists a finite set $\mu \subset (a,b)$ such that

$$\sum_{(a,b)\setminus\mu} \left\| [\mathcal{P}^b(g)_+ \Delta^+ h \mathcal{P}_a(f) - \mathcal{P}^b(g)\Delta^+ h \mathcal{P}_a(f)_-] \right\| < \epsilon \tag{5.47}$$

and

$$\sum_{i\ge 2,\, x_{i-1}^k \in \zeta_k \setminus \mu} \left\| [(G_k)_+ \Delta^+ h(F_k)_+ - G_k \Delta^+ h F_k](x_{i-1}^k) \right\| < \epsilon \tag{5.48}$$

for all $k \ge 1$. As shown at the beginning of this proof, the (LY) integral in (5.23) exists. So by Definition 3.11, there is a finite set $\mu' \subset (a,b)$ such that (5.47) holds for all $\mu \supset \mu'$. To prove (5.48) for some $\mu \supset \mu'$, by (5.30) for f and g and Cauchy's inequality, for each subset $I_k \subset \{2,\ldots,n(k)\}$, we get

$$\sum_{i\in I_k} \left\| [(G_k)_+ \Delta^+ h(F_k)_+ - G_k \Delta^+ h F_k](x_{i-1}^k) \right\|$$
$$\le C(f) \sum_{i\in I_k} \left\| [\Delta^+ G_k \Delta^+ h](x_{i-1}^k) \right\| + C(g) \sum_{i\in I_k} \left\| [\Delta^+ h \Delta^+ F_k](x_{i-1}^k) \right\|$$
$$\le C(f)C(g)\left[\left(\sum_{i\in I_k} \|\Delta_i^{\zeta_k} g\|^2 \right)^{1/2} + \left(\sum_{i\in I_k} \|\Delta_{i-1}^{\zeta_k} f\|^2 \right)^{1/2} \right] \left(\sum_{i\in I_k} \|\Delta^+ h(x_{i-1}^k)\|^2 \right)^{1/2}$$
$$\le C(f)C(g)[V_2(g) + V_2(f)] v_p(h)^{1/2} \max_{i\in I_k} \|\Delta^+ h(x_{i-1}^k)\|^{1-(p/2)}. \tag{5.49}$$

To show that the right side of the above inequality is small if I_k does not contain indices of large jumps of the function h let $\epsilon > 0$. Then by Lemma 2.2, there exists a finite set $\mu \supset \mu'$ such that, if $y \notin \mu$,

$$\|\Delta^+ h(y)\|^{1-(p/2)} < \epsilon / [C(f)C(g)[V_2(g) + V_2(f)]v_p(h)^{1/2}]. \tag{5.50}$$

Therefore by (5.49) and (5.50), it follows that (5.48) holds for all $k \ge 1$. As was noted above, this implies (5.28) and, hence, (5.23). The proof of Theorem 5.5 is complete. $\square$

The following variant of the Duhamel formula will be used to differentiate indefinite product integrals. The proof follows by taking the negative of both sides in (5.23).

PART II. PRODUCT INTEGRALS AND p-VARIATION

Corollary 5.6. *Let* $g, f \in \mathcal{W}_p([a,b]; \mathbb{B})$, $0 < p < 2$. *Then*

$$\prod_a^b (\mathbb{1} + dg) - \prod_a^b (\mathbb{1} + df) = (LY) \int_a^b \mathcal{P}^b(f) \, d(g - f) \, \mathcal{P}_a(g). \qquad (5.51)$$

We finish this subsection by formulating conditions for simpler forms of the Duhamel formula to hold. The following is a special case of Theorem 5.5.

Corollary 5.7. *Let* $g, f \in \mathcal{W}_p([a,b]; \mathbb{B})$, $0 < p < 2$. *If* g *and* f *are right-continuous then*

$$\prod_a^b (\mathbb{1} + dg) - \prod_a^b (\mathbb{1} + df) = (CY) \int_a^b \mathcal{P}^b(g) \, d(g - f) \, \mathcal{P}_a(f)_-^{(a)}.$$

If g *and* f *are left-continuous then*

$$\prod_a^b (\mathbb{1} + dg) - \prod_a^b (\mathbb{1} + df) = (CY) \int_a^b \mathcal{P}^b(g)_+^{(b)} \, d(g - f) \, \mathcal{P}_a(f).$$

5.3. Differentiability. The main result of this subsection is the proof of analyticity of product integral operators. We proceed by recalling and extending some facts concerning differentiability of operators acting between quasinormed spaces.

If (V, α) and (W, β) are two quasinormed spaces, a linear map A from V into W will be called *bounded* if $\gamma(A) := \sup\{\beta(Av) : \alpha(v) \leq 1\} < \infty$. The following is known (cf. Theorems II.1.1 and III.2.10 in Rolewicz, 1972).

Proposition 5.8. *The following are equivalent:*

(1) A *is continuous on* V;
(2) A *is continuous at* 0;
(3) A *is bounded.*

Let $L(V, W)$ be the space of all bounded linear maps of V into W. Then it is easily seen that $L(V, W)$ is a vector space and γ is a quasinorm on it, with $c(\gamma) \leq c(\beta)$.

Let (V, α) be a quasinormed space, U an open subset of V and $u \in U$. Let ψ be a function from U into W where (W, β) is a quasinormed space. Then ψ will be called *Fréchet differentiable* at u if for some $(D\psi)(u)(\cdot) \in L(V, W)$,

$$\beta[\psi(u + v) - \psi(u) - (D\psi)(u)(v)]/\alpha(v) \to 0 \quad \text{as} \quad v \to 0 \quad \text{in} \quad V.$$

This definition agrees with the usual one for normed spaces and with the more general definition in topological vector spaces, in other words differentiability uniformly over bounded sets (cf. Sect. 2.3 in Nashed, 1971 or Averbukh and Smolyanov, 1967).

Let $L^{(1)}(V, W) := L(V, W)$ and $D^1\psi(u) := D\psi(u)$. For $k = 1, 2, \cdots$, let $L^{(k+1)}(V, W) := L(V, L^{(k)}(V, W))$ and $D^{k+1}\psi := D(D^k\psi)$ if this derivative exists (in the same Fréchet sense). Note that for each k, recursively, $L^{(k)}(V, W)$ is a quasinormed space with a quasinorm γ_k and $c(\gamma_k) \leq c(\beta)$ for all k. Here $(D^k\psi)(u)$ will be called the kth *Fréchet derivative* of ψ at u if it exists in $L^{(k)}(V, W)$. So the $(k+1)$st derivative will be the derivative, when it exists, of the kth derivative.

A k-*form* is a k-linear map $B : V \times V \times \cdots \times V \mapsto W$. It will be called *bounded* if

$$\chi(B) := \chi_k(B) := \sup\{\beta(B(u_1, \ldots, u_k)) : \alpha(u_i) \leq 1 \text{ for all } i\} < \infty.$$

Let $L(^kV, W)$ be the set of all bounded k-forms. Then $(L(^kV, W), \chi_k)$ is a quasinormed linear space. The following is straightforward to prove:

Proposition 5.9. *For any quasinormed spaces* (V, α) *and* (W, β) *with* (W, β) *complete, and any* $k = 1, 2, \cdots, (L(^kV, W), \chi_k)$ *is also complete.*

There is a natural linear 1-1 correspondence between $L^{(k)}(V, W)$ and $L(^kV, W)$ which preserves quasinorms: Chae (1985, Secs. 3.11, 3.12) shows this for norms and the proof extends to quasinorms. The element of $L(^kV, W)$ corresponding to $D^k\psi(u)$ will be written $d^k\psi(u)$ and called the *k*th *differential* of ψ at u. $D^k\psi(u)$ also naturally defines an element of $L(V, L(^{k-1}V, W))$ which will be called $(D^1 d^{k-1})\psi(u)$. A *k*th differential $(d^k\psi)(u)$ is always a *symmetric k-form B*, that is a B such that

$$B(u_{\pi(1)}, \ldots, u_{\pi(k)}) \; = \; B(u_1, \ldots, u_k)$$

for any $u_1, \ldots, u_k$ in V and any permutation π of $\{1, \ldots, k\}$: this "Schwarz theorem" is proved in Chae (1985, Secs. 7.4, 7.9) and the proof extends to quasinormed spaces. Any *k*-form B has a *symmetrization* $B^{(s)}$ defined by

$$B^{(s)}(u_1, \ldots, u_k) \; := \; (k!)^{-1} \sum_\pi B(u_{\pi(1)}, \ldots, u_{\pi(k)}),$$

where the sum is over all permutations π of $\{1, \ldots, k\}$. If B is bounded, so is $B^{(s)}$. Let

$$Bu^{\otimes k} := B(u, u, \ldots, u) \qquad \text{and} \qquad Bv^{\otimes i} \otimes w^{\otimes j} := B(v, \ldots, v, w, \ldots, w),$$

where u appears k times, v appears i times and w j times. Let B be a *k*-form and $0 \le l < k$. Define a mapping

$$B(u_1, \ldots, u_l): \underbrace{V \times \cdots \times V}_{k-l \text{ times}} \to W$$

by $(u_{l+1}, \ldots, u_k) \to B(u_1, \ldots, u_l, u_{l+1}, \ldots, u_k)$. For each *symmetric k-form B*, induction on k implies the following binomial formula:

$$B(v + w)^{\otimes k} = Bv^{\otimes k} + \sum_{l=1}^{k-1} \binom{k}{l} Bv^{\otimes k-l} \otimes w^{\otimes l} + Bw^{\otimes k}. \tag{5.52}$$

However, we need such a formula for a possibly asymmetric form.

Let $B \in L(^kV, W)$ and $i_0 = 0 < i_1 < \cdots < i_{l-1} < i_l = k$. Define a partial symmetrization of B by

$$B\{u_1, \ldots, u_{i_1}\} \cdots \{u_{i_{l-1}+1}, \ldots, u_k\} := \sum_\pi B(u_{\pi(1)}, \ldots, u_{\pi(k)}), \tag{5.53}$$

where the sum is over those permutations π of $\{1, \ldots, k\}$ for which $\pi(i_j + 1) < \cdots < \pi(i_{j+1})$ whenever $i_{j+1} > i_j + 1$ and $j = 0, 1, \ldots, l - 1$. Let $d_j := i_j - i_{j-1}$ for $j = 1, \ldots, l$. We note that if $u_{i_{j-1}+1} = \cdots = u_{i_j} = v_j$ for $j = 1, \ldots, l$ then

$$B\{v_1^{\otimes d_1}\} \cdots \{v_l^{\otimes d_l}\} = \frac{m!}{d_1! \cdots d_l!} B^{(s)} v_1^{\otimes d_1} \otimes \cdots \otimes v_l^{\otimes d_l}. \tag{5.54}$$

Formula (5.52) can be extended to possibly asymmetric *k*-forms as follows:

$$B\{(v + w)^{\otimes k}\} = B\{v^{\otimes k}\} + \sum_{l=1}^{k-1} B\{v^{\otimes(k-l)}\}\{w^{\otimes l}\} + B\{w^{\otimes k}\}. \tag{5.55}$$

Definition 5.10. Let V and W be two quasinormed spaces. A mapping $P_k\colon V \mapsto W$, for some $k = 1, 2, \ldots$, is called a *homogeneous polynomial*, or *monomial*, *of degree* k if there exists a k-form $\hat{P}_k\colon V \times \cdots \times V \mapsto W$ such that $P_k(v) = \hat{P}_k(v, \ldots, v)$, $v \in V$.

Remark. In this definition we do not assume the k-form $\hat{P}_k$ to be symmetric. But, since for any k-form B and $v \in V$, $B^{(s)}(v, \ldots, v) = B(v, \ldots, v)$, $\hat{P}_k$ could be taken symmetric if desired.

Definition 5.11. Let (V, α) and (W, β) be two quasinormed spaces and let U be an open subset of V. A mapping $\psi\colon U \to W$ is said to be *analytic* or *holomorphic* on U if, for each $u \in U$ and $k = 1, 2, \ldots$, there is a homogeneous polynomial P_k^u of degree k and an $r > 0$ such that, for all $v \in U$ with $\alpha(v - u) < r$,

$$\psi(v) = \psi(u) + \sum_{k=1}^{\infty} P_k^u(v - u), \tag{5.56}$$

where the series converges in the topology (equivalently the quasinorm or metric) of W.

Remark. Some authors require uniform convergence of the series (5.56) for $\alpha(v - u) \le s < r$ (cf. p. 626 in Nachbin, 1973 or p. 168 in Chae, 1985).

The *radius of uniform convergence* of the series (5.56), say $\rho(\psi(u))$, is the largest ρ such that the series converges uniformly for $\alpha(v - u) \le \rho_0$ for each $\rho_0 < \rho$. Chae (1985, Sec. 11.8) gives an example where a function ψ is analytic on a complex Banach space H with complex values and has a series expansion (5.56) convergent absolutely in $\mathbb{C}$ for all $v \in H$, taking $u = 0$, but where the radius of uniform convergence is finite, specifically where H is the Hilbert space of all square-summable sequences $z := \{z_k\}_{k \ge 1}$ and $\psi(z) := \sum_k z_k^k$.

If the radius of uniform convergence $\rho(\psi(u)) > 0$ for (5.56), then for each $k = 1, 2, \cdots$, $D^k \psi(u)$ exists and $P_k^u = d^k \psi(u)/k!$ for each $u \in U$, as shown by Chae (1985, Sec. 11.12) for normed spaces. We extend this fact to quasinormed spaces as follows:

Theorem 5.12. *Let (V, α) and (W, β) be a quasinormed space and a complete quasinormed space, respectively. Let U be an open subset of V and $\psi\colon U \mapsto W$ be an analytic function on U, with P_k^u as in (5.56). Let $u \in U$ be such that*

$$L := L(u) := \limsup_{k \to \infty} \chi(\hat{P}_k^u)^{1/k} < \infty. \tag{5.57}$$

Then the following two statements hold:

(1) *The radius of uniform convergence of (5.56) $\rho(\psi(u)) \ge (c(\beta)L)^{-1}$.*

(2) *If, in addition, β is a q-subadditive quasinorm then ψ has an nth Fréchet derivative at u and $d^n \psi(u) = n!(\hat{P}_n^u)^{(s)}$ for all $n = 1, 2, \ldots$.*

Proof. (1). Let $v \in U$ be such that $\alpha(u - v) \le r < (c(\beta)L)^{-1}$. Since (W, β) is complete, by Lemma 2.9, we get

$$\beta\Big(\sum_{k=1}^{\infty} P_k^u(v - u)\Big) \le \sum_{k=1}^{\infty} c(\beta)^k \beta\big(P_k^u(v - u)\big) \le \sum_{k=1}^{\infty} c(\beta)^k \chi_k(\hat{P}_k^u) \alpha(v - u)^k.$$

Then the claim (1) follows by the root test on a series of nonnegative numbers.

(2). Let $w \in U$ and $r > 0$ be such that $\alpha(u - w) < r^{1/q} < 1 \wedge (c(\beta)L)^{-1}$. We claim that ψ has an nth Fréchet derivative at w for each $n \geq 1$ and its nth differential is given by

$$d^n \psi(w)(v_1, \ldots, v_n) = \hat{P}^u_n \{v_1\} \cdots \{v_n\} + \sum_{m=1}^{\infty} \hat{P}^u_{m+n} \{(w - u)^{\otimes m}\} \{v_1\} \cdots \{v_n\}$$

$$= n! (\hat{P}^u_n)^{(s)}(v_1, \ldots, v_n) \qquad (5.58)$$

$$+ \sum_{m=1}^{\infty} \frac{(m+n)!}{m!} (\hat{P}^u_{m+n})^{(s)}(w - u)^{\otimes m} \otimes v_1 \otimes \cdots \otimes v_n,$$

where $(v_1, \ldots, v_n) \in V^n$ and the last equality follows from (5.54).

We prove the claim by induction on n. Namely, we show that (5.58) holds for an arbitrary but fixed $n \geq 1$ if $n = 1$, or if $n \geq 2$ and (5.58) holds for $n - 1$. Let

$$R_0 := \begin{cases} \hat{P}^u_{n-1} \underbrace{\{\cdot\} \cdots \{\cdot\}}_{n-1 \text{ times}} = (n-1)! (\hat{P}^u_{n-1})^{(s)}, & \text{for } n \geq 2 \\ \psi(u), & \text{for } n = 1 \end{cases}$$

and let, for any $(v_1, \ldots, v_m) \in V^m$, $m \geq 1$,

$$R_m(v_1, \ldots, v_m) := \begin{cases} \hat{P}^u_{m+n-1} \{v_1, \ldots, v_m\} \underbrace{\{\cdot\} \cdots \{\cdot\}}_{n-1 \text{ times}}, & \text{for } n \geq 2 \\ \hat{P}^u_m(v_1, \ldots, v_m), & \text{for } n = 1. \end{cases}$$

By the induction assumption (cf. (5.58) and (5.56)), for $\alpha(v - u) < r^{1/q}$ the series

$$\phi(v) := R_0 + \sum_{m=1}^{\infty} R_m(v - u)^{\otimes m} = \begin{cases} d^{n-1} \psi(v), & \text{for } n \geq 2 \\ \psi(v), & \text{for } n = 1 \end{cases}$$

converges in $L(^{n-1}V, W)$, where $L(^0 V, W) := W$. First, we show that, for all v with $\alpha(v - w) < [r - \alpha(w - u)^q]^{1/q}$, the series

$$Q_k(v - w)^{\otimes k} := R_k(v - w)^{\otimes k} + \sum_{m=k+1}^{\infty} R_m \{(w - u)^{\otimes (m-k)}\} \{(v - w)^{\otimes k}\}$$

for each $k \geq 1$ and $\sum_{k=1}^{\infty} Q_k(v - w)^{\otimes k}$ converge in $L(^{n-1}V, W)$. Then we show that

$$\phi(v) = \phi(w) + \sum_{k=1}^{\infty} Q_k(v - w)^{\otimes k}. \qquad (5.59)$$

Since the quasinorm β is q-subadditive, by (5.53), for each $g, h, v_1, \ldots, v_{n-1} \in V$ and each $1 \leq k < m$, we have

$$\beta \left(\hat{P}^u_{m+n-1} \{g^{\otimes (m-k)}\} \{h^{\otimes k}\} \{v_1\} \cdots \{v_{n-1}\} \right)$$

$$\leq \left(\sum_\pi \beta(\hat{P}^u_{m+n-1}(d_{\pi(1)}, \ldots, d_{\pi(m+n-1)}))^q \right)^{1/q} \leq \left(\frac{(m+n-1)!}{(m-k)!k!} \right)^{1/q} \times$$

$$\times \alpha(g)^{m-k}\alpha(h)^k\alpha(v_1)\cdots\alpha(v_{n-1}) \sup_{\alpha(d_i)\leq 1} \beta(\hat{P}^u_{m+n-1}(d_1, \ldots, d_{m+n-1})),$$

where $d_1 = \cdots = d_{m-k} = g$, $d_{m-k+1} = \cdots = d_m = h$, $d_{m+1} = v_1, \ldots, d_{m+n-1} = v_{n-1}$ and the sum is over those permutations π of $\{1, \ldots, m+n-1\}$ for which $\pi(1) < \cdots < \pi(m-k)$ and $\pi(m-k+1) < \cdots < \pi(m)$. Hence

$$\chi_{n-1}\left(\hat{P}^u_{m+n-1}\{g^{\otimes(m-k)}\}\{h^{\otimes k}\}\right)$$

$$\leq \left((m+n-1)\cdots(m+1)\binom{m}{k} \right)^{1/q} \chi(\hat{P}^u_{m+n-1})\alpha(g)^{m-k}\alpha(h)^k$$

for $n \geq 2$. Similarly, we have

$$\beta\left(\hat{P}^u_m\{g^{\otimes(m-k)}\}\{h^{\otimes k}\}\right) \leq \binom{m}{k}^{1/q} \chi(\hat{P}^u_m)\alpha(g)^{m-k}\alpha(h)^k$$

for $n = 1$. In the case $n \geq 2$, since $c(\chi_{n-1}) \leq c(\beta)$, for each v such that $\alpha(v-w) < [r - \alpha(w-u)^q]^{1/q}$, we get by Lemma 2.9 and Proposition 5.9,

$$\chi_{n-1}\left(\sum_{k=1}^\infty \sum_{m=k+1}^\infty \hat{P}^u_{m+n-1}\{(w-u)^{\otimes(m-k)}\}\{(v-w)^{\otimes k}\}\right)$$

$$\leq \sum_{k=1}^\infty c(\chi_{n-1})^k \sum_{m=k+1}^\infty c(\chi_{n-1})^{m-k}\chi_{n-1}\left(\hat{P}^u_{m+n-1}\{(w-u)^{\otimes(m-k)}\}\{(v-w)^{\otimes k}\}\right)$$

$$\leq \sum_{m=2}^\infty c(\beta)^m (m+n-1)^{(n-1)/q}\chi(\hat{P}^u_{m+n-1}) \sum_{k=1}^{m-1} \binom{m}{k}^{1/q}\alpha(w-u)^{m-k}\alpha(v-w)^k$$

$$\leq \frac{1}{c(\beta)^{n-1}r^{(n-1)/q}} \sum_{l=n+1}^\infty c(\beta)^l l^{(n-1)/q}\chi(\hat{P}^u_l)r^{l/q}. \tag{5.60}$$

By the root test on a series of nonnegative numbers, assumption (5.57) and choice of r, the last series converges. In the case $n = 1$, similarly we get

$$\beta\left(\sum_{k=1}^\infty \sum_{m=k+1}^\infty \hat{P}^u_m\{(w-u)^{\otimes(m-k)}\}\{(v-w)^{\otimes k}\}\right)$$

$$\leq \sum_{k=1}^\infty c(\beta)^k \sum_{m=k+1}^\infty c(\beta)^{m-k}\beta\left(\hat{P}^u_m\{(w-u)^{\otimes(m-k)}\}\{(v-w)^{\otimes k}\}\right)$$

$$\leq \sum_{m=2}^\infty c(\beta)^m\chi(\hat{P}^u_m) \sum_{k=1}^{m-1} \binom{m}{k}^{1/q}\alpha(w-u)^{m-k}\alpha(v-w)^k$$

$$\leq \sum_{m=2}^{\infty} c(\beta)^m \chi(\hat{P}_m^u) r^{m/p} < \infty. \tag{5.61}$$

Then, (5.60) and (5.61) imply that the series

$$S := \sum_{k=1}^{\infty} \sum_{m=k+1}^{\infty} R_m\{(w-u)^{\otimes(m-k)}\}\{(v-w)^{\otimes k}\}$$

converges in $L(^{n-1}V, W)$ absolutely and uniformly for all v with $\alpha(v-w) < [r - \alpha(w-u)^q]^{1/q}$. Moreover, by (5.60) and (5.61) again, we get

$$S = \lim_{M\to\infty} \sum_{k=1}^{M-1} \sum_{m=k+1}^{M} R_m\{(w-u)^{\otimes(m-k)}\}\{(v-w)^{\otimes k}\}$$

$$= \sum_{m=2}^{\infty} \sum_{k=1}^{m-1} R_m\{(w-u)^{\otimes(m-k)}\}\{(v-w)^{\otimes k}\}, \tag{5.62}$$

where the last equality is a result of an interchange of the order of summation. Since $\alpha(w-u) \vee \alpha(v-w) < r^{1/q} < (c(\beta)L)^{-1}$ the sums $\sum_m R_m(w-u)^{\otimes m}$ and $\sum_m R_m(v-w)^{\otimes m}$ converge too. Then, by (5.55) and (5.62), we get

$$\phi(v) = R_0 + R_1(w-u) + R_1(v-w) + \sum_{m=2}^{\infty} R_m[(w-u)+(v-w)]^{\otimes m}$$

$$= \left(R_0 + R_1(w-u) + \sum_{m=2}^{\infty} R_m(w-u)^{\otimes m} \right) + R_1(v-w)$$

$$+ \sum_{m=2}^{\infty} R_m(v-w)^{\otimes m} + \sum_{m=2}^{\infty} \sum_{k=1}^{m-1} R_m\{(w-u)^{\otimes(m-k)}\}\{(v-w)^{\otimes k}\}$$

$$= \phi(w) + \left(R_1(v-w) + \sum_{m=2}^{\infty} R_m\{(w-u)^{\otimes(m-1)}\}\{v-w\} \right)$$

$$+ \sum_{k=2}^{\infty} \left(R_k(v-w)^{\otimes k} + \sum_{m=k+1}^{\infty} R_m\{(w-u)^{\otimes(m-k)}\}\{(v-w)^{\otimes k}\} \right)$$

$$= \phi(w) + \sum_{k=1}^{\infty} Q_k(v-w)^{\otimes k}.$$

Therefore (5.59) holds.

Now we are ready to prove that ϕ is Fréchet differentiable at w with $D^1\phi(w) = Q_1$. If $n \geq 2$ then, by (5.59), for each v with $\alpha(v-w) < [r - \alpha(w-u)^q]^{1/q}$, it follows as in (5.60) that

$$\chi_{n-1}\big(\phi(v) - \phi(w) - Q_1(v-w)\big) = \chi_{n-1}\left(\sum_{k=2}^{\infty} Q_k(v-w)^{\otimes k} \right)$$

$$\leq \frac{\alpha(v-w)^2}{c(\beta)^2 r^{2/q}} \sum_{m=2}^{\infty} c(\beta)^m (m+n-1)^{(n-1)/q} \chi(\hat{P}^u_{m+n-1})r^{m/q}.$$

By the root test on a series of nonnegative numbers again, the last series converges. Hence, ϕ is differentiable at w and $D^1\phi(w) = D^1 d^{n-1}\psi(w) = Q_1$ in the case when $n \geq 2$. If $n = 1$ then similarly, for each v with $\alpha(v - w) < [r - \alpha(w - u)^q]^{1/q}$, it follows that

$$\beta\left(\psi(v) - \psi(w) - Q_1(v-w)\right) \leq \frac{\alpha(v-w)^2}{r^{2/p}} \sum_{m=2}^{\infty} m^{2/p} c(\beta)^m \chi(\hat{P}^u_m)r^{m/p} < \infty.$$

This bound implies that ψ is differentiable at w. Hence, by induction, (5.58) holds for each $n \geq 1$ and the proof of Theorem 5.12 is complete. □

Let H be a linear subspace of $\mathcal{W}_p := \mathcal{W}_p([a,b]; \mathbb{B})$. Recall that, by Proposition 5.3 and Lemma 2.3, the indefinite product integrals $\mathcal{P}_a(f)$ and $\mathcal{P}^b(g)$ belong to $\mathcal{W}_p$ for each $f, g \in H \subset \mathcal{W}_p$, $0 < p < 2$. So, the indefinite product integrals $\mathcal{P}_a(f)$ and $\mathcal{P}^b(g)$ induce product integral operators $\mathcal{P}_a$ and $\mathcal{P}^b$, respectively, from H into $\mathcal{W}_p$ for $0 < p < 2$. We will prove that the product integral operators $\mathcal{P}_a$ and $\mathcal{P}^b$ are Fréchet differentiable as operators acting on $\mathcal{W}_p$ and are analytic on the subspaces of right- or left-continuous elements of $\mathcal{W}_p$. Gill and Johansen (1990, Theorem 8; cf. also Section 4 in Gill, 1994) proved a related result for product integrals with respect to functions of bounded variation, restated as (1.5) above. An integral defined by integration by parts was used in Gill and Johansen's paper to express the derivative of their product integral operator, while we use the (LY) integral in (5.72) below.

We start with three auxiliary statements. Given $f \in \mathcal{W}_p([a,b]; \mathbb{B})$, $0 < p < 2$, by Theorem 4.23, for each $a \leq y \leq z \leq b$, the product integral with respect to f over the interval $[y, z]$ exists. So we can define

$$U_f(z, y) := \prod_y^z (\mathbb{1} + df), \quad y \leq z, \ y, z \in [a, b].$$

Lemma 5.13. *For any $V < \infty$ there exists a finite constant $C(V)$ such that if $f \in \mathcal{W}_p([a,b]; \mathbb{B})$, $0 < p < 2$, and $V_p(f) \leq V$ then*

$$Q_p(U_f) := Q_p(U_f; [a,b])$$

$$:= \sup\left\{\left(\sum_{i=1}^{n} \|U_f(x_i, x_{i-1}) - \mathbb{1}\|^p\right)^{1/p}: \{x_i: i = 0, \dots, n\} \in Q([a,b])\right\}$$

$$\leq C(V), \tag{5.63}$$

and

$$\|U_f\|_{(\infty)} := \sup\left\{\|U_f(z,y)\|: a \leq y \leq z \leq b\right\} < \infty.$$

Proof. Let $\kappa^* = \{y_j: j = 0, 1, \dots, N\}$ be the partition of $[a, b]$ constructed in Lemma 2.20 using the number $c = 1/\left[4c_p^2(2 + \zeta(2/p))\right]^p$, where $c_p = 1 \vee 2^{(1/p)-1}$. Therefore, for each $j \notin J_{\pm}(\kappa^*)$, we have

$$2c_p^2(2 + \zeta(2/p))V_p\left(f; [\Delta_j^{\kappa^*}]\right) \leq 1/2. \tag{5.64}$$

Let $\kappa = \{x_i\colon i = 0, 1, \ldots, n\}$ be any partition of $[a, b]$ and let

$$I(0) := \{i = 1, \ldots, n\colon y_j \in [x_{i-1}, x_i] \text{ for some } j = 0, \ldots, N\},$$

so that $\mathrm{card} I(0) \leq 2N \leq C_1(V)$ for some finite constant $C_1(V)$. For all $i \notin I(0)$, $[x_{i-1}, x_i]$ is included in the interior of some $[y_{j-1}, y_j]$, which is then not an interval $[x-, x]$ or $[x, x+]$. Let $I(j) := \{i = 1, \ldots, n\colon [x_{i-1}, x_i] \subset (y_{j-1}, y_j)\}$ for $j = 1, \ldots, N$. By (5.64), we can apply Proposition 4.27 with

$$\gamma_{cd} = 2c_p^2(2 + \zeta(2/p))V_p(f; [\Delta_i^\kappa]) \leq 1/2.$$

Then by (4.77) and Theorem 4.29 we get

$$\sum_{j=1}^{N} \sum_{i \in I(j)} \|U_f(x_i, x_{i-1}) - 1\!\!1\|^p = \sum_{j=1}^{N} \sum_{i \in I(j)} \|G(f; x_{i-1}, x_i) - 1\!\!1\|^p$$

$$\leq 2^p \sum_{j=1}^{N} \sum_{i \in I(j)} v_p(f; [\Delta_i^\kappa]) \leq 2^p v_p(f; [a, b]).$$

We apply Lemma 4.19 and Definition 4.20 to conclude that there exists a finite constant $C_2(V)$ such that, for all $i = 1, \ldots, n$,

$$\|U_f(x_i, x_{i-1}) - 1\!\!1\| \leq 1 + \|U_f\|_{(\infty)} \leq 1 + C_2(V).$$

Therefore using inequality (2.21) (letting $v_i = 0$ for $i \in I(0)$ and $u_i = 0$ for $i \notin I(0)$) it follows that

$$\left(\sum_{i=1}^{n} \|U_f(x_i, x_{i-1}) - 1\!\!1\|^p\right)^{1/p} \leq c_p\Big[\Big(\sum_{i \in I(0)} \|U_f(x_i, x_{i-1}) - 1\!\!1\|^p\Big)^{1/p}$$

$$+ \Big(\sum_{j=1}^{N} \sum_{i \in I(j)} \|U_f(x_i, x_{i-1}) - 1\!\!1\|^p\Big)^{1/p}\Big]$$

$$\leq c_p\big[C_1(V)^{1/p}(1 + C_2(V)) + 2V_p(f; [a, b])\big].$$

Since κ is an arbitrary partition this implies (5.63) with

$$C(V) := c_p\big[(1 + C_2(V))C_1(V)^{1/p} + 2V\big].$$

The proof of Lemma 5.13 is complete.   □

By the previous lemma, for each $f \in \mathcal{W}_p([a, b]; \mathbb{B})$, $0 < p < 2$,

$$\Theta_p(f) := 2c_p^2(1 + \zeta(2/p))\|U_f\|_{(\infty)}(Q_p(U_f) + 1)^2 < \infty, \qquad (5.65)$$

where $c_p = 1 \vee 2^{(1/p)-1}$. This quantity will appear in the next lemma in upper bounds for the p-variation of certain integrals and in Theorem 5.17 in lower bounds for the radius of uniform convergence. Moreover, from Lemma 5.13 it follows that

$$V_p(\mathcal{P}_a(f)) \leq C(V)\|\mathcal{P}_a(f)\|_\infty. \qquad (5.66)$$

This can be compared with an analogous bound for $V_p^*(\mathcal{P}_a(f))$ given by Proposition 5.3.

Lemma 5.14. *Let $0 < p < 2$ and let $f, h, \phi \in \mathcal{W}_p([a, b]; \mathbb{B})$. Then the function Φ_a defined by*

$$\Phi_a := \Phi_a(\phi)(\cdot) := (CY) \int_a^{(\cdot)} \mathcal{P}^{(\cdot)}(f)^{(a+)} \, dh \, \phi_-^{(a)}$$

exists and

$$V_{p,\infty}(\Phi_a(\phi)) \le \Theta_p(f) V_p(h) V_{p,\infty}(\phi) < \infty. \tag{5.67}$$

Likewise the function Φ^b defined by

$$\Phi^b := \Phi^b(\phi)(\cdot) := (CY) \int_{(\cdot)}^b \phi_+^{(b)} \, dh \, \mathcal{P}_{(\cdot)}(f)^{(b-)}$$

exists and

$$V_{p,\infty}(\Phi^b) \le \Theta_p(f) V_p(h) V_{p,\infty}(\phi) < \infty. \tag{5.68}$$

Proof. We prove (5.67) only and note that a proof of (5.68) is similar, so it is omitted. By Theorem 4.23, Proposition 5.3 and Lemmas 2.3 and 2.5, $\mathcal{P}^x(f)^{(a+)}$ exists and belongs to $\mathcal{W}_p([a, x], \mathbb{B})$ for each $x \in [a, b]$. Therefore by Theorem 3.27 and the relation (2.15), the function Φ_a is well defined. To bound the p-variation of $\mathcal{P}^x(f)^{(a+)}$ let $x \in [a+, b]$ and let $\kappa^x = \{y_j : j = 0, \ldots, k\}$ be a partition of $[a+, x]$. Then we have

$$\left(\sum_{j=1}^k \|\mathcal{P}^x(f)(y_j) - \mathcal{P}^x(f)(y_{j-1})\|^p \right)^{1/p} = \left(\sum_{j=1}^k \|U_f(x, y_j)[\mathbb{1} - U_f(y_j, y_{j-1})]\|^p \right)^{1/p}$$

$$\le \sup_{a \le y \le x} \|U_f(x, y)\| Q_p(U_f; [a, x]) \le \|U_f\|_{(\infty)} Q_p(U_f).$$

Since κ^x is an arbitrary partition of $[a+, x]$, by Lemmas 2.5 and 5.13, it follows that

$$\sup_{x \in [a,b]} V_{p,\infty}(\mathcal{P}^x(f)^{(a+)}; [a, x]) \le \sup_{x \in [a,b]} V_p(\mathcal{P}^x(f)^{(a+)}; [a, x]) + \sup_{x \in [a,b]} \|\mathcal{P}^x(f)^{(a+)}\|_\infty$$

$$\le \|U_f\|_{(\infty)} Q_p(U_f) + \|U_f\|_{(\infty)} < \infty. \tag{5.69}$$

To show (5.67) let $\kappa = \{x_i : i = 0, \ldots, n\}$ be a partition of $[a, b]$. Using inequality (2.21), then applying twice the integral inequality (3.94) and the relation (2.15), we get

$$\left(\sum_{i=1}^n \|\Delta_i^\kappa \Phi_a(\phi)\|^p \right)^{1/p} \le c_p \left(\sum_{i=1}^n \left\| (CY) \int_{x_{i-1}}^{x_i} \mathcal{P}^{x_i}(f)^{(a+)} \, dh \, \phi_-^{(a)} \right\|^p \right)^{1/p}$$

$$+ c_p \left(\sum_{i=1}^n \left\| [U_f(x_i, x_{i-1}) - \mathbb{1}](CY) \int_a^{x_{i-1}} \mathcal{P}^{x_{i-1}}(f)^{(a+)} \, dh \, \phi_-^{(a)} \right\|^p \right)^{1/p}$$

$$\le c_p^2 (1 + \zeta(2/p)) \left(\sum_{i=1}^n v_p(h; [\Delta_i^\kappa]) V_{p,\infty}(\mathcal{P}^{x_i}(f); [\Delta_i^\kappa])^p V_{p,\infty}(\phi_-^{(a)}; [\Delta_i^\kappa])^p \right)^{1/p}$$

$$+ c_p \|\Phi_a(\phi)\|_\infty Q_p(U_f)$$

$$\leq c_p^2(1+\zeta(2/p))V_p(h)V_{p,\infty}(\phi_-^{(a)}) \sup_{x\in[a,b]} V_{p,\infty}(\mathcal{P}^x(f);[a,x])$$

$$+c_p^2(1+\zeta(2/p))V_p(h)V_{p,\infty}(\phi_-^{(a)})Q_p(U_f) \sup_{x\in[a,b]} V_{p,\infty}(\mathcal{P}^x(f);[a,x])$$

$$\leq \big[c_p^2(1+\zeta(2/p))(1+Q_p(U_f)) \sup_{x\in[a,b]} V_{p,\infty}(\mathcal{P}^x(f);[a,x])\big]V_p(h)V_{p,\infty}(\phi).$$

Since κ is an arbitrary partition of $[a,b]$ and since $V_{p,\infty}(\Phi_a)\leq 2V_p(\Phi_a)$ this in conjunction with (5.69) implies (5.67). The proof of Lemma 5.14 is complete. □

Lemma 5.15. *Let* $f,\psi_1,\psi_2\in\mathcal{W}_p([a,b];\mathbb{B})$, $0<p<2$, $\phi\in\ell_\infty([a,b];\mathbb{B})$ *and let* Δ_j, $j=1,2$, *denote either* Δ^+ *or* Δ_-. *Then*

$$V_{p,\infty}\Big(\sum_{(a,\cdot)}\mathcal{P}^{(\cdot)}(f)_+\Delta_1\psi_1\Delta_2\psi_2\phi\Big)$$

$$\leq 2c_p\|U_f\|_{(\infty)}(1+Q_p(U_f))\mathfrak{S}_p(\psi_1)\mathfrak{S}_p(\psi_2)\|\phi\|_\infty \quad (5.70)$$

and

$$V_{p,\infty}\Big(\sum_{(\cdot,b)}\phi\Delta_1\psi_1\Delta_2\psi_2\mathcal{P}_{(\cdot)}(f)_-\Big)$$

$$\leq 2c_p\|U_f\|_{(\infty)}(1+Q_p(U_f))\mathfrak{S}_p(\psi_1)\mathfrak{S}_p(\psi_2)\|\phi\|_\infty, \quad (5.71)$$

where $c_p=1\vee 2^{(1/p)-1}$. *The same inequalities hold when* $\mathcal{P}^{(\cdot)}(f)_+$ *and* $\mathcal{P}_{(\cdot)}(f)_-$ *are replaced by* $\mathcal{P}^{(\cdot)}(f)$ *and* $\mathcal{P}_{(\cdot)}(f)$, *respectively.*

Proof. We will prove (5.70) only and note that a proof of (5.71) is similar. Let $\kappa=\{x_i: i=0,\ldots,n\}$ be a partition of $[a,b]$. Using inequalities (2.21) and (3.94) we get

$$\Big(\sum_{i=1}^n\Big\|\Delta_i^\kappa\Big(\sum_{(a,\cdot)}\mathcal{P}^{(\cdot)}(f)_+\Delta_1\psi_1\Delta_2\psi_2\phi\Big)\Big\|^p\Big)^{1/p}$$

$$\leq c_p\Big(\Big\|\sum_{(a,x_1)}\mathcal{P}^{x_1}(f)_+\Delta_1\psi_1\Delta_2\psi_2\phi\Big\|^p+\sum_{i=2}^n\Big\|\sum_{[x_{i-1},x_i)}\mathcal{P}^{x_i}(f)_+\Delta_1\psi_1\Delta_2\psi_2\phi\Big\|^p\Big)^{1/p}$$

$$+c_p\Big(\sum_{i=2}^n\big\|U_f(x_i,x_{i-1})-\mathbb{1}\big\|^p\Big(\sum_{(a,x_{i-1})}\|\mathcal{P}^{x_{i-1}}(f)_+\Delta_1\psi_1\Delta_2\psi_2\phi\|\big)^p\Big)^{1/p}$$

$$\leq c_p\|U_f\|_{(\infty)}\|\phi\|_\infty\Big(\Big(\sum_{(a,x_1)}\|\Delta_1\psi_1\|\,\|\Delta_2\psi_2\|\Big)^p+\sum_{i=2}^n\Big(\sum_{[x_{i-1},x_i)}\|\Delta_1\psi_1\|\,\|\Delta_2\psi_2\|\Big)^p\Big)^{1/p}$$

$$+c_p\|U_f\|_{(\infty)}\|\phi\|_\infty\Big(\sum_{i=1}^n\|U_f(x_i,x_{i-1})-\mathbb{1}\|^p\Big)^{1/p}\sum_{(a,b)}\|\Delta_1\psi_1\|\,\|\Delta_2\psi_2\|$$

$$\leq c_p\|U_f\|_{(\infty)}(1+Q_p(U_f))\mathfrak{S}_p(\psi_1)\mathfrak{S}_p(\psi_2)\|\phi\|_\infty.$$

The right side is finite by Lemma 5.13 and (2.12). Since κ is an arbitrary partition of $[a,b]$ and since $V_{p,\infty}(F)\leq 2V_p(F)$ whenever $F(a)=0$, this implies (5.70). The same arguments hold when $\mathcal{P}_+^{(\cdot)}$ is replaced by $\mathcal{P}^{(\cdot)}$. The proof of Lemma 5.15 is complete. □

Now we are ready to prove that product integral operators on $\mathcal{W}_p$ are Fréchet differentiable.

Theorem 5.16. *Let $\mathcal{P}_a$ and $\mathcal{P}^b$ be the product integral operators on $\mathcal{W}_p([a,b];\mathbb{B})$ with $0 < p < 2$. Then both of them have values in $\mathcal{W}_p([a,b];\mathbb{B})$ and are Fréchet differentiable at each point $f \in \mathcal{W}_p$ with derivatives $D\mathcal{P}_a(f)$ and $D\mathcal{P}^b(f)$ defined by, for each $h \in \mathcal{W}_p$,*

$$DP_a(f)h = (LY) \int_a^{(\cdot)} \mathcal{P}^{(\cdot)}(f)\,dh\,\mathcal{P}_a(f) \tag{5.72}$$

and

$$DP^b(f)h = (RY) \int_{(\cdot)}^b \mathcal{P}^b(f)\,dh\,\mathcal{P}_{(\cdot)}(f). \tag{5.73}$$

Proof. The operators take $\mathcal{W}_p([a,b];\mathbb{B})$ into itself by Lemma 5.2, Proposition 5.3 and Lemma 2.3. We will prove differentiability of $\mathcal{P}_a(f)$ only since the other case is similar. Let $f, h \in \mathcal{W}_p$. Then the (LY) integral is well defined by Corollary 3.28. By (5.3), for all $a \le x < y \le b$, we have

$$\Delta^+(\mathcal{P}^y(f))(x) = \prod_{x+}^y (\mathbb{1} + df) - \prod_x^y (\mathbb{1} + df) = -[\mathcal{P}^y(f)_+(\Delta^+f)](x). \tag{5.74}$$

Moreover, by the Duhamel formula (5.51) and relation (3.75), for all $a < x \le b$, we get

$$\Delta_-[\mathcal{P}_a(f+h) - \mathcal{P}_a(f)](x) = (CY) \int_{x-}^x \mathcal{P}^x(f)\,dh\,\mathcal{P}_a(f+h)_-$$

$$= [\Delta_- h\mathcal{P}_a(f+h)_-](x).$$

Hence by the Duhamel formula (5.51) again, it follows that, for each $y \in [a,b]$,

$$R(f,h)(y) := [\mathcal{P}_a(f+h) - \mathcal{P}_a(f) - D\mathcal{P}_a(f)h](y)$$

$$= (LY) \int_a^y \mathcal{P}^y(f)\,dh\,[\mathcal{P}_a(f+h) - \mathcal{P}_a(f)]$$

$$= (CY) \int_a^y \mathcal{P}^y(f)^{(a+)}\,dh\,[\mathcal{P}_a(f+h) - \mathcal{P}_a(f)]_-^{(a)}$$

$$- \sum_{(a,y)} \mathcal{P}^y(f)_+\Delta^+f\Delta^+h[\mathcal{P}_a(f+h) - \mathcal{P}_a(f)]_- + \sum_{(a,y)} \mathcal{P}^y(f)_+\Delta^+h\Delta_-h\mathcal{P}_a(f+h)_-.$$

By (2.20) and Lemmas 5.14 and 5.15, it follows that

$$V_{p,\infty}(R(f,h)) \le c_p V_{p,\infty}\big(\Phi_a(\mathcal{P}_a(f+h) - \mathcal{P}_a(f))\big)$$

$$+ c_p^2 V_{p,\infty}\Big(\sum_{(a,\cdot)} \mathcal{P}^{(\cdot)}(f)_+\Delta^+f\Delta^+h[\mathcal{P}_a(f+h) - \mathcal{P}_a(f)]_-\Big)$$

$$+ c_p^2 V_{p,\infty}\Big(\sum_{(a,\cdot)} \mathcal{P}^{(\cdot)}(f)_+\Delta^+h\Delta_-h\mathcal{P}_a(f+h)_-\Big)$$

$$\le c_p \Theta_p(f) V_p(h) V_{p,\infty}(\mathcal{P}_a(f+h) - \mathcal{P}_a(f))$$

$$+ 2c_p^3 \|U_f\|_{(\infty)} (1 + P_p(U_f)) V_p(h) V_p(f) \|\mathcal{P}_a(f+h) - \mathcal{P}_a(f)\|_\infty$$

$$+ 2c_p^3 \|U_f\|_{(\infty)} (1 + P_p(U_f)) V_p(h)^2 \|\mathcal{P}_a(f+h)\|_\infty. \tag{5.75}$$

By (5.4), for all $a < x \leq b$, we have

$$\Delta_-\big(\mathcal{P}_a(f+h)\big)(x) = \prod_a^x (\mathbb{1} + d(f+h)) - \prod_a^{x-} (\mathbb{1} + d(f+h))$$
$$= \big[\Delta_-(f+h)\mathcal{P}_a(f+h)_-\big](x).$$

Hence by (5.74) and by the Duhamel formula (5.51) once again, we get

$$[\mathcal{P}_a(f+h) - \mathcal{P}_a(f)](y) = (CY)\int_a^y \mathcal{P}^y(f)^{(a+)}\, dh\, \mathcal{P}_a(f+h)_-^{(a)}$$

$$- \sum_{(a,y)} \mathcal{P}^y(f)_+ \Delta^+ f \Delta^+ h \mathcal{P}_a(f+h)_- + \sum_{(a,y)} \mathcal{P}^y(f)_+ \Delta^+ h \Delta_-(f+h)\mathcal{P}_a(f+h)_-.$$

Thus by (2.20) and Lemmas 5.14 and 5.15 again, we get

$$V_{p,\infty}\big(\mathcal{P}_a(f+h) - \mathcal{P}_a(f)\big) \leq c_p\Theta_p(f)V_p(h)V_{p,\infty}(\mathcal{P}_a(f+h))$$

$$+2c_p^3\|U_f\|_{(\infty)}(1 + P_p(U_f))V_p(h)\|\mathcal{P}_a(f+h)\|_\infty \times \times \big[V_p(f) + V_p(f+h)\big].$$

Plugging this bound into (5.75) we arrive at

$$V_{p,\infty}(R(f,h)) \leq c_p^2 V_p(h)^2 V_{p,\infty}(\mathcal{P}_a(f+h))\{\Theta_p(f)^2 + 2c_p\|U_f\|_{(\infty)}(1 + P_p(U_f))\times$$

$$\times\big[1 + c_p(2V_p(f) + V_p(f+h)) \times \times [\Theta_p(f) + 2c_p^2\|U_f\|_{(\infty)}(1 + P_p(U_f))V_p(f)]\big]\}.$$

Using (5.66) and Lemma 4.19 one can conclude that

$$V_{p,\infty}(\mathcal{P}_a(f+h)) \leq C \quad \text{as } V_p(h) \to 0 \tag{5.76}$$

for some finite constant C which does not depend on h. Therefore

$$V_{p,\infty}(R(f,h)) = O(V_p(h)^2) \quad \text{as } V_p(h) \to 0.$$

The proof of Theorem 5.16 is complete.   □

For simplicity, in higher order derivatives of product integral operators we will assume one-sided continuity of the function f. Namely, we will prove analyticity of the operators $\mathcal{P}_a$ and $\mathcal{P}^b$ restricted to the subspace $\mathcal{W}_p^r$ of right-continuous elements of $\mathcal{W}_p$ and to the subspace $\mathcal{W}_p^l$ of left-continuous elements of $\mathcal{W}_p$, respectively. Moreover, we show that the kth differential of the product integral operator $\mathcal{P}_a$ at $f \in \mathcal{W}_p^r$ is defined by

$$d^k \mathcal{P}_a(f)\, h^{\otimes k} := (LY)\int_a^{(\cdot)} \mathcal{P}^{(\cdot)}(f)\, dh\, d^{k-1}\mathcal{P}_a(f)\, h^{\otimes k-1}$$

$$= (CY)\int_a^{(\cdot)} \mathcal{P}^{(\cdot)}(f)\, dh\, \big(d^{k-1}\mathcal{P}_a(f)\, h^{\otimes k-1}\big)_-^{(a)}, \tag{5.77}$$

where $d^0 \mathcal{P}_a(f) h^{\otimes 0} := \mathcal{P}_a(f)$, and the kth differential of the product integral operator $\mathcal{P}^b$ at $f \in W_p^l$ is defined by

$$d^k \mathcal{P}^b(f) h^{\otimes k} := (RY) \int_{(\cdot)}^b d^{k-1} \mathcal{P}^b(f) h^{\otimes k-1} \, dh \, \mathcal{P}_{(\cdot)}(f)$$

$$= (CY) \int_{(\cdot)}^b \left(d^{k-1} \mathcal{P}^b(f) h^{\otimes k-1} \right)_+^{(b)} dh \, \mathcal{P}_{(\cdot)}(f), \qquad (5.78)$$

where $d^0 \mathcal{P}^b(f) h^{\otimes 0} := \mathcal{P}^b(f)$. In these cases the sums in the (LY) integral (5.72) and in the (RY) integral (5.73) (from Definition 3.11) are equal to zero, since the jumps of h in them are. Therefore the formulas (5.77) and (5.78) for $k = 1$ coincide with the derivatives (5.72) and (5.73), respectively.

Recalling the definition (5.65) of $\Theta_p(f)$ we have:

Theorem 5.17. *Let $\mathcal{P}_a$ and $\mathcal{P}^b$ be the product integral operators respectively from $W_p^r([a,b]; \mathbb{B})$ and from $W_p^l([a,b]; \mathbb{B})$ into $W_p([a,b]; \mathbb{B})$ with $0 < p < 2$. Then $\mathcal{P}_a$ and $\mathcal{P}^b$ are analytic on $W_p^r([a,b]; \mathbb{B})$ and $W_p^l([a,b]; \mathbb{B})$, respectively, with the same lower bound for the radius of uniform convergence,*

$$\rho(\mathcal{P}_a(f)) \ge \left(c_p \Theta_p(f) \right)^{-1} \quad \text{and} \quad \rho(\mathcal{P}^b(f)) \ge 1 / \left(c_p \Theta_p(f) \right)^{-1},$$

where $c_p = 1 \vee 2^{(1/p)-1}$.

Proof. We will prove the analyticity of the operator $\mathcal{P}_a$ and note that the same arguments work for the operator $\mathcal{P}^b$. Let $f \in W_p^r([a,b]; \mathbb{B})$. For each $k \ge 1$ and any $h_1, \ldots, h_k \in W_p^r$, define the function

$$A_k(h_1, \ldots, h_k) := (LY) \int_a^{(\cdot)} \mathcal{P}^{(\cdot)}(f)(x_1) \, dh_1(x_1) (LY) \int_a^{x_1} \mathcal{P}^{x_1}(f)(x_2) \, dh_2(x_2) \cdots$$

$$\cdots (LY) \int_a^{x_{k-1}} \mathcal{P}^{x_{k-1}}(f)(x_k) \, dh_k(x_k) \mathcal{P}_a(f)(x_k).$$

By Theorem 4.23, Proposition 5.3 and Lemma 2.3, $\mathcal{P}_a(f)$ exists and belongs to W_p. Therefore using Lemma 5.14 recursively one can conclude that the function $A_k(h_1, \ldots, h_k)$ is well defined and that A_k is a map from $W_p^r \times \cdots \times W_p^r$ into W_p. By Proposition 3.21, for each $k \ge 1$, A_k is a k-form. Applying (5.67) recursively, for each $k \ge 1$, we arrive at the inequality

$$V_{p,\infty}(A_k(h_1, \ldots, h_k)) \le \Theta_p(f) V_p(h_1) V_{p,\infty}(A_{k-1}(h_2, \ldots, h_k))$$

$$\cdots$$

$$\le \Theta_p(f)^k V_{p,\infty}(\mathcal{P}_a(f)) V_p(h_1) \cdots V_p(h_k). \qquad (5.79)$$

Therefore A_k is a bounded k-form from $W_p^r \times \cdots \times W_p^r$ into W_p and a mapping $P_k^f : W_p^r \mapsto W_p$ defined by

$$P_k^f(h) := A_k h^{\otimes k} = (LY) \int_a^{(\cdot)} \mathcal{P}^{(\cdot)}(f) \, dh \, A_{k-1} h^{\otimes k-1}$$

is a homogeneous polynomial of degree k. To prove the convergence of the series (5.56) with $\psi = \mathcal{P}_a$, by the Duhamel formula (5.51), for each $n \geq 1$ and all $h \in \mathcal{W}_p^r$, we have

$$R_n(f, h) := \mathcal{P}_a(f + h) - \mathcal{P}_a(f) - \sum_{k=1}^{n} P_k^f(h)$$

$$= (LY) \int_a^{(\cdot)} \mathcal{P}^{(\cdot)}(f) \, dh \, \mathcal{P}_a(f + h) - \sum_{k=1}^{n} P_k^f(h)$$

$$= (LY) \int_a^{(\cdot)} \mathcal{P}^{(\cdot)}(f) \, dh \, R_{n-1}(f, h).$$

Again applying (5.67) recursively, for each $n \geq 1$, we arrive at the inequality

$$V_{p,\infty}(R_n(f, h)) \leq \Theta_p(f) V_p(h) V_{p,\infty}(R_{n-1}(f, h))$$

$$\cdots$$

$$\leq (\Theta_p(f) V_p(h))^n V_{p,\infty}((CY) \int_a^{(\cdot)} \mathcal{P}^{(\cdot)}(f) \, dh \, (\mathcal{P}_a(f + h))_-^{(a)})$$

$$\leq (\Theta_p(f) V_p(h))^{n+1} V_{p,\infty}(\mathcal{P}_a(f + h)). \tag{5.80}$$

The right side of (5.80) converges to 0 for all $h \in \mathcal{W}_p^r$ with $V_{p,\infty}(h) < 1/\Theta_p(f)$. Therefore, for all such h, we have

$$\mathcal{P}_a(f + h) = \mathcal{P}_a(f) + \sum_{k=1}^{\infty} P_k^f(h). \tag{5.81}$$

Since f is an arbitrary function in $\mathcal{W}_p^r$ this implies that the product integral operator $\mathcal{P}_a$ is analytic on $\mathcal{W}_p^r$.

Let

$$\chi_k(\mathcal{P}_a) := \sup \{V_{p,\infty}(A_k(h_1, \ldots, h_k)): V_{p,\infty}(h_i) \leq 1 \quad \text{for all } i\}.$$

Then by (5.79), it follows that $L(f) := \limsup_{k \to \infty} \chi_k(\mathcal{P}_a)^{1/k} \leq \Theta_p(f)$. Therefore by part (1) of Theorem 5.12, Lemma 2.6 and Proposition 2.10, it follows that the radius of uniform convergence of the series in (5.81) is $\rho(\mathcal{P}_a(f)) \geq (c_p \Theta_p(f))^{-1}$. The proof of Theorem 5.17 is complete. $\qquad\square$

The following statement is a consequence of Theorem 5.17 and of part (2) of Theorem 5.12.

Corollary 5.18. *Let $\mathcal{P}_a$ and $\mathcal{P}^b$ be the product integral operators respectively from $\mathcal{W}_p^r([a, b]; \mathbb{B})$ and from $\mathcal{W}_p^l([a, b]; \mathbb{B})$ into $\mathcal{W}_p([a, b]; \mathbb{B})$ with $0 < p < 2$. Assume also that $\mathcal{W}_p$ is equipped with p-subadditive quasinorm $\tilde{V}_{p,\infty}$ whenever $0 < p < 1$, or its norm if $1 \leq p < 2$. Then for each $n \geq 1$, both operators are n times differentiable at each point f of their domain of definition, with k-th differentials given by (5.77) and (5.78), respectively.*

5.4. Linear integral equations. We consider linear integral equations for integrals in the left or right Young sense. It follows from the Duhamel formula that an indefinite product integral provides a solution. Here we establish the uniqueness of this solution. Under a continuity assumption the existence and uniqueness of a solution of a forward linear integral equation was proved by Freedman (1983a, Theorem 4.1) using the Banach fixed point theorem.

We start with a definition:

Definition 5.19. Let $f, F \in \mathcal{R}(\tau_a; \mathbb{B})$ be such that the integral $(LY) \int_{\tau_a} df\, F$ exists. We say that F *satisfies the forward linear integral equation with respect to* f if, for all $y \in \tau_a$,

$$F(y) = \mathbb{1} + (LY) \int_a^y df\, F. \qquad (5.82)$$

Let $g, G \in \mathcal{R}(\tau^b; \mathbb{B})$ be such that the integral $(RY) \int_{\tau^b} G\, dg$ exists. We say that G satisfies the *backward linear integral equation with respect to* g if, for all $y \in \tau^b$,

$$G(y) = \mathbb{1} + (RY) \int_y^b G\, dg. \qquad (5.83)$$

Remark. Hildebrandt (1959) solved the integral equation (5.82) for matrix-valued functions f, F, of bounded 1-variation, using the (RYS) integral. Similar results for functions having finite 1-variation were established by Hinton (1966) using Cauchy left and right integrals. For later results we refer to Hönig (1980, §4.2) and references there. Analogous equations when the integrals are stochastic have been treated for stochastic processes, specifically local martingales, with continuous sample paths by Maisonneuve (1968), local semimartingales with right-continuous sample paths by Doléans-Dade (1970) and semimartingales with regulated sample paths by Gal'chuk (1984, Theorem 5.1).

The next statement implies that the left and right Young integrals in the above equations reduce to the corresponding (MPS) integrals if f and g are either right-continuous or left-continuous. The statement follows from Lemma 3.26.

Corollary 5.20. *If F satisfies the forward linear integral equation with respect to some $f \in \mathcal{R}(\tau_a; \mathbb{B})$ then, for each $y \in \tau_a$ which is not the right endpoint of τ_a,*

$$\Delta^+ F(y) = (\Delta^+ f)(y)\, F(y)$$

and, for each $y > a$, $y \in \tau_a$,

$$\Delta_- F(y) = (\Delta_- f)(y)\, F(y-).$$

Likewise, if G satisfies the backward linear integral equation with respect to some $g \in \mathcal{R}(\tau^b; \mathbb{B})$ then, for each $y < b$, $y \in \tau^b$,

$$\Delta^+ G(y) = -G(y+)(\Delta^+ g)(y)$$

and, for each $y \in \tau^b$ which is not the left endpoint of τ^b,

$$\Delta_- G(y) = -G(y)(\Delta_- g)(y).$$

Now we are ready to prove the existence of a unique solution of the forward linear integral equation.

Theorem 5.21. *Let $\tau = \tau_a$ and $f \in \mathcal{W}_p(\tau; \mathbb{B})$, $0 < p < 2$. Then the indefinite product integral $\mathcal{P}_a(f)$ exists, is in $\mathcal{W}_p(\tau; \mathbb{B})$ and is the unique solution in $\mathcal{W}_r(\tau; \mathbb{B})$, for any $r \geq p$ such that $1/p + 1/r > 1$, of the forward integral equation (5.82).*

Proof. The indefinite product integral $\mathcal{P}_a(f)$ exists by Theorem 4.23 and is in $\mathcal{W}_p(\tau; \mathbb{B})$ by Proposition 5.3 and Lemma 2.3. Taking $g = 0$ in (5.23) we get that $\mathcal{P}_a(f)$ satisfies (5.82). Let F be another solution of (5.82) in $\mathcal{W}_p(\tau; \mathbb{B})$ and $H := \mathcal{P}_a(f) - F$. Then by Proposition 3.21 $H(y) = (LY) \int_a^y df\, H$ for $a \leq y \in \tau$. So $H(a) = 0$. Then by Corollary 5.20, $H(a+) = 0$. Let $\Psi_a(y) := (CY) \int_a^y df\, H_-^{(a)}$. Then by (3.108), (3.93) and (2.15), it follows that

$$V_{p,\infty}(\Psi_a) \leq C_{p,r} V_p(f) V_{r,\infty}(H), \qquad (5.84)$$

where $C_{p,r} = (1 + c_p)(1 + \zeta(1/p + 1/r))$. Let $S_a(y) := \sum_{(a,y)} \Delta^+ f \Delta_- H$. Then

$$V_{p,\infty}(S_a) \leq 2\mathfrak{S}_p(f)\mathfrak{S}_r(H) \qquad (5.85)$$

by (3.109) and (3.95). Since $V_p(H)$ is a decreasing function of p, by (5.84), (5.85) and Definition 3.11 of the (LY) integral, it follows that

$$V_{r,\infty}(H) \leq V_{p,\infty}(H) \leq D_{p,r} V_p(f) V_{r,\infty}(H) \qquad (5.86)$$

for a constant $D_{p,r}$ depending only on p and r. Then (5.86) holds for quasinorms over any interval included in $[a, b]$. Taking an interval $(a+, y)$ for y close enough to a, we will have $V_p(f) < 1/D_{p,r}$ by Lemma 2.18, and thus $H \equiv 0$ on $[a, y)$. There is a largest y for which this holds, with $y \leq b$ where b is the right endpoint of τ. If $y = b \notin \tau$ or if $y = b \in \tau$ and $H(y) = 0$ we are done. If $y < b$ and $H(y) = 0$ then we can repeat the argument with y in place of a and contradict maximality of y. So suppose $H(y) \neq 0$ and $y \in \tau$. Then $H(y) = (LY) \int_a^y df\, H$ but from the form (3.44) of the (LY) integral we see that there is no term at the right endpoint (even if $\Delta_- f(y) \neq 0$), so $H(y) = 0$, again a contradiction. So uniqeness is proved. $\square$

The next statement concerns uniqueness of the solution of the backward linear integral equation and is analogous to Theorem 5.21. The proof is based on symmetric arguments and, hence, is omitted.

Theorem 5.22. *Let $\tau = \tau^b$ and $g \in \mathcal{W}_p(\tau; \mathbb{B})$, $0 < p < 2$. Then the indefinite product integral $\mathcal{P}^b(g)$ exists, is in $\mathcal{W}_p(\tau; \mathbb{B})$ and is the unique solution in $\mathcal{W}_r(\tau; \mathbb{B})$, for any $r \geq p$ such that $1/p + 1/r > 1$, of the backward integral equation (5.83).*

It was mentioned in subsection 4.1 that almost all sample paths of certain stochastic processes have finite p-variation for some $p < 2$. These sample paths provide non-trivial examples of functions which satisfy our linear integral equations. The next statement is a consequence of Theorems 4.4 and 5.21, and of Theorem 2 of Fristedt and Taylor (1973).

Corollary 5.23. *Let $X = \{X(t): t \in [0, T]\}$, $T < \infty$, be a real-valued stable process of index $\alpha \in (0, 2)$ with right-continuous sample paths, which has no drift if $\alpha < 1$*

and has a Lévy measure symmetric about 0 if $\alpha = 1$. Then, for almost all sample paths of X, the indefinite product integral

$$\mathcal{P}_0(X)(t) := \prod_0^t (1 + dX) = \exp\left\{X(t) - X(0)\right\} \prod_{[0,t]} (1 + \Delta X)e^{-\Delta X} \quad \text{for } t \in [0,T]$$

exists and is the unique solution of the forward linear integral equation

$$\mathcal{P}_0(X)(t) = 1 + (MPS) \int_0^t (\mathcal{P}_0(X))_-^{(0)} dX \quad \text{for } t \in [0,T].$$

The last statement of this section is a consequence of Theorems 4.4 and 5.21, of the result of Fernique (1964), and of Theorem 3 of Kawada and Kôno (1973).

Corollary 5.24. *Let $B_H = \{B_H(t): t \in [0,T]\}$, $T < \infty$, be a fractional Brownian motion with index H, $1/2 < H < 1$. Then, for almost all sample paths of B_H, the indefinite product integral*

$$\mathcal{P}_0(B_H)(t) := \prod_0^t (1 + dB_H) = \exp\left\{B_H(t) - B_H(0)\right\} \quad \text{for } t \in [0,T]$$

exists and is the unique solution of the forward linear integral equation

$$\mathcal{P}_0(B_H)(t) = 1 + (MPS) \int_0^t \mathcal{P}_0(B_H) \, dB_H \quad \text{for } t \in [0,T].$$

6. THE LOGARITHM OPERATOR

Throughout this section $\mathbb{B} = (\mathbb{B}, \| \cdot \|)$ will continue to denote a Banach algebra with identity $\mathbb{I}$. Here τ will stand for a non-open, non-degenerate interval, $[a, b)$, $(a, b]$ or $[a, b]$. Given a suitable $\mathbb{B}$-valued function f on τ we define a $\mathbb{B}$-valued function $\mathcal{L}(f)$ on τ such that $\mathcal{L}$ is an inverse operation to the indefinite product integral. We call $\mathcal{L}$ the logarithm operator.

6.1. Reciprocals of indefinite integrals. Consider a $\mathbb{B}$-valued function f on τ. We will say that f is *invertible* if $f(x)$ is invertible for each $x \in \tau$ and f^{inv} is the *reciprocal function* of f if $f^{\text{inv}}(x) := [f(x)]^{-1}$ for each $x \in \tau$. The next statement is preparatory to a definition of the logarithm operator.

Lemma 6.1. *Let $f \in \mathcal{R}(\tau; \mathbb{B})$ be invertible and let the reciprocal function $f^{\text{inv}} \in \ell_\infty(\tau; \mathbb{B})$. Then $f^{\text{inv}} \in \mathcal{R}(\tau; \mathbb{B})$ and f_+, f_- are invertible. Moreover, we have*

$$(f^{\text{inv}})_+(x) = (f_+)^{\text{inv}}(x) =: f_+^{\text{inv}}(x)$$

for each $x \in \bar{\tau}$ which is not the right endpoint of τ and

$$(f^{\text{inv}})_-(x) = (f_-)^{\text{inv}}(x) =: f_-^{\text{inv}}(x)$$

for each $x \in \bar{\tau}$ which is not the left endpoint of τ.

Proof. For each $x, y \in \tau$, we have:

$$f^{\text{inv}}(y) - f^{\text{inv}}(x) = f^{\text{inv}}(x) [f(x) - f(y)] f^{\text{inv}}(y). \tag{6.1}$$

Therefore since f is regulated and $f^{\text{inv}} \in \ell_\infty(\tau; \mathbb{B})$ it follows that f^{inv} is regulated. Let $x \in \bar{\tau}$ where x is not the right endpoint of τ. Then

$$f^{\text{inv}}(x+) f(x+) = \lim_{y \downarrow x} f^{\text{inv}}(y) f(y) = \mathbb{I}.$$

Therefore $f_+(x)$ is invertible with the inverse $f(x+)^{-1}$. Likewise it follows that $f_-(x)$ is invertible at each $x \in \bar{\tau}$ which is not the left endpoint of τ and the inverse of $f_-(x)$ is equal to $f(x-)^{-1}$. This completes the proof of Lemma 6.1. $\square$

Turning to the definition of the logarithm operator it is worthwile to recall a definition of the logarithm function. Namely, the logarithm of y is

$$\int_1^y \frac{dx}{x} \quad \left(= \int_1^y dx\, x^{-1} = \int_1^y x^{-1}\, dx \right)$$

for $y > 0$. We wish to extend this definition, replacing x and x^{-1} by regulated functions f and f^{inv}, respectively. Since $\mathbb{B}$ may not be commutative, depending on which side we multiply by the inverse, two variants of the logarithm operator will be defined. Also, it is convenient to treat these two variants separately on a left closed interval $\tau_a := [a, b)$ or $\tau_a := [a, b]$ or on a right closed interval $\tau^b := (a, b]$ or $\tau^b := [a, b]$, respectively, for some $-\infty < a < b < +\infty$.

Definition 6.2. Let $f \in \mathcal{R}(\tau; \mathbb{B})$ be invertible and let the reciprocal function $f^{\mathrm{inv}} \in \ell_\infty(\tau; \mathbb{B})$. We say that the function $\mathcal{L}_a(f)$ on τ exists and equals

$$\mathcal{L}_a(f)(y) := (LY) \int_a^y df \, f^{\mathrm{inv}} \qquad (6.2)$$

if $a \in \tau$, so $\tau = \tau_a$, and the (LY) integral exists for all $y \in \tau_a$. Likewise we say that the function $\mathcal{L}^b(f)$ on τ exists and equals

$$\mathcal{L}^b(f)(y) := (RY) \int_y^b f^{\mathrm{inv}} \, df \qquad (6.3)$$

if $b \in \tau$ and the (RY) integral exists for all $y \in \tau = \tau^b$.

Next we describe a subset of $f \in \mathcal{W}_p$ on which the functions $\mathcal{L}_a(f)$ and $\mathcal{L}^b(f)$ will be shown to exist whenever $0 < p < 2$.

Definition 6.3. Let $0 < p < \infty$. Given a $\mathbb{B}$-valued function f on τ we write $f \in \mathcal{I}_p(\tau; \mathbb{B})$ if $f \in \mathcal{W}_p(\tau; \mathbb{B})$, f is invertible and the reciprocal function $f^{\mathrm{inv}} \in \ell_\infty(\tau; \mathbb{B})$.

The following statement is a consequence of identity (6.1), so we omit the proof.

Lemma 6.4. If $f \in \mathcal{I}_p(\tau; \mathbb{B})$ for some $0 < p < \infty$ then $f^{\mathrm{inv}} \in \mathcal{W}_p(\tau; \mathbb{B})$.

Lemma 6.5. Let τ be an interval, $\mathbb{B}$ a Banach algebra and $0 < p < \infty$. Then $\mathcal{I}_p := \mathcal{I}_p(\tau, \mathbb{B})$ is an open subset of $\mathcal{W}_p(\tau; \mathbb{B})$ and $\psi \colon f \mapsto f^{\mathrm{inv}}$ is analytic from $\mathcal{I}_p$ into itself.

Proof. Let $f \in \mathcal{I}_p$, so that by Lemma 6.4, for some $M < \infty$, $V_{p,\infty}(f^{\mathrm{inv}}) \leq M$. If $g \in \mathcal{W}_p(\tau; \mathbb{B})$ and $V_{p,\infty}(g) < 1/(2c_p^3 M)$, then by Lemma 2.8, $V_{p,\infty}(f^{\mathrm{inv}} g) < 1/(2c_p^2)$. By induction and Lemma 2.8, $V_{p,\infty}((f^{\mathrm{inv}} g)^k) \leq 1/(2^k c_p^{k+1})$ for all k. So, in

$$h := [\mathbb{I} - f^{\mathrm{inv}} g + (f^{\mathrm{inv}} g)^2 - \cdots] f^{\mathrm{inv}},$$

the series converges in $\mathcal{W}_p(\tau; \mathbb{B})$ by Lemma 2.9 and $h = (f + g)^{\mathrm{inv}}$ by a well-known telescoping sum. Since h is bounded, $f + g \in \mathcal{I}_p$ and $\mathcal{I}_p$ is open in $\mathcal{W}_p$. Thus the inverse operator ψ has an expansion (5.56) with the homogeneous polynomials P_k^u defined by $P_k^u g = (-1)^k (f^{\mathrm{inv}} g)^k f^{\mathrm{inv}}$ for each $k = 1, 2, \ldots$. It follows from Definition 5.11 that ψ is analytic on $\mathcal{I}_p$. $\qquad \square$

Next we show that $\mathcal{L}_a$ and $\mathcal{L}^b$ are (nonlinear) operators from $\mathcal{I}_p(\tau; \mathbb{B})$ into $\mathcal{W}_p(\tau; \mathbb{B})$ for $0 < p < 2$. We call $\mathcal{L}_a$ and $\mathcal{L}^b$ *logarithm operators*.

Lemma 6.6. Let $0 < p < 2$ and let $\mathcal{L}$ be either $\mathcal{L}_a$ or $\mathcal{L}^b$. The operator $\mathcal{L}$ maps each function $f \in \mathcal{I}_p(\tau; \mathbb{B})$ into $\mathcal{W}_p(\tau; \mathbb{B})$ where τ is either τ_a or τ^b, respectively.

Proof. This is a consequence of Lemma 6.4 and Proposition 3.32. $\qquad \square$

Lemma 6.7. Let $\mathcal{L}_a(f)$ exist for some $f \in \mathcal{R}(\tau_a; \mathbb{B})$ and let $u \in \mathbb{B}$ be invertible. Then $\mathcal{L}_a(fu)$ exists and $\mathcal{L}_a(fu) = \mathcal{L}_a(f)$. Likewise if $\mathcal{L}^b(f)$ exists for some $f \in \mathcal{R}(\tau^b; \mathbb{B})$ then $\mathcal{L}^b(uf)$ exists and $\mathcal{L}^b(uf) = \mathcal{L}^b(f)$.

Proof. Since $(vu)^{-1} = u^{-1} v^{-1}$ the statement follows from Proposition 3.21. $\qquad \square$

The following theorem is an analogue of Theorems 1.3.1 and 3.5.4 of Dollard and Friedman (1979) to which they refer as "the fundamental theorem of product integration".

Theorem 6.8. *Let $f \in \mathcal{I}_p(\tau_a; \mathbb{B})$, $0 < p < 2$. Then the product integral with respect to $\mathcal{L}_a(f)$ over τ_a exists and, for all $y \leq z$, $y, z \in \tau_a$,*

$$f(z)f(y)^{-1} = \prod_{y}^{z} (\mathbb{1} + d\mathcal{L}_a(f)). \tag{6.4}$$

In particular, if $y = a$ then $f(z)f(a)^{-1} = \mathcal{P}_a(\mathcal{L}_a(f))(z)$, $\forall z \in \tau_a$, where $\mathcal{P}_a(g)$ denotes the indefinite product integral with respect to g over τ_a.

Proof. By Lemma 6.6, the function $\mathcal{L}_a(f) \in \mathcal{W}_p(\tau_a; \mathbb{B})$. Thus by Theorem 4.23, there exists a family

$$\{\prod_{y}^{z} (\mathbb{1} + d\mathcal{L}_a(f))\colon a \leq y \leq z, \, z \in \tau_a\}$$

of product integrals with respect to $\mathcal{L}_a(f)$. To show that (6.4) holds take any $a \leq y \leq z$, $z \in \tau_a$. By Lemma 4.19, there exists a finite constant $C(\mathcal{L}_a(f))$ such that, for any partition κ of any subinterval of τ_a,

$$\left\| \prod_{\kappa} [\mathbb{1} + (\Delta^{\kappa})(\mathcal{L}_a(f))] \right\| \leq C(\mathcal{L}_a(f)).$$

Then by the algebraic identity (4.49), for each partition $\kappa = \{x_i \colon i = 0, \dots, n\}$ of $[y, z]$, we have

$$\left\| \left(\prod_{i=1}^{n} [\mathbb{1} + (\Delta_i^{\kappa})(\mathcal{L}_a(f))] \right) - f(z)f(y)^{-1} \right\|$$

$$= \left\| \left(\prod_{i=1}^{n} [\mathbb{1} + (\Delta_i^{\kappa})(\mathcal{L}_a(f))] \right) - \left(\prod_{i=1}^{n} [f(x_i)f(x_{i-1})^{-1}] \right) \right\|$$

$$= \left\| \sum_{i=1}^{n} \left(\prod_{j>i} [\mathbb{1} + (\Delta_j^{\kappa})(\mathcal{L}_a(f))] \right) [\mathbb{1} + (\Delta_i^{\kappa})(\mathcal{L}_a(f)) - f(x_i)f(x_{i-1})^{-1}] f(x_{i-1})f(y)^{-1} \right\|$$

$$\leq C(\mathcal{L}_a(f)) \|f\|_{\infty} \|f^{\text{inv}}\|_{\infty} \sum_{i=1}^{n} \left\| \mathbb{1} + (\Delta_i^{\kappa})(\mathcal{L}_a(f)) - f(x_i)f(x_{i-1})^{-1} \right\|. \tag{6.5}$$

Using the additivity property of the (LY) integral (cf. Proposition 3.25) and the (Y_2) definition of (CY) integral, for each $i = 1, \dots, n$, we get

$$\mathbb{1} - f(x_i)f(x_{i-1})^{-1} + (\Delta_i^{\kappa})(\mathcal{L}_a(f)) - \sum_{(x_{i-1}, x_i)} [(\Delta^{+}f)(\Delta_{-}f^{\text{inv}})]$$

$$= \mathbb{1} - f(x_i)f(x_{i-1})^{-1} + (CY) \int_{x_{i-1}}^{x_i} df\,(f_{-}^{\text{inv}})^{(x_{i-1})}$$

$$= \mathbb{1} - f(x_i)f(x_{i-1})^{-1} + (MPS) \int_{x_{i-1}}^{x_i} df_{+}^{(x_i)} (f^{\text{inv}})_{-}^{(x_{i-1})} + [\Delta^{+}f\,f^{\text{inv}}](x_{i-1})$$

$$= (MPS)\int_{x_{i-1}}^{x_i} df_+^{(x_i)} (f_-^{\text{inv}})^{(x_{i-1})} - [f(x_i) - f(x_{i-1}+)]f(x_{i-1})^{-1}.$$

Thus by the integral inequality (3.97), for each $i = 1, \ldots, n$ and any $p_1 \in (p, 2)$, we get

$$\|\mathbb{1} + (\Delta_i^\kappa)(\mathcal{L}_a(f)) - f(x_i)f(x_{i-1})^{-1}\| \leq \sum_{(x_{i-1}, x_i)} \|(\Delta^+ f)(\Delta_- f^{\text{inv}})\|$$

$$+(1 + \zeta(2/p_1))V_{p_1}(f_+^{(x_i)}; [\Delta_i^\kappa])V_{p_1}((f_-^{\text{inv}})^{(x_{i-1})}; [\Delta_i^\kappa]). \qquad (6.6)$$

Let $\epsilon > 0$. By Lemmas 2.2, 2.4 and 6.4, the set D_1 of all $x \in \tau_a$ such that

$$\max\{\|\Delta^+ f_-^{\text{inv}}(x)\|, \|\Delta_- f_+(x)\|\} \geq \epsilon$$

is finite. Using Lemma 2.15 one can find a number

$$0 < \delta < \min\{|x - y|: x, y \in D_1, \ x \neq y\}$$

such that the oscillation on any interval $[c, d] \subset \tau_a \setminus D_1$ of length $< \delta$, of either of the two functions f_-^{inv} and f_+, is less than 2ϵ. By existence of left and right limits, for each $x \in D_1$, one can find $c_x \in [a, x)$, with $\Delta_- f^{\text{inv}}(c_x) = 0$, and $d_x > x$, $d_x \in \tau_a$, with $\Delta^+ f(d_x) = 0$, such that, for all $x \in D_1$,

$$Osc_{[c_x, x]}(f_-^{\text{inv}}) < 2\epsilon \quad \text{and} \quad Osc_{[x, d_x]}(f_+) < 2\epsilon.$$

By Corollary 3.28 the sum in Definition 3.11 of the (LY) integral $\mathcal{L}_a(f)(b)$ converges absolutely. Hence, there exists a finite set D_2 of points of τ_a such that

$$\sum_{\tau_a \setminus D_2} \|(\Delta^+ f)(\Delta_- f^{\text{inv}})\| < \epsilon.$$

Let κ_1 be a partition of $[y, z]$ which includes the set $D_1 \cup \{c_x, d_x: x \in D_1\} \cup D_2$ and such that $|\kappa_1| < \delta$. Then summing the bounds (6.6) with any $\kappa \supset \kappa_1$ and using Lemma 3.31 it follows that

$$\sum_{i=1}^n \|\mathbb{1} + (\Delta_i^\kappa)(\mathcal{L}_a(f)) - f(x_i)f(x_{i-1})^{-1}\|$$

$$\leq (2\epsilon)^{(p_1 - p)/p_1}(1 + \zeta(2/p_1))[V_p(f; \tau_a)V_p(f^{\text{inv}}; \tau_a)]^{p/p_1} \times$$

$$\times [Osc_{\tau_a}(f_+) \vee Osc_{\tau_a}(f_-^{\text{inv}})]^{(p_1 - p)/p_1} + \epsilon =: (2\epsilon)^{(p_1 - p)/p_1}C(f) + \epsilon \qquad (6.7)$$

for some finite constant $C(f)$. By Theorem 4.23, there exists a partition κ_2 of $[y, z]$ such that

$$\left\|\prod_y^z(\mathbb{1} + d\mathcal{L}_a(f)) - \prod_\kappa[\mathbb{1} + (\Delta^\kappa)(\mathcal{L}_a(f))]\right\| < \epsilon$$

for all refinements κ of κ_2. We can and do assume that $\kappa_2 \supset \kappa_1$. Therefore, by (6.5), (6.6) and (6.7), we arrive at

$$\left\|\prod_y^z(\mathbb{1} + d\mathcal{L}_a(f)) - f(z)f(y)^{-1}\right\|$$

$$\leq \epsilon + [(2\epsilon)^{(p_1 - p)/p_1}C(f) + \epsilon]C(\mathcal{L}_a(f))\|f\|_\infty\|f^{\text{inv}}\|_\infty.$$

Since $\epsilon > 0$ is arbitrary this proves (6.4) and, hence, Theorem 6.8. $\qquad \square$

Next is an analogous statement for the logarithm operator L^b.

Theorem 6.9. *Let* $g \in \mathcal{I}_p(\tau^b; \mathbb{B})$, $0 < p < 2$. *Then the product integral with respect to* $L^b(g)$ *over* τ^b *exists and, for all* $y \leq z$, $y, z \in \tau^b$,

$$g(z)^{-1}g(y) = \prod_y^z (\mathbb{1} + dL^b(g)).$$

In particular, if $z = b$ *then* $g(b)^{-1}g(y) = \mathcal{P}^b(L^b(g))(y)$, *for all* $y \in \tau^b$, *where* $\mathcal{P}^b(f)$ *denotes the indefinite product integral with respect to* f *over* τ^b.

Proof. The proof is based on arguments symmetric to those already used to prove Theorem 6.8. We omit the details.   □

Theorem 6.8 and the following statement show that the operators $\mathcal{P}_a$ and $\mathcal{L}_a$ are inverses of each other.

Theorem 6.10. *Let* $f \in \mathcal{W}_p(\tau_a, \mathbb{B})$, $0 < p < 2$, *be such that the indefinite product integral* $\mathcal{P}_a(f) \in \mathcal{I}_p(\tau_a; \mathbb{B})$. *Then the function* $\mathcal{L}_a(\mathcal{P}_a(f))$ *on* τ_a *exists and, for all* $y \in \tau_a$,

$$\mathcal{L}_a(\mathcal{P}_a(f))(y) = f(y) - f(a).$$

Remark. Proposition 4.30 in conjunction with Proposition 5.3 and Lemma 2.3 provides a sufficient condition for f to satisfy the assumption of the above theorem. Namely, the desired indefinite product integral exists if $f \in \mathcal{W}_p(\tau_a; \mathbb{B})$, $0 < p < 2$, is such that for every $x \in \tau_a$, $\mathbb{1} + (\Delta_- f)(x)$ and $\mathbb{1} + (\Delta^+ f)(x)$ are invertible, except that if x is an endpoint of τ_a, only the jump of f on the side toward τ_a is restricted.

Proof. Let $f \in \mathcal{W}_p(\tau_a, \mathbb{B})$ be such that the indefinite product integral $\mathcal{P}_a(f) \in \mathcal{I}_p(\tau_a; \mathbb{B})$. Hence, by Lemma 6.6, the function $\mathcal{L}_a(\mathcal{P}_a)$ is well defined. Let $P_a(x) := \mathcal{P}_a(f)(x)$ and $P_a^{\mathrm{inv}}(x) := [\mathcal{P}_a(f)]^{\mathrm{inv}}(x)$ for all $x \in \tau_a$.

If f is a step function then it is easily seen that $\mathcal{P}_a(f)$ and $\mathcal{L}_a(\mathcal{P}_a(f))$ are also step functions, with jumps at the same points. The jumps are the same by Lemmas 3.26 and 5.1, so the conclusion holds.

In general, let $p < r < 2$. For any $\delta > 0$, by Lemma 2.16 take a step function ϕ with $V_{r,\infty}(f - \phi) < \delta$. The operator $\mathcal{P}_a$, being Fréchet differentiable by Theorem 5.16, is continuous from $\mathcal{W}_r$ into $\mathcal{W}_r$. The operator $g \mapsto g^{\mathrm{inv}}$ is analytic, thus continuous from $\mathcal{I}_r$ into itself by Lemma 6.5. The bilinear operator $(g,h) \mapsto (LY) \int_a^{(\cdot)} dgh$ is jointly continuous from $\mathcal{W}_r \times \mathcal{W}_r$ into $\mathcal{W}_r$ by Proposition 3.32. It follows that the operator $\mathcal{L}_a(\cdot)$ is continuous from $\mathcal{I}_r$ into $\mathcal{W}_r$. Letting $\phi \to f$ in $\mathcal{W}_r$, Theorem 6.10 follows.   □

Corollary 6.11. *Let* $f \in \mathcal{I}_p(\tau_a; \mathbb{B})$ *with* $0 < p < 2$. *Then* $\mathcal{L}_a(f)$ *is a function in* $\mathcal{W}_p(\tau_a; \mathbb{B})$, *unique up to an additive constant, such that*

$$f(y)f(a)^{-1} = \mathcal{P}_a(\mathcal{L}_a(f))(y), \quad \forall y \in \tau_a. \tag{6.8}$$

Proof. By Lemma 6.6 and Theorem 6.8, the function $\mathcal{L}_a(f)$ belongs to $\mathcal{W}_p(\tau_a; \mathbb{B})$ and satisfies (6.8). Suppose that $\tilde{L} \in \mathcal{W}_p(\tau_a; \mathbb{B})$ also satisfies

$$f(y)f(a)^{-1} = \mathcal{P}_a(\tilde{L})(y), \quad \forall y \in \tau_a.$$

Then $\left[\mathcal{P}_a(\tilde{L})\right]^{\mathrm{inv}} = f(a)f^{\mathrm{inv}} \in \ell_\infty(\tau_a; \mathbb{B})$. By Proposition 5.3 and Lemma 2.3, $\mathcal{P}_a(\tilde{L}) \in \mathcal{W}_p(\tau_a; \mathbb{B})$. Therefore $\mathcal{P}_a(\tilde{L}) \in \mathcal{I}_p(\tau_a; \mathbb{B})$ and, by Theorem 6.10 and Lemma 6.7, we have

$$\tilde{L}(y) - \tilde{L}(a) = \mathcal{L}_a(\mathcal{P}_a(\tilde{L}))(y) = L_a(f\, f(a)^{-1})(y) = \mathcal{L}_a(f)(y), \quad \forall y \in \tau_a$$

This proves Corollary 6.11.   □

We finish this subsection by formulating results for the logarithm operator $\mathcal{L}^b$ analogous to those contained in Theorem 6.10 and Corollary 6.11 for $\mathcal{L}_a$. The proofs are based on symmetric arguments and, hence, are omitted.

Theorem 6.12. *Let $g \in \mathcal{W}_p(\tau^b; \mathbb{B})$, $0 < p < 2$, be such that the indefinite product integral $\mathcal{P}^b(g) \in \mathcal{I}_p(\tau^b; \mathbb{B})$. Then the function $\mathcal{L}^b(\mathcal{P}^b(g))$ on τ^b exists and, for all $y \in \tau^b$, $\mathcal{L}^b(\mathcal{P}^b(g))(y) = g(y) - g(b)$.*

Corollary 6.13. *Let $g \in \mathcal{I}_p(\tau^b; \mathbb{B})$, $0 < p < 2$. Then $\mathcal{L}^b(g)$ is a function in $\mathcal{W}_p(\tau^b; \mathbb{B})$, unique up to an additive constant, such that $g(b)^{-1}g(y) = \mathcal{P}^b(\mathcal{L}^b(g))(y)$ for all $y \in \tau^b$.*

6.2. The evolution representation. A family $U = \{U(y, x): x \le y, \, x, y \in \tau\} \subset \mathbb{B}$ is called an *evolution* in $\mathbb{B}$ defined on τ if $U(x, x) = \mathbb{I}$ for each $x \in \tau$ and

$$U(z, y)\, U(y, x) = U(z, x), \quad \forall x \le y \le z, \quad x, z \in \tau.$$

The notion of evolution generalizes the concept of a one-parameter semigroup of bounded linear operators on a Banach space. The classical Hille-Yosida theorem describes any strongly continuous, contractive semigroup in terms of its generator. In this way the Hille-Yosida theorem provides a one-to-one correspondence between semigroups and their generators. Similar pairing results have been established for evolutions under certain conditions on the function $[a, b] \ni t \mapsto U(\cdot, a)$. For example, $U(\cdot, a)$ is analytic, or almost differentiable, or of bounded variation in one type of conditions. Other conditions imply that the generators will be unbounded operators in some situations, bounded in other. Freedman (1983a) extended earlier results on the evolution representation problem to the case when $U(\cdot, a)$ is continuous and of bounded p-variation with $1 \le p < 2$. In Freedman (1983b), he solved this problem under different assumptions, which allow the generator to be unbounded and $U(\cdot, a)$ to be discontinuous.

Let $f \in \mathcal{W}_p(\tau; \mathbb{B})$, $0 < p < 2$, and let $\tau = [a, b]$. By Theorem 4.23, there exists a family

$$U_f := \Big\{ \prod_y^z (\mathbb{I} + df): y \le z, \, y, z \in \tau \Big\}$$

of product integrals with respect to f. By Corollary 4.24, U_f is an evolution in $\mathbb{B}$. If, in addition, $\mathbb{I} + (\Delta_- f)(x)$, $x \in (a, b]$, and $\mathbb{I} + (\Delta^+ f)(x)$, $x \in [a, b)$, are invertible then, by Propositions 4.30 and 5.3, we have $U_f(\cdot, a) = \mathcal{P}_a(f) \in \mathcal{I}_p(\tau; \mathbb{B})$. We show that each evolution U such that $U_a := U(\cdot, a) \in \mathcal{I}_p(\tau; \mathbb{B})$, $0 < p < 2$, can be represented in such a way via the family of product integrals with respect to the function $\mathcal{L}_a(U_a) \in \mathcal{W}_p(\tau; \mathbb{B})$. Freedman (1983a, Theorem 8.1) proved the following under an additional continuity assumption, which implies our invertibility assumption:

Theorem 6.14. *Let U be an evolution in $\mathbb{B}$ defined on τ_a such that $U_a \in \mathcal{I}_p(\tau_a; \mathbb{B})$, $0 < p < 2$. Then the product integral with respect to $\mathcal{L}_a(U_a)$ over τ_a exists and $\mathcal{L}_a(U_a)$ is a function in $\mathcal{W}_p(\tau_a; \mathbb{B})$, unique up to an additive constant, such that*

$$U(z,y) = \prod_y^z (\mathbb{1} + d\mathcal{L}_a(U_a)), \quad \forall a \le y \le z, \ z \in \tau_a. \tag{6.9}$$

Proof. Since $U_a \in \mathcal{I}_p(\tau_a; \mathbb{B})$, by Theorem 6.8, the product integral with respect to $\mathcal{L}_a(U_a)$ over τ_a exists and, for all $a \le y \le z$, $z \in \tau_a$,

$$\prod_y^z (\mathbb{1} + d\mathcal{L}_a(U_a)) = U_a(z)U_a(y)^{-1} = U(z,y).$$

This implies (6.9). Suppose that $\widetilde{L} \in \mathcal{W}_p(\tau_a; \mathbb{B})$ also satisfies

$$U(z,y) = \prod_y^z (\mathbb{1} + d\widetilde{L}), \quad \forall a \le y \le z, \quad z \in \tau_a. \tag{6.10}$$

Since $U_a \in \mathcal{I}_p(\tau_a; \mathbb{B})$ it follows that $\mathcal{P}_a(\widetilde{L})$ is invertible and the inverse function $[\mathcal{P}_a(\widetilde{L})]^{\text{inv}} \in \ell_\infty(\tau_a; \mathbb{B})$. By Proposition 5.3 and Lemma 2.3, $\mathcal{P}_a(\widetilde{L}) \in \mathcal{W}_p(\tau_a; \mathbb{B})$. Therefore $\mathcal{P}_a(\widetilde{L}) \in \mathcal{I}_p(\tau_a; \mathbb{B})$ and by Theorem 6.10 and (6.10), we have

$$\begin{aligned}
\widetilde{L}(z) - \widetilde{L}(y) &= \mathcal{L}_a(\mathcal{P}_a(\widetilde{L}))(z) - L_a(\mathcal{P}_a(\widetilde{L}))(y) \\
&= \mathcal{L}_a(U_a)(z) - \mathcal{L}_a(U_a)(y), \quad \forall a \le y \le z, z \in \tau_a.
\end{aligned}$$

This proves Theorem 6.14.   □

7. REFERENCES

Averbukh, V. I. and Smolyanov, O. G. (1967). The theory of differentiation in linear topological spaces. *Uspekhi Mat. Nauk*, **22**(6), 201-260 (in Russian). Translated in *Russian Math. Surveys*, **22**(6), 201–258.

Birkhoff, Garrett (1937). On product integration. *J. Math. and Phys.* (MIT, Cambridge, Mass), **16**, 104-132.

Bonsall, F. F. and Duncan, J. (1973). *Complete normed algebras.* Springer-Verlag, Berlin.

Bruneau, M. (1974). *Variation totale d'une fonction. Lect. Notes in Math.* (Springer-Verlag), **721**.

Chae, S. B. (1985). *Holomorphy and calculus in normed spaces.* Dekker, New York.

Dobrushin, R. L. (1953). Generalization of Kolmogorov's equations for Markov processes with a finite number of possible states. *Mat. Sb.*, **33** (3), 567-596 (in Russian).

Doléans-Dade, C. (1970). Quelques applications de la formule de changement de variables pour les semimartingales. *Z. Wahrsch. Verw. Gebiete*, **16**, 181-194.

Dollard, J. D. and Friedman, C. N. (1979). *Product integration with applications to differential equations. Encyclopedia of Math. and its Appl.*, Vol. 10, Addison-Wesley, Reading, Massachusetts.

Doran, R. S. and Belfi, V. A. (1986). *Characterizations of C^*-algebras: the Gelfand-Naimark theorems.* Dekker, New York.

Dudley, R. M. (1973). Sample functions of the Gaussian process. *Ann. Probab.*, **1**, 66-103.

Dudley, R. M. (1992). Fréchet differentiability, p-variation and uniform Donsker classes. *Ann. Probab.*, **20**, 1968-1982.

Dudley, R. M. (1997). Empirical processes and p-variation. In *Festschrift for Lucien Le Cam*, Eds. D. Pollard, E. Torgersen, G. L. Yang. Springer-Verlag, New York, pp. 219-233.

Dvoretzky, A. and Rogers, C. A. (1950). Absolute and unconditional convergence in normed linear spaces. *Proc. Nat. Acad. Sci. U.S.A.*, **36**, 192-197.

Emery, M. (1978). Stabilité des solutions des équations différentielles stochastiques application aux intégrales multiplicatives stochastiques. *Z. Wahrsch. Verw. Gebiete*, **41**, 241-262.

Fernique, X. (1964). Continuité des processus gaussiens. *C. R. Acad. Sci. Paris*, **258**, 6058-6060.

Fernique, X. (1971). Régularité de processus gaussiens. *Invent. Math.*, **12**, 304-320.

Freedman, M. A. (1983a). Operators of p-variation and the evolution representation problem. *Trans. Amer. Math. Soc.*, **279**, 95-112.

Freedman, M. A. (1983b). Necessary and sufficient conditions for discontinuous evolutions with applications to Stieltjes integral equations. *J. Integral Equations*, **5**, 237-270.

Fristedt, B. and Taylor, S. J. (1973). Strong variation for the sample functions of a stable process. *Duke Math. J.*, **40**, 259-278.

Gal'chuk, L. I. (1984). Stochastic integrals with respect to optional semimartingales and random measures. *Theory Probab. Appl.*, **29**, 93-108.

Gantmacher, F. R. (1960). *The Theory of Matrices*. Vol. 2. Chelsea, New York.

Gill, R. D. (1994). Lectures on survival analysis. In: *Ecole d'Eté de Probabilités de Saint Flour XXII*, (ed. P. Bernard). *Lect. Notes in Math.* (Springer-Verlag), **1581**, 115-241.

Gill, R. D. and Johansen, S. (1990). A survey on product-integration with a view toward application in survival analysis. *Ann. Statist.*, **18**, 1501-1555.

Gochman, E. Ch. (1958). *The Stieltjes integral and its applications*. Moscow, GIFML (in Russian).

Hildebrandt, T. H. (1959). On systems of linear differentio-Stieltjes-integral equation. *Illinois J. Math.*, **3**, 352-373.

Hildebrandt, T. H. (1963). *Introduction to the theory of integration*. Academic Press, New York.

Hinton, D. B. (1966). A Stieltjes-Volterra integral equation theory. *Canad. J. Math.*, **18**, 314-331.

Hönig, C. S. (1975). *Volterra Stieltjes-Integral Equations*. Mathematics Studies 16. North-Holland, Amsterdam.

Hönig, C. S. (1980). *Volterra Stieltjes-Integral Equations*. In: *Functional Differential Equations and Bifurcation*, (ed. A. F. Izé). *Lect. Notes in Math.* (Springer-Verlag), **799**, 173-216.

Jarník, J. and Kurzweil, J. (1987). A general form of the product integral and linear ordinary differential equations. *Czechoslovak Math. J.*, **37**, 642-659.

Kawada, T. and Kôno, N. (1973). On the variation of Gaussian processes. In: *Proc. of the Second Japan-USSR Symposium on Probability Theory, Lect. Notes in Math.* (Springer-Verlag), **330**, 176-192.

Köthe, G. (1969). *Topological Vector Spaces I*. Translated by D. J. H. Garling. Springer, New York.

Krabbe, G. L. (1961a). Integration with respect to operator-valued functions. *Bull. Amer. Math. Soc.*, **67**, 214-218.

Krabbe, G. L. (1961b). Integration with respect to operator-valued functions. *Acta Sci. Math.* (Szeged), **22**, 301-319.

Lebesgue, H. (1973). *Leçons sur l'intégration et la recherche des fonctions primitives.* 3-rd Edition. Chelsea, New York.

Leśniewicz, R. and Orlicz, W. (1973). On generalized variations (II). *Studia Math.*, **45**, 71-109.

Love, E. R. (1951). A generalization of absolute continuity. *J. London Math. Soc.*, **26**, 1-13.

Love, E. R. (1993). The refinement-Ross-Riemann-Stieltjes (R^3S) integral. In *Analysis, geometry and groups: a Riemann legacy volume*, Eds. H. M. Srivastava, Th. M. Rassias, Part I. Hadronic Press, Palm Harbor, FL, pp. 289-312.

Love, E. R. and Young, L. C. (1938). On fractional integration by parts. *Proc. London Math. Soc.*, **44**, 1-35.

Lyons, Terry (1994). Differential equations driven by rough signals (I): an extension of an inequality of L. C. Young. *Math. Res. Lett.*, **1**, 451-464.

MacNerney, P. R. (1955). Stieltjes integrals in linear spaces. *Ann. of Math.*, **61**, 354-367.

Maisonneuve, B. (1968). Quelques martingales remarquables associées à une martingale continue. *Publ. Inst. Statist. Univ. Paris*, **17**, no. 3, 13-27.

Manturov, O. V. (1990). A multiplicative integral. In: *Itogi Nauki i Tekhniki*, Ser. Probl. Geom., Vol. 22, R. V. Gamkrelidze ed., VINITI, Moscow, pp. 167-215 (in Russian). Translated in *J. Soviet Math.* **55** (1991), no. 5, 2042-2076.

Marcus, M. B. and Shepp, L. A. (1971). Sample behavior of Gaussian processes. *Proc. Sixth Berkeley Symp. math. Statist. Prob.* (1970) **2**, 423-442. Univ. of Calif. Press, Berkeley and Los Angeles.

Masani, P. R. (1947). Multiplicative Riemann integration in normed rings. *Trans. Amer. Math. Soc.*, **61**, 147-192.

Moore, E. H. (1915). Definition of limit in general integral analysis. *Proc. Nat. Acad. Sci. U.S.A.*, **1**, 628.

Nachbin, L. (1973). Recent developments in infinite dimensional holomorphy. *Bull. Amer. Math. Soc.*, **79**, 625-640.

Nashed, M. Z. (1971). Differentiability and related properties of non-linear operators: some aspects of the role of differentials in nonlinear analysis. In: *Nonlinear Functional Analysis and Applications*, L. B. Rall (ed.), Academic Press, New York, 103-309.

Peano, G. (1886-7). Integrazione per serie delle equazioni differenziali lineari. *Atti Accad. sci. Torino*, **22**, 437-466. Engl. transl. in "Selected works of Giuseppe Peano", ed. H. C. Kennedy, Univ. Toronto Press, 1973, pp. 58-66.

Pollard, S. (1923). The Stieltjes' integral and its generalizations. *Quart. J. of Pure and Appl. Math.*, **49**, 73-138.

Rolewicz, S. (1985). *Metric Linear Spaces.* Reidel, Dordrecht; PWN, Warsaw.

Schlesinger, L. (1931). Neue Grundlagen für einen Infinitesimalkalkul der Matrizen. *Math. Z.*, **33**, 33-61.

Schlesinger, L. (1932). Weitere Beiträge zum Infinitesimalkalkul der Matrizen. *Math. Z.*, **35**, 485-501.

Schmidt, G. (1971). On multiplicative Lebesgue integration and families of evolution operators. *Math. Scand.*, **29**, 113-133.

Schwabik, Š. (1990). The Perron product integral and generalized linear differential equations. *Časopis Pěst. Mat.*, **115**, 368-404.

Schwabik, Š. (1994). Bochner product integration. *Math. Bohem.*, **119**, 305-335.

Segal, I. E. and Kunze, R. A. (1968). *Integrals and Operators.* McGraw-Hill, New York.

Taylor, S. J. (1972). Exact asymptotic estimates of Brownian path variation. *Duke Math. J.*, **39**, 219-241.

Volterra, V. (1887). Sui fondamenti della teoria delle equazioni differenziali lineari. *Memorie della Società Italiana delle Scienze (detta dei XL)*, serie III, vol. VI, n. 8. In: *Opere Matematiche I*, 1954, pp. 209-289. Accad. Nazionale dei Lincei, Roma.

Volterra, V. and Hostinsky, B. (1938). *Opérations infinitesimales linéaires.* Gauthier - Villars, Paris.

Wiener, N. (1924). The quadratic variation of a function and its Fourier series. *J. Math. and Phys.* (MIT, Cambridge, Mass), **3**, 73-94.

Young, L. C. (1936). An inequality of the Hölder type, connected with Stieltjes integration. *Acta Math.*, **67**, 251-282.

Young, L. C. (1938). General inequalities for Stieltjes integrals and the convergence of Fourier series. *Math. Ann.*, **115**, 581-612.

PART III

DIFFERENTIABILITY OF THE COMPOSITION AND QUANTILE OPERATORS FOR REGULATED AND A. E. CONTINUOUS FUNCTIONS

BY R. M. DUDLEY AND R. NORVAIŠA

ABSTRACT. If F is a function defined on the range of a function G, let $(F{\circ}G)(x) = F(G(x))$ for all x. Let (Ω, μ) be a finite measure space. The paper treats differentiability of the two-function composition operator $f, g \mapsto (F + f){\circ}(G + g)$ into $L^q(\Omega, \mu)$. where $g \to 0$ in L^s and $1 \leq q < s$. The case where $f = 0$, namely $g \mapsto F{\circ}(G + g)$, for suitable F, G, is a special case of the so-called Nemytskii or superposition operator, which has been extensively studied, as in the book by J. Appell and P. P. Zabrejko, Nonlinear Superposition Operators, Cambridge University Press, 1990, Chapter 3. The remainder R_0 in differentiating the two-function composition operator splits as $R_0 = R_1 + R_2$, where $R_1 := f{\circ}(G+g) - f{\circ}G$ and $R_2 := F{\circ}(G+g) - F{\circ}G - (F'{\circ}G) \cdot g$. Then R_2 is the remainder for the Nemytskii operator. Thus, this paper concentrates on R_1. For suitable G, the question then is, for what f, and uniformly over what classes of f, is $\|t f{\circ}(G + g) - t f{\circ}G\|_q = o(|t| + \|g\|_s)$ as $|t| + \|g\|_s \to 0$, or equivalently $\|f{\circ}(G + g) - f{\circ}G\|_q = o(1)$ as $\|g\|_s \to 0$. This is a question of continuity or equicontinuity of Nemytskii operators at points. Previously, for the most part, global continuity had been treated. The individual f are shown to be exactly those which are continuous almost everywhere, suitably measurable, and such that $|f(x)|/(1 + |x|^{s/q})$ is bounded in x. Large classes of f, called "uniformly Riemann," are given over which the differentiability is uniform. These give in particular Fréchet differentiability $\mathcal{W}_\Phi \times L^s \mapsto L^q$ for an arbitrary Φ-variation space $\mathcal{W}_\Phi$, e.g. any p-variation space $\mathcal{W}_p$. Very similar results are found for the quantile operator $g \mapsto (G + g)^{\leftarrow}$ for functions G and g from an interval J into $\mathbb{R}$, where $H^{\leftarrow}(y) := \inf\{x \in J : H(x) \geq y\}$. Also, a theorem is given on composition of Banach-valued functions with supremum norms, where again f need not be differentiable.

CONTENTS

1 Introduction.

Let $(X, |\cdot|)$ and $(Y, \|\cdot\|)$ be normed real linear spaces and let T be a function mapping an open subset U of X into Y. Then T is said to be **Gateaux differentiable** at $F \in U$ if and only if there is a bounded linear operator $T'(F)(\cdot)$ from X into Y such that for all $g \in X$,

$$(1.1) \qquad \|T(F + tg) - T(F) - tT'(F)(g)\| = o(|t|)$$

as $t \to 0$ in $\mathbb{R}$. Early work on differentiable statistical functionals by von Mises [55, p. 203]; [56, p. 323] and Filippova [27] treated Gateaux differentiability with some additional probabilistic conditions, where F is a probability distribution function.

Both analysts and statisticians have realized that differentiability is useful when the $o(|t|)$ holds uniformly for g in sets in some large enough class. If

$$\|T(F + h) - T(F) - T'(F)(h)\| = o(|h|)$$

as $|h| \to 0$, then T is called **Fréchet differentiable** at F. Equivalently, the $o(|t|)$ in (1.1) holds uniformly for g in any bounded set in X.

If $\mathcal{C}$ is a class of bounded subsets of X, containing all singletons $\{x\}$, $x \in X$, then T is called $\mathcal{C}$-**differentiable** at F if, for each $C \in \mathcal{C}$, in (1.1) the $o(|t|)$ holds uniformly for $g \in C$ (Sebastião e Silva [44]). Specifically, T is called **compactly** differentiable if it is $\mathcal{C}$-differentiable for the class $\mathcal{C}$ of compact subsets of X (for the norm $|\cdot|$). The usefulness of $\mathcal{C}$-differentiability was emphasized in two surveys by Averbukh and Smolyanov [6, 7].

Let $F^{\leftarrow}(y) := F^{\leftarrow}_{[0,1]}(y) := \inf(\{1\} \cup \{x: 0 \le x \le 1, F(x) \ge y\})$ and $(F \circ G)(x) := F(G(x))$. Then we have two operators, the **quantile operator** $F \mapsto F^{\leftarrow}$ and the **composition operator** $(F, G) \mapsto F \circ G$.

For example, consider the identity function $U(x) = x$, $0 \le x \le 1$. Let $D[0, 1]$ be the space of right-continuous functions on $[0, 1]$ with left limits, equipped with the supremum norm, and consider the L^p norms

$$\|f\|_p := \left(\int_0^1 |f(x)|^p dx\right)^{1/p}, \quad 1 \le p < \infty.$$

Then the quantile operator is not Fréchet differentiable at U from $D[0, 1]$ into L^p, but Reeds [41, Theorem 6.4.2] showed that it is compactly differentiable. Extending functions in $D[0, 1]$ to be constant on $(-\infty, 0]$ and on $[1, \infty)$, we get a space $D_\infty[0, 1]$ of functions on $\mathbb{R}$. Consider the composition operator $(f, g) \mapsto (F + f) \circ (G + g)$ for $f \in D_\infty[0, 1]$ with sup norm and $g \in L^p[0, 1]$, where for example $G(x) = x$, $0 \le x \le 1$, and $F(x) \equiv x$ on $\mathbb{R}$. This operator is not Fréchet differentiable at $f = g = 0$ into L^r for any p, $r \ge 1$, not even for $p = \infty$, $r = 1$, e.g. Dudley [22, Proposition 2.8(b)]; in its proof, $\|f_n \circ G\|_1 = 1/(4n)$. But Reeds [41, proof of Theorem 6.4.3] showed that the operator is compactly differentiable there for $1 \le r = p < \infty$ (and thus for $1 \le r \le p < \infty$). Fernholz [26, Proposition 6.1.6] proved an overlapping fact.

The aims of the present paper are to find *all* directions f in which the two operators (composition and quantile) are differentiable at suitable F (and G) and to show uniformity over classes $\mathcal{C}$ of sets C as large as possible. As one special case, it will be shown that one can take $\mathcal{C}$ to be the class of compact sets for the Skorohod topology (defined after Lemma 2.4 below). This class strictly and by far includes the class of compact sets for the sup norm. There are various probability limit theorems holding in the Skorohod topology but not in the supremum norm, e.g. Skorohod [49], Bingham [11, Section 9]. Possible applications of the present paper's differentiability facts to such cases are here left for the future.

The chain rule $(F \circ G)'(x) = (F'(G(x)))(G'(x))$ holds for Fréchet derivatives, as Fréchet [28, p. 311] proved in what was, according to Taylor [51, p. 376], the first paper on Fréchet derivatives in full generality. The chain rule also holds for compact derivatives. Note that chain rules give derivatives of the function $x \mapsto F(G(x))$, while this paper treats differentiability of the operator $(F, G) \mapsto F \circ G$.

The partial derivatives of $(F, G) \mapsto F \circ G$ are easy to find, at least formally. For fixed G, the operator $F \mapsto F \circ G$ is linear and so will be its own derivative (if it is a bounded operator), with $(F + f)(G) - F(G) \equiv f(G)$. For fixed F, the operator $G \mapsto F \circ G$ is one case of the Nemytskii or superposition operator which in general is given by $T(G)(x) := F(G(x), x)$. Such operators and their differentiability have been studied in detail Appell [3], Appell and Zabrejko [4]. If G is real valued and F is differentiable, we would have

$$(1.2) \qquad F(G + g) - F(G) = (F' \circ G)g + o(|g|),$$

in other words the derivative of $G \mapsto F \circ G$ in the direction g is multiplication of g by $F' \circ G$. Note that derivatives of G do not appear in these partial derivatives of the composition operator.

For example, the function on $\mathbb{R}^2$, $f(x, y) := xy/(x^2 + y^2)$, with $f(0, 0) := 0$, is everywhere differentiable in y for fixed x and in x for fixed y, but is not jointly (Fréchet) differentiable or even jointly continuous at $(0, 0)$. For composition the gap between joint and separate differentiability is perhaps even wider. Still, composition has been shown to be jointly differentiable in a number of cases, as mentioned in Part I after Corollary 2.3. For example, Gray [32] considered the derivative of composition for C^1 functions between subsets of normed spaces. A result related to his is given in Sec. 6 below. On nonlinear functions between (non-normed) topological vector spaces see e.g. Bücher [14, Thme. 4.27].

When we have non-supremum norms such as L^p norms, then the remainder $o(|g(x)|)$ is not $o(\|g\|)$ pointwise everywhere, but only "outside of sets of small measure", and differentiability is less straightforward to prove. Until recently, the best available results on differentiability of composition for L^p norms on g and on $(F + f) \circ (G + g)$ were apparently those of Reeds [41] and Fernholz [26]. Brokate and Colonius [13, Prop. 2.5] prove joint differentiability of the composition operator $(f, g) \mapsto (F + f) \circ (G + g)$ into L^p, $1 \le p < \infty$, under uniform Lipschitz conditions on both f and g, which are much stronger than the conditions of Reeds or the present paper, and with a stronger conclusion of *continuous* Fréchet differentiability.

One can ask whether compact differentiability for the supremum norm on $\mathbb{R}$ follows from Fréchet differentiability for some other norm (such as a p-variation norm) when Fréchet differentiability fails for the sup norm. Such a state of affairs is ruled out by the following, presumably known fact (in Dunford and Schwartz [24, II.1.16] for $(X, |\cdot|)$ complete):

1.3 Proposition.   If $(X, |\cdot|)$ is a normed linear space and $\|\cdot\|$ is another norm on X such that all compact sets for $|\cdot|$, or just all sequences with $|x_n| \to 0$, are bounded for $\|\cdot\|$, then for some $K < \infty$, $\|x\| \le K|x|$ for all $x \in X$.

Proof.   For some $\delta > 0$ and $M < \infty$, $|x| \le \delta$ implies $\|x\| \le M$, since otherwise take x_n with $|x_n| \le 1/n$ and $\|x_n\| > n$ for all n. Then by homogeneity, $\|x\| \le (M/\delta)|x|$ for all x. $\square$

To extend the Reeds-Fernholz results the following definitions will be helpful. A real function f on a bounded interval $[a, b]$ will be called a **Riemann function** if it is bounded and continuous Lebesgue almost everywhere (a.e.). A bounded function f is continuous a.e. if and only if it is Riemann integrable (e.g. Shilov and Gurevich [45, Sec. 1.7, Theorem 4]; Cohn [16, p. 76] or Riesz and Sz.-Nagy [43, Sec. 13]). Riemann [42, Section 5] showed that an integral $\int_0^1 f(x) \, dx$ exists in his sense if and only if both (a) f is bounded and (b) for every $\epsilon > 0$ there exist decompositions of $[0, 1]$ into finitely many intervals I_j such that the total length of the I_j on which f oscillates by more than ϵ is arbitrarily small. It can be shown that (b) is equivalent to f being continuous a.e. (Riemann (-Stieltjes) integrals are not used in this paper, but were in Dudley [21].) For Riemann integrals $\int_a^b f(x)\,dx$, existence in the "mesh $\to 0$" sense is equivalent to existence under refinements of partitions, e.g. Hildebrandt [33, Theorem 10.9 p. 51]

Let $R[a, b]$ denote the set of all Riemann functions on $[a, b]$. A set $\mathcal{F} \subset R[a, b]$ will be called **uniformly Riemann** if

(a) For every $\varepsilon > 0$ there is a $\delta > 0$ such that for each $f \in \mathcal{F}$ there is a set $B = B_f \subset [a, b]$ such that $\lambda(B) < \varepsilon$ and if $x \notin B$ and $|y - x| < \delta$, then $|f(x) - f(y)| < \varepsilon$, and

(b) $\mathcal{F}$ is **bounded in diameter**, meaning that

$$\sup\{|f(x) - f(y)| : \ a \leq x \leq y \leq b, \ f \in \mathcal{F}\} < \infty.$$

For $0 < \tau < \infty$, let $R_\tau(\mathbb{R})$ be the set of all real-valued functions f on $\mathbb{R}$ which are universally measurable (u.m.), continuous almost everywhere for Lebesgue measure and satisfy $\|f\|_{\{\tau\}} := \sup_x |f(x)|/(1 + |x|^\tau) < \infty$. (On u.m. sets and functions, see the Appendix; we note here only that Borel implies universal measurability, which implies Lebesgue measurability.) Functions in $R_\tau(\mathbb{R})$ will be called τ-**Riemann**. A set $\mathcal{F} \subset R_\tau(\mathbb{R})$ will be called **uniformly τ-Riemann** if the restrictions of functions in $\mathcal{F}$ to any bounded interval are uniformly Riemann and $\sup\{\|f\|_{\{\tau\}} : \ f \in \mathcal{F}\} < \infty$. A main result, Theorem 5.1 below, says that for a finite measure space (Ω, μ), the composition operator $(F, G) \mapsto F \circ G$ is jointly differentiable from $R_{s/q}(\mathbb{R}) \times \mathcal{L}^s(\Omega, \mu)$ into $\mathcal{L}^q(\Omega, \mu)$ whenever $1 \leq q < s$, for Fréchet differentiability in G and $\mathcal{C}$-differentiability in F, where $\mathcal{C}$ is the class of uniformly (s/q)-Riemann sets, at F, G such that F is Lipschitz, $G \in \mathcal{L}^s$, and $\mu \circ G^{-1}$ is absolutely continuous with respect to Lebesgue measure. For composition to be jointly differentiable into L^q at such G with G varying in $\mathcal{L}^s$, and F in the direction of one function f, taking $F + tf$ as $t \downarrow 0$, Theorem 5.3 shows that the (s/q)-Riemann condition on f is necessary. Under the same conditions on F, G and g, Theorem 5.1 also gives joint Fréchet differentiability of the composition operator for any Φ-variation norm on f. This was apparently unknown until [22] even for the usual total variation norm on f.

For the quantile operator, the main result (Theorem 4.5 below) gives $\mathcal{C}$-differentiability, for the class $\mathcal{C}$ of uniformly Riemann sets uniformly bounded on $[0, 1]$, into $L^p[0, 1]$ at any diffeomorphism F. Also, Corollary 4.15 shows that at $F = U$ with $U(x) = x$, $0 \leq x \leq 1$, differentiability of the quantile operator along f and h, with $h(x) \equiv f(1 - x)$, implies the a.e. continuity and boundedness of f.

Both main theorems improve considerably on those of Reeds [41] and Fernholz [26] for the given operators, since compact sets in D for the supremum norm are very small compared to uniformly Riemann sets.

Sec. 3 gives some equivalent conditions for a function to be a Riemann function, and for $\mathcal{F}$ to be a uniformly Riemann set of functions.

A function f from an interval into $\mathbb{R}$ is called **regulated** if it has left and right limits everywhere. A regulated function on a bounded, closed interval is bounded, and continuous except for a countable set of jumps, so it is a Riemann function. Sec. 2 collects a number of (known) conditions equivalent to being regulated. Then, "uniformly regulated" collections of functions are defined and again equivalent conditions are given. Proposition 3.3 shows that uniformly regulated classes are uniformly Riemann. This will yield Fréchet differentiability for many norms (Φ-variation norms) which is, broadly speaking, better than compact differentiability for the sup norm, although for each specific Φ there are compact sets in D for the sup norm not bounded in Φ-variation (cf. Prop. 1.3). Specializing again, taking $\Phi(x) = x^p$, $0 \leq x < \infty$, $1 \leq p < \infty$, we get classes of functions of bounded p-variation, treated elsewhere, specifically for the Riemann-Stieltjes bilinear functional $(F, G) \mapsto \int F dG$: Dudley [21, 23].

2 Regulated functions.

The space $D[0, 1]$ of right-continuous functions with left limits on $[0, 1]$ ("cadlag" functions) is familiar in probability and statistics. The right continuity is sometimes very useful, for example in the theory of stochastic processes. But for present purposes the restriction to right-continuous functions is somewhat arbitrary, and unnecessary, so larger spaces will be treated.

For any interval J (open or closed, bounded or not), let $E(J)$ denote the set of all real-valued functions on J which have left and right limits at each point in the interior of J, a right limit at the left endpoint and a left limit at the right endpoint (even when those endpoints may be infinite). Functions in $E(J)$ are called **regulated functions** on J. Note that if F is a distribution function (right-continuous, as usual) then F^{-1} is left-continuous. Thus $E(\mathbb{R})$ contains both F and F^{-1}.

By a (somewhat arbitrary) choice of increasing homeomorphism, we can transform the interval J to $[0, 1]$, or a subinterval of it without one or both endpoints, preserving the space E except possibly for the endpoints.

Definitions. Let f be a real-valued function on an interval J and $\varepsilon > 0$. For a subinterval $[c, d] \subset J$ let $|f([c, d])| := |f(d) - f(c)|$. Subintervals of J will be called **nonoverlapping** if they are disjoint except possibly for their endpoints. Let $\sup_{\{J_i\}}$ denote the supremum over all sequences $\{J_i\}$ of nonoverlapping subintervals of J. Then let

$$n(f, \varepsilon) := \sup_{\{J_i\}} \{n : |f(J_i)| > \epsilon \quad \text{for } i = 1, \ldots, n\}.$$

For $n = 1, 2, \ldots$, let

$$v(f, n) := \sup_{\{J_i\}} \{ \textstyle\sum_{i=1}^n |f(J_i)| \}.$$

For a sequence $\Lambda := \{\lambda_i\}_{i \geq 1}$ of positive numbers, the Λ-**variation** of a function f on J is $\sup_{\{J_i\}} \sum_i |f(J_i)|/\lambda_i$. Then f is of **bounded Λ-variation**, or $f \in \Lambda BV$, iff its Λ-variation is finite.

An M-**function** will be a strictly increasing, convex function Φ on $[0, \infty[$ with $\Phi(0) = 0$. Then for a real function f on J, the Φ-**variation** $v_\Phi(f)$ is $\sup_{\{J_i\}} \sum_i \Phi(|f(J_i)|)$.

A **step function** on an interval J is a finite linear combination of indicator functions of subintervals of J (which may be open or closed at either end).

Among M-functions are, of course, the functions $\Phi(x) = x^r$ on $0 \leq x < \infty$, where $1 \leq r < \infty$.

The following equivalences are essentially known, but in scattered references. It seems worthwhile to collect and prove them since most of the steps are rather short and none is too long. Also, the equivalences for an individual function will prepare the way for equivalences on suitably "bounded" families of functions.

2.1 Theorem. For a real-valued function f on a closed, bounded interval $J = [a, b]$, the following are equivalent:

(I) f is regulated on J ($f \in E(J)$).

(II) For any $\varepsilon > 0$, $n(f,\varepsilon) < \infty$.

(III) $v(f,n) = o(n)$ as $n \to \infty$.

(IV) For some $\{\lambda_i\}_{i\geq 1}$ such that $\lambda_i > 0$ for all i and $\sum_i 1/\lambda_i = +\infty$, $f \in \Lambda BV$.

(V) $v_\Phi(f) < \infty$ for some M-function Φ.

(VI) $f \equiv g \circ h$ where g is continuous on $\mathbb{R}$ and h is nondecreasing and real valued on J.

(VI') $f \equiv g \circ h$ where g is continuous on $\mathbb{R}$ and h is strictly increasing and real valued on J.

(VII) f is a uniform limit of step functions on J.

Proof.   First, (I) implies (II): suppose that for some $\varepsilon > 0$, $n(f,\varepsilon) = +\infty$. Dividing J into two halves, we have $n(f,\varepsilon) = +\infty$ on at least one half. Continuing the process, we get a sequence of nested intervals $J_k = [a_k, b_k]$ converging to a point x, such that $n(f,\varepsilon) = +\infty$ on each J_k. Then we have $n(f,\varepsilon) = +\infty$ either on infinitely many intervals $[a_k, x]$, contradicting existence of $f(x-)$, or on infinitely many intervals $[x, b_k]$, contradicting existence of $f(x+)$. So (II) follows.

Next, each of the eight conditions implies that f is bounded: this is clear for (VI), (VI') and (VII). If f is unbounded on J, take a sequence $x_n \in J$ with $|f(x_{n+1})| \geq |f(x_n)| + 1$ for all n. For any given $n = 1, 2, \ldots$, let $y_1 \leq y_2 \leq \ldots \leq y_n$ be $x_1, x_2, \ldots, x_n$ arranged in order. Then $|f(y_j) - f(y_{j-1})| \geq 1$ for $j = 2, \ldots, n$. Letting $n \to \infty$, this clearly contradicts any of (II) through (V), so it also contradicts (I). So suppose for some $K < \infty$, $|f(x)| \leq K$ for all $x \in J$.

(II) implies (III): for any $\delta > 0$, with $m := n(f,\delta)$, for $n > m$ we have $v(f,n) \leq 2Km + (n-m)\delta \leq 2n\delta$ for $n \geq 2Km/\delta$. Let $\delta \downarrow 0$.

(III) implies (IV): let $\varepsilon_n := v(f,n)/n$. It follows easily from the definition of $v(f,n)$ that $\varepsilon_{n+1} \leq \varepsilon_n$ for all n. So $\varepsilon_n \downarrow 0$ as $n \to \infty$. It is well known that for any sequence of positive numbers $\varepsilon_n \downarrow 0$, there is a sequence $c_n \downarrow 0$ with $\sum_n c_n = +\infty$ and $\sum_n \varepsilon_n c_n < \infty$. (Proof: we can assume $\varepsilon_1 < 1$. If $\sum_n \varepsilon_n < \infty$, let $c_n \equiv 1/n$. Otherwise, for $k = 1, 2, \ldots$, let n_k be the number of elements in the set A_k of values of n such that $2^{-k} \leq \varepsilon_n < 2^{1-k}$. Let $c_n := d_k := \min(1, 1/n_1, 1/n_2, \ldots, 1/n_k)$ for each $n \in A_k$; then the given properties hold.) Let $\lambda_n := 1/c_n$ for all n. Then $\lambda_i \uparrow +\infty$ and $\sum_i 1/\lambda_i = +\infty$. For any nonoverlapping intervals J_i, a sum $\sum_i |f(J_i)|/\lambda_i$ is maximized when $|f(J_i)|$ is nonincreasing in i (by induction and since if $a \geq b$ and $c \geq d$, then $ac + bd \geq ad + bc$). Then $|f(J_i)| \leq \varepsilon_i$ and (IV) follows.

Clearly, (IV) implies (II), so (II), (III) and (IV) are equivalent.

(IV) implies (V): given $\{\lambda_i\}_{i\geq 1}$ such that $\lambda_i > 0$ for all i and $\sum_i 1/\lambda_i = +\infty$, and given $M < \infty$ such that $\sup_{\{J_i\}} \sum_i |f(J_i)|/\lambda_i \leq M$, we also have $\varepsilon_n = v(f,n)/n \downarrow 0$ as $n \to \infty$. We can assume $\lambda_i \geq 1$ and that λ_i is nondecreasing in i. Then let $g(\varepsilon_i) := \varepsilon_i/\lambda_k$ for the largest value of $k = k(i)$ such that $\varepsilon_k = \varepsilon_i$ ($k = i$ if ε_i decrease strictly). Then g is well-defined, positive, and strictly increasing on the ε_i. For any sequence $\{i(k)\}_{k\geq 1}$, $g(\varepsilon_{i(k)}) \downarrow 0$ if and only if $i(k) \uparrow +\infty$. For $x \geq 0$, let $\Phi(x)$ be the supremum of all affine functions $f(x) \equiv ax + b$ such that $f(\varepsilon_i) \leq g(\varepsilon_i)$ for all i. Any supremum of affine functions is convex.

For each i, there are finitely many affine functions f_j which equal 0 at $\varepsilon_{k(i)+1}$ and $g(\varepsilon_j)$ at ε_j, for $j = 1, \ldots, k(i)$. All these functions are strictly positive at ε_i. The one with smallest slope (which is smallest at ε_i) is $\leq g$ at ε_k for all k. So $\Phi(\varepsilon_i) > 0$ for all i, while $\Phi(0) = 0$, and the convex function Φ is strictly increasing on $0 \leq x < \infty$.

Any sequence $\{J_i\}_{i\geq 1}$ of nonoverlapping intervals J_i can be arranged in order so that $|f(J_i)|$ is nonincreasing in i. We then have $|f(J_i)| \leq \varepsilon_i$ for all i, so by convexity $\Phi(|f(J_i)|) \leq |f(J_i)|\Phi(\varepsilon_i)/\varepsilon_i \leq |f(J_i)|/\lambda_i$, so $\sum_i \Phi(|f(J_i)|) \leq M$ as desired.

(V) implies (VI): for an M-function Φ with $v_\Phi(f) < \infty$, let ϕ be its inverse, a strictly increasing function from $[0, \infty[$ onto itself. Then ϕ is concave, and subadditive: $\phi(x + y) \leq \phi(x) + \phi(y)$ for all x, $y \geq 0$. Let $h(x)$ be the Φ-variation of f on the interval $[a, x]$. Then h is nondecreasing and has finite values on $[a, b]$. For $a \leq x \leq y \leq b$, we have $h(y) \geq h(x) + \Phi(|f(y) - f(x)|)$, so $\phi(h(y) - h(x)) \geq |f(y) - f(x)|$. Since ϕ vanishes only at 0, it follows that f is a function of h, $f(x) \equiv g(h(x))$ for a unique function g on the range of h. $d_\phi(u, v) := \phi(|u - v|)$ defines a metric on $\mathbb{R}$ and g is a Lipschitz function for d_ϕ, $|g(s) - g(t)| \leq d_\phi(s, t)$ for s, t in the domain of g. Then g extends to a function on all of $\mathbb{R}$ satisfying the same Lipschitz condition, by the (Helly-Hahn-Banach-Kirszbraun-McShane) extension theorem, e.g. Dudley [20, Theorem 6.1.1]. Since d_ϕ metrizes the usual topology, g is continuous, proving (VI).

(VI) implies (VI$'$): let h be a nondecreasing function and $h_1(x) \equiv x + h(x)$. Then h_1 is strictly increasing, and for any x, $y \geq 0$, $|h(x) - h(y)| \leq |h_1(x) - h_1(y)|$. So, as in the previous step, $h \equiv \gamma \circ h_1$ for a Lipschitz and, hence, continuous function γ, and $f \equiv (g \circ \gamma) \circ h_1$, proving (VI$'$).

(VI$'$) implies (VII): a continuous real function g on a closed, bounded interval, here $[h(a), h(b)]$, is a uniform limit of step functions. A step function composed with an increasing function h is a step function, so (VII) follows.

(VII) implies (I): step functions are clearly in $E(J)$, so it suffices to show that $E(J)$ is complete for uniform convergence. Suppose $f_n \in E(J)$ and $f_n \to f$ uniformly on J. Then for $a < x \leq b$, $f_n(x-)$ converge to some $f(x)-$, and by the proof that a uniform limit of continuous functions is continuous, it follows that f indeed has a left limit at x and $f(x-) \equiv f(x)-$. Likewise, f has right limits $f(x+) = \lim_{n \to \infty} f_n(x+)$ for $a \leq x < b$, and (I) follows. □

Bibliographic remarks.   That (I) implies (II) follows e.g. from Taberski [50, Theorem 3]. Dellacherie and Meyer [18, Theorem IV-22 p. 94-IV] prove that for f bounded (I) is equivalent to (II$'$): for any $c < d$, f has only finitely many upcrossings (or downcrossings) of the interval $[c, d]$. Clearly, (II) implies (II$'$). The equivalences of (I) with the other conditions have been given: (III) announced in Chanturia [15, Theorem 5]; (IV) proved in Perlman [40]; (V) in Goffman, Moran and Waterman [30]; (VI) in Sierpiński [48] and Tsereteli [52, 53]; (VII) in Dieudonné [19, (7.6.1)] and Bourbaki [12, II, §1 #3]. In citation indices (*Science Citation Index* and *Compumath Index*, 1945 through 1996) the only citation of Sierpiński [48] we found was that of Perlman [40]. Also, the only citation of Tsereteli [52, 53] we found was that of Asatiani and Chanturia [5]. L. C. Young [58, p. 583], who first defined Φ-variation, showed that (V) implies both (VI) and (I).

Next, some subsets of $E(J)$ will be characterized in the same ways. A family $\mathcal{F}$ of functions is bounded in diameter, as defined in §1, iff $\sup\{v(f, 1): f \in \mathcal{F}\} < \infty$. This does not follow from $\sup\{n(f, \varepsilon): f \in \mathcal{F}\} \leq 1$ (even for all $\varepsilon > 0$): let $J := [a, b]$, $\mathcal{F} := \{f: f(x) = 0, a \leq x < b\}$ (but with arbitrary values at b). So "bounded in diameter" needs to be assumed separately in (a) below:

2.2 Theorem.   For a subset $\mathcal{F} \subset E(J)$, where J is a closed, bounded interval, the following are equivalent:

(a) $\mathcal{F}$ is bounded in diameter and for every $\varepsilon > 0$, $\sup\{n(f, \varepsilon): f \in \mathcal{F}\} < \infty$.

(b) $\lim_{n \to \infty} \sup\{v(f, n): f \in \mathcal{F}\}/n = 0$.

(c) $\sup\left\{\sup_{\{J_i\}}\{\sum_i |f(J_i)|/\lambda_i\}: f \in \mathcal{F}\right\} < \infty$ for some $\{\lambda_i\}_{i \geq 1}$ such that $\lambda_i > 0$ for all i and $\sum_i 1/\lambda_i = +\infty$.

(d) For some M-function Φ, $\sup\{v_\Phi(f): f \in \mathcal{F}\} < \infty$.

Proof.   Clearly, (b), (c) or (d) implies that $\mathcal{F}$ is bounded in diameter. Then, the chain of implications (a) $\to$ (b) $\to$ (c) $\to$ (d) follows by a straightforward extension of the corresponding steps of the proof of Theorem 2.1, and (d) $\to$ (a) also follows directly. $\square$

A class $\mathcal{F}$ of real functions on J will be called **uniformly regulated** iff any of the equivalent conditions in Theorem 2.2 holds.

Now, for simplicity, J will be taken to be $[0,1]$. The following development of metrics for $E := E[0,1]$ is much the same as that for $D := D[0,1]$ given in Billingsley [9, Chap. 3]. Some of the proofs will be omitted except for remarks on what changes may be needed for E. Recall that each f in E is bounded.

Two known metrics for D are also defined on E. A metric d on E is defined by, for any f and g in E, $d(f,g) := \inf\{\varepsilon > 0 : $ for some increasing homeomorphism α of $[0,1]$, $|\alpha(x) - x| < \varepsilon$ and $|f(\alpha(x)) - g(x)| < \varepsilon$ for $0 \le x \le 1\}$. For any increasing homeomorphism α of $[0,1]$, let

$$m(\alpha) := \sup\left\{\left|\log \frac{\alpha(t) - \alpha(s)}{t - s}\right| : 0 \le s < t \le 1\right\}.$$

Then another metric on $E[0,1]$ is $d_0(f,g) := \inf\{\varepsilon > 0 : $ for some increasing homeomorphism α of $[0,1]$, $m(\alpha) < \varepsilon$ and $|f(\alpha(x)) - g(x)| < \varepsilon$ for $0 \le x \le 1\}$.

For any interval J and function f, let $w_f(J)$, called the **oscillation** of f on J, be

$$\mathrm{osc}_J f := w_f(J) := (\sup - \inf)_J f = \sup\{|f(x) - f(u)| : x, u \in J\}.$$

For $\delta > 0$ and $f \in E$, let

$$w_f^*(\delta) := \inf\{\max_{1 \le i \le r} w_f(]t_{i-1}, t_i[) : 0 = t_0 < t_1 < \ldots < t_r = 1,$$

$$t_i - t_{i-1} > \delta \quad \text{for } i - 1, \ldots, r; \ r = 1, 2, \ldots\}.$$

(For half-open intervals $[t_{i-1}, t_i[$ and D, rather than open intervals $]t_{i-1}, t_i[$ and E, Billingsley [9, p. 110] called the corresponding quantity $w_f'(\delta)$.)

2.3 Lemma.   If $f, g \in E$ and $d(f,g) < \delta^2$, where $0 < \delta < 1/4$, then $d_0(f,g) \le 4\delta + w_f^*(\delta)$.

Proof.   From the definition of $w_f^*(\delta)$, take points t_i with $w_f(]t_{i-1}, t_i[) < w_f^*(\delta) + \delta$, $i = 1, 2, \ldots, r$. Then by definition of $d(f,g)$, take an increasing homeomorphism α with $\|f - g \circ \alpha\|_\infty = \|f \circ \alpha^{-1} - g\|_\infty < \delta^2$ and $\sup_t |\alpha(t) - t| < \delta^2$. Let β be the increasing homeomorphism of $[0,1]$ which agrees with α at the t_i and is linear on each interval $J_i := [t_{i-1}, t_i]$. Then the composition $\alpha^{-1} \circ \beta$ leaves each t_i fixed and is increasing, so t and $\alpha^{-1}(\beta(t))$ are either both in the same open interval $]t_{i-1}, t_i[$ or both equal some t_i. So for $0 \le t \le 1$,

$$|f(t) - g(\beta(t))| \le |f(t) - f(\alpha^{-1}(\beta(t)))| + |f(\alpha^{-1}(\beta(t))) - g(\beta(t))|$$

$$< w_f^*(\delta) + \delta + \delta^2 < 4\delta + w_f^*(\delta).$$

Now it will be enough to show that $m(\beta) \le 4\delta$. Since $\beta = \alpha$ at each t_i, $|\beta(t_i) - \beta(t_{i-1}) - (t_i - t_{i-1})| < 2\delta^2 < 2\delta(t_i - t_{i-1})$. Since β is piecewise linear, it follows that

$$|\beta(s) - \beta(t) - (s - t)| \le 2\delta|s - t|$$

for any s, t in $[0,1]$. So since $\delta < 1/4$,

$$-4\delta \; < \; \log(1 - 2\delta) \; \leq \; \log\frac{\beta(t) - \beta(s)}{t - s} \; \leq \; \log(1 + 2\delta) \; < \; 2\delta,$$

so $m(\beta) \leq 4\delta$. $\square$

2.4 Lemma. For any $f \in E$, $\lim_{\delta\downarrow 0} w_f^*(\delta) = 0$.

Proof. For any $\varepsilon > 0$ and t with $0 \leq t < 1$, since f has a right limit $f(t+)$, there is an s with $t < s \leq 1$ such that $w_f(]t, s[) \leq \varepsilon/2$. Let $g(t, \varepsilon)$ be the supremum of such s. Then $w_f(]t, g(t,\varepsilon)[) \leq \varepsilon/2 < \varepsilon$. Let $t_0 := 0$ and $t_{i+1} := g(t_i, \varepsilon)$, $i = 0, 1, \ldots$. Then $t_r = 1$ for some r, since if t_j are defined for all j, $t_j \uparrow t_\infty \leq 1$, and f oscillates by at least $\varepsilon/2$ on each interval $]t_{2j}, t_{2j+2}[$, contradicting existence of the left limit $f(t_\infty -)$. Then if $\delta < \min_{1 \leq i \leq r}(t_i - t_{i-1})$, $w_f^*(\delta) < \varepsilon$ as claimed. $\square$

Just as in D, if $d_0(f, g) < 1/4$, we have $d(f, g) \leq 2d_0(f, g)$. In the other direction, Lemmas 2.3 and 2.4 show that convergence for d implies convergence for d_0. So the metrics d and d_0 define the same topology on $E[0, 1]$. This topology will be called the **Skorohod** topology, as in D. The metric d_0 makes E complete (which is proved just as for D), while d does not.

No doubt, the Skorohod topology has some drawbacks, such as the fact that addition of functions is not continuous. But the point here is that the metric d will give another characterization of uniformly regulated sets, and in turn, C-differentiability for the class C of uniformly regulated sets is stronger than sup norm-compact differentiability.

For the metrics d and d_0, $E[0, 1]$ is separable: a countable dense set consists of step functions having only rational values, on finitely many intervals separated by rational endpoints. For E, some of the intervals may reduce to single points, even in the interior of $[0, 1]$.

If a sequence of functions in E converges for d or d_0 to a continuous function, then the convergence is uniform.

2.5 Theorem. A set $A \subset E[0, 1]$ has compact closure (equivalently, is totally bounded for d_0) if and only if A is uniformly bounded and $\lim_{\delta\downarrow 0}\sup\{w_f^*(\delta) : f \in A\} = 0$.

Proof. Compactness and closure are the same for d and d_0. Since d_0 is complete, having compact closure is equivalent to being totally bounded for d_0. It is also equivalent to the assertion that every sequence f_n in A has a subsequence converging to some f in E.

To prove "only if", if $\overline{A}$ is compact, it is bounded for d, and for any f, $d(f, 0) = \|f\|_\infty$ (supremum norm), so A is uniformly bounded.

For a fixed $\delta > 0$, one can check that the function $f \mapsto w_f^*(\delta)$ is upper semicontinuous on E. Then by Dini's theorem for upper semicontinuous functions, the convergence $w_f^*(\delta) \downarrow 0$ as $\delta \downarrow 0$, given by Lemma 2.4, holds uniformly for $f \in \overline{A}$, proving "only if".

Now to prove "if", let $A \subset E$ be uniformly bounded and $w_f^*(\delta) \downarrow 0$ as $\delta \downarrow 0$ uniformly on A. First it will be proved that A is totally bounded for d. Given $\varepsilon > 0$, take a positive integer k such that $1/k < \varepsilon$ and $w_f^*(1/k) < \varepsilon$ for all f in A. Let $\alpha := \sup\{\|f\|_\infty : f \in A\}$. A set F is called an ε-net for a set G and metric d if for any $y \in G$, there is an $x \in F$ with $d(x, y) < \varepsilon$. (F is not necessarily included in G.) Let H be a finite ε-net for $[-\alpha, \alpha]$. Let B be the finite set of functions g from $[0, 1]$ into H such that for $j = 1, \ldots, k$, g is constant on the open interval $](j - 1)/k, j/k[$. Then $B \subset E$. It will be shown that B is a 2ε-net for A. Given $f \in A$, since $w_f^*(1/k) < \varepsilon$, take points t_i with $0 = t_0 < t_1 < \ldots < t_r = 1$

with $t_i - t_{i-1} > 1/k$ and $w_f(]t_{i-1}, t_i[) < \varepsilon$, $i = 1, \ldots, r$. Let j_i be the integers such that $j_i/k \le t_i < (j_i+1)/k$, $i = 0, 1, \ldots, r$. Then $0 = j_0 < j_1 < \ldots < j_r = k$ since $t_i - t_{i-1} > 1/k$.

Let γ be an increasing homeomorphism of $[0, 1]$ with $\gamma(j_i/k) = t_i$, $i = 0, 1, \ldots, r$, and such that γ is linear on each interval $[j_{i-1}/k, j_i/k]$. Choose $g \in B$ such that $|g(j/k) - f(\gamma(j/k))| < \epsilon$ for $j = 0, 1, \ldots, k$ and $|g((j/k)-) - f(\gamma(j/k)-)| < \epsilon$ for $j = 1, \ldots, k$.

Any interval $]\gamma((j-1)/k), \gamma(j/k)[$ must be included in one of the intervals $]\gamma(j_{i-1}/k), \gamma(j_i/k)[=]t_{i-1}, t_i[$. So the function $f \circ \gamma$ oscillates by at most ε on any interval $](j-1)/k, j/k[$. Since g is constant over each such interval, and is within ε of $f \circ \gamma$ near the right endpoint, we have $|g(t) - f(\gamma(t))| < 2\varepsilon$ for all t. By the definitions, $|\gamma(j_i/k) - j_i/k| = |t_i - j_i/k| < 1/k < \varepsilon$. Since γ is piecewise linear, $\sup_t |\gamma(t) - t| < \varepsilon$. So $d(f, g) < 2\varepsilon$, and B is indeed a 2ε-net for A in the metric d.

Now given $\eta > 0$, take δ with $0 < \delta < 1/4$ such that $4\delta + w_f^*(\delta) < \eta$ for all $f \in A$. Then let $\varepsilon = \delta^2/3$. Take the set B as above for this ε. Then by Lemma 2.3, B is an η-net for A in the metric d_0. So A is totally bounded for d_0. $\qquad\square$

Now, uniformly regulated and bounded sets can be characterized as follows:

2.6 Theorem.   A set $\mathcal{F} \subset E[0, 1]$ is totally bounded for d if and only if it is both uniformly bounded and uniformly regulated.

Proof.   If $\mathcal{F}$ is totally bounded for d, it is uniformly bounded. Given $\varepsilon > 0$, there are some $m < \infty$ and $f_1, \ldots, f_m \in \mathcal{F}$ such that for all $f \in \mathcal{F}$, $d(f, f_j) < \varepsilon$ for some j. For any fixed $g \in E$, there are some $r < \infty$ and $t_0 = 0 < t_1 < \ldots < t_r = 1$ such that $w_g(]t_{i-1}, t_i[) < \varepsilon$ for $i = 1, \cdots, r$, by Lemma 2.4. If $d(g, h) < \varepsilon$, $h \in E$, then for some increasing homeomorphism α, $\|g - h \circ \alpha\|_\infty < \varepsilon$. Let $s_i := \alpha(t_i)$. Then $w_h(]s_{i-1}, s_i[) < 3\varepsilon$ for $i = 1, \ldots, r$, so $n(h, 3\varepsilon) \le 3r$. Letting $g = f_j$, $h = f \in \mathcal{F}$ and taking the maximum over m values of j, then letting $\varepsilon \downarrow 0$, shows that $\mathcal{F}$ is uniformly regulated.

Conversely, suppose that for some $M < \infty$, $\|f\|_\infty \le M$ for all $f \in \mathcal{F}$ and for each $\varepsilon > 0$, there is a $k = k(\varepsilon) < \infty$ such that $n(f, \varepsilon) \le k$ for all $f \in \mathcal{F}$. Let H be a finite ε-net in $[-M, M]$. Define a distance between two finite sequences $0 = t_0 < t_1 < \ldots < t_r = 1$ and $0 = s_0 < s_1 < \ldots < s_r = 1$ by $e(\{s_i\}_{1 \le i \le r}, \{t_i\}_{1 \le i \le r}) := \max_i |s_i - t_i|$. Then there is a finite ε-net G for e in the set of such sequences for $r = 1, \ldots, 2k$. For each $t := \{t_i\} \in G$, there is a finite set F_t of functions in E which have values in H and are constant on each open interval $]t_{i-1}, t_i[$.

For $f \in \mathcal{F}$, let $s_0 := 0$ and given $s_0, \ldots, s_i$ with $s_i < 1$, let $s_{i+1} := \inf\{s \le 1: w_f(]s_i, s[) > \varepsilon\}$, or 1 if there is no such s. Then $w_f(]s_i, s_{i+1}[) \le \varepsilon$ and for some $r \le 2k$, $s_r = 1$. Then for some $t \in G$, $e(\{s_i\}_{i=1}^r, t) < \varepsilon$ and there is an increasing, piecewise linear homeomorphism h with $h(s_i) = t_i$, $i = 1, \ldots, r$, and $|h(x) - x| < \varepsilon$, $0 \le x \le 1$. Take $g \in F_t$ with $|g(t_i) - f(s_i)| < \varepsilon$, $i = 0, 1, \ldots, r$, and $|g(t_i-) - f(s_i-)| < \varepsilon$, $i = 1, \ldots, r$. Then $\|g \circ h - f\|_\infty < 2\varepsilon$, so $d(f, g) \le 2\varepsilon$. Thus the union of all F_t, $t \in G$, gives a finite 2ε-net for d and $\mathcal{F}$. $\qquad\square$

It follows from Theorem 2.6 that any compact set in $E[0, 1]$ for the Skorohod topology is uniformly regulated.

Theorem 2.6 suggests that totally bounded sets in E for d are of some interest, so one might ask about their compact closures in the completion $\overline{E}^d$ of E for d. But $\overline{E}^d$ seems to be a rather pathological object. Let $f_n := 1_{A(n)}$ where for $n = 2, 3, \ldots$, $A_{2n-1} := [\frac{1}{2} - \frac{1}{n}, \frac{1}{2})$, $A_{2n} := [\frac{1}{2}, \frac{1}{2} + \frac{1}{n})$. Then $\{f_n\}_{n \ge 3}$ is a Cauchy sequence in $D := D[0, 1] \subset E[0, 1]$ for d, but not convergent. For $g := 1_{[1/2, 1]} \in D$, $\{f_n + g\}$ is not a Cauchy sequence. The subsequence $\{f_{2n-1} + g\}$ converges to g, but $\{f_{2n-1}\}$ does not converge in D or E. Also, $\{f_{2n} + g\}$ is a Cauchy sequence but not convergent. As these examples show, addition in D

is not continuous for d. The procedure for defining addition of elements of the completion, which works for metrics e satisfying $e(x, y) = e(x + u, y + u)$ for all x, y and u, fails for $\overline{E}^d$. Also, since d and d_0 metrize the same topology, and d_0 is complete, E is a G_δ (a countable intersection of open sets) in $\overline{E}^d$ [20, Theorem 2.5.4]. Thus, $\overline{E}^d \setminus E$ is a countable union of nowhere dense sets. This also never happens when an incomplete abelian group (such as a vector space) is completed for a metric with the property of e above.

$E = E[0, 1]$ can also be considered as a subset of the space $\overline{\mathbb{R}}^I$ of all functions from $I := [0, 1]$ into $\overline{\mathbb{R}} := [-\infty, \infty]$ with the product topology (pointwise convergence).

2.7 Theorem (Nelson [39, Theorem 3.4]).   In $\overline{\mathbb{R}}^I$, E is a $K_{\sigma\delta}$, that is, a countable intersection of countable unions of compact sets.

Proof.   For m, $n = 1, 2, \ldots$, let K_{mn} be the set of all functions f from I into $\mathbb{R}$ such that $|f(x)| \le m$ for all $x \in I$ and such that there are no more than m nonoverlapping intervals $J_i \subset I$ with $|f(J_i)| > 1/n$. Then it is easy to see that each K_{mn} is compact (its complement is open in the product topology). It follows from Theorem 2.1 that $E = \bigcap_n \bigcup_m K_{mn}$.   □

We also have the following sequential compactness property in $E[0, 1]$ for pointwise convergence:

2.8 Proposition.   Any uniformly regulated and uniformly bounded sequence $\{f_k\}$ in $E[0, 1]$ has a subsequence converging pointwise (everywhere) to a function in $E[0, 1]$.

Proof.   By Theorem 2.6, taking a subsequence, we can assume that $\{f_i\}$ is a Cauchy sequence for d, and that $f_i(x)$ converges for each rational $x \in [0, 1]$ to some $f(x)$. For each $\varepsilon > 0$, $\sup_i n(f_i, \varepsilon) < \infty$ by Theorem 2.2. Thus $n(f, \varepsilon) < \infty$ (for f as a function on $\mathbb{Q} \cap [0, 1]$). So (as in Theorem 2.1) right limits $f(x+)$ (as $y \downarrow x$ through rationals) exist for $0 \le x < 1$, and left limits $f(x-)$ for $0 < x \le 1$. There are at most countably many points where $f(x+) \ne f(x-)$. Taking another subsequence, we can assume f_i converges at these points.

Call x an ε-**oscillation point** for $\{f_i\}$ iff for all $\delta > 0$, for infinitely many i, $\mathrm{osc}_{[x-\delta, x+\delta]} f_i > \varepsilon$. Then by the d-Cauchy property, $\mathrm{osc}_{[x-2\delta, x+2\delta]} f_i > \varepsilon/2$ for all i large enough. It follows that there can only be finitely many ε-oscillation points for a given $\varepsilon > 0$, and countably many for all $\varepsilon > 0$. We can assume that f_i converges at all such points x to some $f(x)$. If x is not an ε-oscillation point for any $\varepsilon > 0$, then convergence $f_i \to f$ at rationals in the neighborhood of x implies f_i converges at x to a limit we call $f(x)$. Then f is defined on $[0, 1]$ and is the pointwise limit of f_i. For each $n = 1, 2, \ldots$, there is an m such that all f_i are in K_{mn} as in the last proof, so f is also in K_{mn}. It follows that $f \in E$.   □

Let K be the set consisting of points $0, 0+, x-, x, x+$ for $0 < x < 1$, and of $1-, 1$, with the natural linear ordering: $x- < x < x+ < y-$ if $x < y$. On K we have the "interval topology" with a base given by the set of all sets $\{v : u < v < w\}$, $\{v : 0 \le v < w\}$, and $\{v : u < v \le 1\}$ for u, $w \in K$. Then $E[0, 1]$ can be identified with the set of all continuous real-valued functions on K (which is compact, not metrizable and is totally disconnected). This identification also has a Banach algebra aspect: Berberian [8, Theorem 5].

3   Almost everywhere continuous functions.

Let $I := [0, 1]$. Let λ be Lebesgue measure on $\mathbb{R}$. For any interval J and real-valued function f on J, recall that $w_f(J) := \mathrm{osc}_J f := \sup_J f - \inf_J f$.

For bounded functions f, the next fact follows from proofs of the well-known equivalence of a.e. continuity with Riemann integrability, e.g. [16, p. 76]. For unbounded functions it is also presumably known, but I do not have a reference for it; the rather short proof will be given for completeness and for use in the next proof.

3.1 Theorem.   For any (bounded or unbounded) interval $J \subset \mathbb{R}$ and function f from J into $\mathbb{R}$, the following are equivalent:

(a) f is continuous on J almost everywhere (for λ), i.e. at each $x \in D$ where $\lambda(J \backslash D) = 0$.

(b) For any bounded subinterval $B \subset J$ and any $\varepsilon > 0$, there are finitely many intervals $I_j \subset B$ whose union has Lebesgue measure at least $\lambda(B) - \varepsilon$ and such that for each j, $w_f(I_j) < \varepsilon$.

(c) For any bounded subinterval $B \subset J$ and $\varepsilon > 0$, there is a $\delta > 0$ such that for some measurable set $A \subset B$ with $\lambda(A) > \lambda(B) - \varepsilon$, for all x in A and all $y \in J$ such that $|y - x| < \delta$, we have $|f(y) - f(x)| \leq \varepsilon$.

Proof.   From the form of the statement, we can assume $J = I$. Then (a) implies (b): given $\varepsilon > 0$, there is a compact $K \subset I$ with $\lambda(K) > 1 - \varepsilon$ such that f is continuous at each point of K. Each $x \in K$ is in some open interval U_x such that $|f(y) - f(x)| < \varepsilon/3$ for all $y \in U_x$. The covering of K by the U_x has a finite subcover, so (b) follows.

(b) implies (c): apply (b) for $\varepsilon/2$ to get intervals $I_j := [a_j, b_j]$, $j = 1, \ldots, n$, whose union has Lebesgue measure larger that $1 - \varepsilon/2$, with $w_f(I_j) < \varepsilon/2$ for each j. Let $\delta := \varepsilon/(4n)$. Let A be the union of the intervals $[a_j + \delta, b_j - \delta]$, $j = 1, \ldots, n$ (some of which may be empty or degenerate). Then $\lambda(A) > 1 - \varepsilon$ and (c) holds.

(c) implies (a): for $\varepsilon = 1/2^n$, $n = 1, 2, \ldots$, (c) gives sets A_n and numbers $\delta_n > 0$. Let $C := \liminf A_n := \bigcup_m \bigcap_{n \geq m} A_n$. Then $\lambda(C) = 1$ and f is continuous at each point of C. $\square$

3.2 Theorem.   For any (bounded or unbounded) interval $J \subset \mathbb{R}$ and class $\mathcal{F}$ of real-valued functions on J, the following are equivalent:

(I) For any bounded interval $H \subset J$ and $\varepsilon > 0$, there is an $n(\varepsilon) < \infty$ such that for every $f \in \mathcal{F}$, there are $n(\varepsilon)$ intervals $J_i \subset H$, whose union has Lebesgue measure at least $\lambda(H) - \varepsilon$, with $w_f(J_i) < \varepsilon$ for all i.

(II) For every bounded interval $H \subset J$ and $\varepsilon > 0$, there is a $\delta > 0$ such that for all $f \in \mathcal{F}$ there is a set $A \subset H$ with $\lambda(A) > \lambda(H) - \varepsilon$ such that whenever $x \in A$ and $y \in J$, $|x - y| < \delta$ implies $|f(x) - f(y)| < \varepsilon$.

(III) For any bounded interval $H \subset J$ and $\varepsilon > 0$, there is a $\delta > 0$ such that for each $f \in \mathcal{F}$ there are finitely many intervals $J_i \subset H$ of lengths $\lambda(J_i) > \delta$ whose union has Lebesgue measure larger than $\lambda(H) - \varepsilon$ and such that $w_f(J_i) < \varepsilon$ for each i.

Proof.   Again, we can take $H = J = I = [0, 1]$.

(I) implies (II): given $\varepsilon > 0$ let $\delta := \varepsilon/(4N)$ where $N := n(\varepsilon/2)$. The rest follows as in the proof that (b) implies (c) in Theorem 3.1.

(II) implies (III): given $\varepsilon > 0$, apply (II) for $\varepsilon/4$ to get a $\delta > 0$, where we can take $\delta < 1/2$. Then for $f \in \mathcal{F}$, (II) gives a set A which can be taken not to contain 0 or 1 and to be compact. For each $x \in A$, let $I_x :=]\max(0, x - \delta), \min(1, x + \delta)[$. Then the I_x are open intervals covering A, with $w_f(I_x) \leq \varepsilon/2 < \varepsilon$ for each x. Taking a finite subcover, we get finitely many intervals of lengths larger than δ, proving (III).

(III) implies (I): given $\varepsilon > 0$, suppose $\delta > 0$ from (III), $f \in \mathcal{F}$, and J_i, $i = 1, \ldots, n$ from (III), where we can and do take each J_i to be open. Delete J_1 from the list of intervals J_i if it is included in the union of the other J_i and, recursively for $i = 2, \ldots, n$, delete J_i if it is included in the union of the others which have not previously been deleted. We are left with a list of intervals, say $I_1, \ldots, I_r$, with the same union as that of $J_1, \ldots, J_n$, such that no I_j is included in the union of the I_i for $i \neq j$. Claim: for any three open intervals A, B and C, such that the intersection of all three is nonempty, at least one is included in the union of the other two. Proof: we can assume that B has the smallest left endpoint, say b. If B also has the largest right endpoint, both A and C are included in B, so assume C has the largest right endpoint, say c. Let $x \in A \cap B \cap C$. Then $B \supset]b, x]$ and $C \supset [x, c[$, so $A \subset]b, c[=]b, x] \cup [x, c[\subset B \cup C$ as claimed.

So, each point of $[0, 1]$ belongs to at most two of the intervals I_j. So $2 \geq \int_0^1 \sum_{j=1}^r 1_{I_j} \geq r\delta$, so $r \leq 2/\delta$ and we can set $n(\varepsilon) = 2/\delta(\varepsilon)$ (not depending on f), proving (I). $\square$

A Riemann function on I, as defined in §1, is thus a bounded universally measurable function satisfying any of the equivalent conditions in Theorem 3.1. Likewise, a uniformly Riemann set as defined in §1 is a set $\mathcal{F}$ of Riemann functions on a bounded interval, bounded in diameter and satisfying any of the equivalent conditions of Theorem 3.2.

3.3 Proposition. Any function f in $E[0, 1]$ is a Borel measurable Riemann function. Also, any uniformly regulated set in $E[0, 1]$ is uniformly Riemann.

Proof. A function in E is bounded, and continuous except on a countable set. So it is Borel measurable Riemann function. For $\mathcal{F}$ satisfying Theorem 2.2(a), applying the construction in the proof of Lemma 2.4 to each $f \in \mathcal{F}$ yields at most $2n(f, \varepsilon/2) + 1$ intervals $J_i \subset I$ with $w_f(J_i) \leq \varepsilon/2$ for all i, and whose union has total length $1 > 1 - \varepsilon$ as desired, so $\mathcal{F}$ is uniformly Riemann. $\square$

Proposition 3.3 extends directly to any bounded interval $[a, b]$ in place of I. It follows that a regulated real function on $\mathbb{R}$ is Riemann on each bounded interval, and that a set of functions on $\mathbb{R}$, uniformly regulated on each bounded interval, is uniformly Riemann on each such interval.

We recall the following:

3.4 Proposition. Any real-valued function f on an interval $J \subset \mathbb{R}$, continuous almost everywhere for Lebesgue measure λ, is Lebesgue measurable.

Proof. We can assume J is bounded and, taking $\tan^{-1} f$, that f is bounded. Then, see e.g. [16, Theorem 2.5.1] $\square$

The following will not be used below but may be of interest. Let (S, d) be a separable metric space with Borel σ-algebra $\mathcal{A}$. Let $\mathcal{B}(S)$ be the set of all bounded Borel measurable real-valued functions on S. Let P be a law (probability measure) on $(S, \mathcal{A})$. Billingsley and Topsøe [10] call a class $\mathcal{F} \subset \mathcal{B}(S)$ a P-**uniformity class** if whenever a sequence of laws P_n converges weakly to P, $\sup_{f \in \mathcal{F}} |\int f d(P_n - P)| \to 0$ as $n \to \infty$. For $r > 0$ and $x \in S$

let $B(x,r) := \{y \in S: d(x,y) < r\}$. For a set A and real-valued function f on A let $\omega_f(A) := \sup\{|f(x) - f(y)|: x,y \in A\}$. Let $\omega_{\mathcal{F}}(S) := \sup_{f \in \mathcal{F}} \omega_f(S)$. Billingsley and Topsøe [10, Theorem 1] prove that a class $\mathcal{F} \subset B(S)$ is a P-uniformity class if and only if both $\omega_{\mathcal{F}}(S) < \infty$ and for every $\varepsilon > 0$

$$\lim_{r \downarrow 0} \sup_{f \in \mathcal{F}} P\{x: \omega_f(B(x,r)) > \varepsilon\} = 0.$$

So, from Theorem 3.2 above, if S is an interval $[a,b]$, $a < b$, and P is the uniform distribution on $[a,b]$, then $\mathcal{F} \subset B[a,b]$ is a P-uniformity class if and only if it is uniformly Riemann and bounded in diameter.

4   The quantile operator.

Let H be any function from an interval $[a,b]$ into $\mathbb{R}$. For any $y \in \mathbb{R}$, let

$$H^{\leftarrow}(y) := H^{\leftarrow}_{[a,b]}(y) := \inf\{x: a \leq x \leq b, H(x) \geq y\}$$
$$:= b \quad \text{if } H(x) < y \text{ for } a \leq x \leq b.$$

If G is a strictly increasing, continuous function on $[a,b]$ with range $[c,d]$ then $G^{\leftarrow}$ on $[c,d]$ is the inverse function G^{-1} of G, but functions H to be considered will not necessarily be non-decreasing. It is known ([41, Theorem 6.4.2], [26, Prop. 6.1.1]) that $G \mapsto G^{\leftarrow}$ from $D[0,1]$ into L^q, $1 \leq q < \infty$, is continuous and sup-norm-compactly differentiable at the identity function U on $[0,1]$. For differentiation at U, some of the formulas for values of derivatives simplify, but it appears that Reeds and Fernholz's results extend also to differentiation at other functions such as the diffeomorphisms to be considered here, in Theorem 4.5 below. So the main difference here is a wider differentiability condition.

Some technical facts will be needed. In the following, $f^{\leftarrow}$ will mean $f^{\leftarrow}_{[a,b]}$, e.g. for $f = G$ or H, unless specified otherwise.

4.1 Lemma.   Let G and H be functions from $[a,b]$ into $\mathbb{R}$ where G is non-decreasing, $-\infty < a < b < +\infty$. Let $c := G(a)$ and $d := G(b)$. Let $y = y_0$ be a point in $[c,d]$, and $t \geq 0$ a fixed number. Let H also be defined on $[A,B]$, $A \leq a < b \leq B$.

(a) Assume that G is continuous from the right and for some $\beta > 0$, $G(x) - G(u) \geq \beta(x-u)$ for $a \leq u < x \leq b$. Then

   (a1) Let $x_0 := G^{\leftarrow}(y) - t/\beta$. If $H(x) \leq G(x) + t$ for all x with $a \leq x < x_0$, then $H^{\leftarrow}(y) \geq x_0$. If also $H(x) < y$ for $A \leq x < a$, then $H^{\leftarrow}_{[A,B]}(y) \geq x_0$.

   (a2) Let $x_1 := G^{\leftarrow}(y) + t/\beta$. Assume that $x_1 > b$ or, if $x_1 \leq b$, that $H(x_1) \geq G(x_1) - t$. Then $H^{\leftarrow}(y) \leq x_1$. If $x_1 \leq b$, then $H^{\leftarrow}_{[A,B]}(y) \leq x_1$.

(b) Assume that $\eta < \infty$ is such that $0 < G(x) - G(u) \leq \eta(x-u)$ whenever $a \leq u < x \leq b$.

   (b1) Let $x_2 := G^{\leftarrow}(y) + t/\eta$ and $x_3 := \min(b, x_2)$. If $H(x) \leq G(x) - t$ whenever $a \leq x < x_3$, then $H^{\leftarrow}(y) \geq x_3$. If also $H(x) < y$ for $A \leq x < a$, then $H^{\leftarrow}_{[A,B]}(y) \geq x_3$.

   (b2) Let $x_4 := G^{\leftarrow}(y) - t/\eta$ and $x_5 := \max(a, x_4)$. If $H(x_5) \geq G(x_5) + t$ then $H^{\leftarrow}_{[A,B]}(y) \leq H^{\leftarrow}(y) \leq x_5$.

Proof. In (a), for any $x \in [a, x_0]$ if $x_0 > a$, $H(x) \le G(x) + t < G(x_0) + t$. If $t > 0$, then for any v with $x_0 < v < G^{\leftarrow}(y)$,

$$G(x_0) \le G(v) - \beta(v - x_0) < y - \beta(v - x_0).$$

Letting $v \uparrow G^{\leftarrow}(y)$ gives $H(x) < y - t + t = y$, so $H^{\leftarrow}(y) \ge x_0$ as claimed. If $t = 0$ the result is clear.

Otherwise, $x_0 \le a$ (so the interval for x is empty). Then $H^{\leftarrow}(y) \ge a \ge x_0$, proving (a1) for $[a, b]$.

If $H(x) < y$ for $A \le x < a$, then clearly also $H^{\leftarrow}_{[A,B]}(y) \ge x_0$.

For (a2), if $x_1 \le b$, and if $t = 0$, then since G is right-continuous, $H(x_1) \ge G(x_1) \ge y$, so $H^{\leftarrow}(y) \le x_1$. If $t > 0$, then for any v with $G^{\leftarrow}(y) < v < x_1$, $G(x_1) - G(v) \ge \beta(x_1 - v)$, so $G(x_1) \ge \beta(x_1 - v) + y$. Letting $v \downarrow G^{\leftarrow}(y)$ gives $G(x_1) - t \ge y + t - t = y$, so $H^{\leftarrow}(y) \le x_1$ and a fortiori $H^{\leftarrow}_{[A,B]}(y) \le x_1$. Or if $x_1 > b$, then $H^{\leftarrow}(y) \le b < x_1$, proving (a2).

In case (b), $G(G^{\leftarrow}(y)) = y$ for $c \le y \le d$ by continuity and $G^{\leftarrow} \equiv G^{-1}$ on $[c, d]$. For (b1), consider two cases. Case 1: if $x_2 \le b$, then for $a \le x < x_3 = x_2$, we have

$$H(x) \le G(x) - t < G(x_2) - t = G(G^{-1}(y) + t/\eta) - t \le y + t - t = y,$$

so $H^{\leftarrow}(y) \ge x_3$.

Case 2: $x_3 = b < x_2 = G^{-1}(y) + t/\eta$, so $b - t/\eta < G^{-1}(y)$. If $b - t/\eta \ge a$, then $y > G(b - t/\eta) \ge G(b) - t$, and

$$H(x) \le G(x) - t \le G(b) - t < y \quad \text{for } a \le x < b,$$

so $H^{\leftarrow}(y) = b = x_3$. Otherwise, we have $b - t/\eta < a$, and for $a \le x < b$,

$$H(x) \le G(x) - t \le G(b) - t < G(b) - \eta(b - a) \le G(a) = c \le y,$$

so $H^{\leftarrow}(y) = b = x_3$, proving (b1) for $[a, b]$.

If $H(x) < y$ for $A \le x < a$, then clearly $H^{\leftarrow}_{[A,B]}(y) \ge x_3$.

For (b2), if $x_4 \ge a$, then $x_5 = x_4$ and

$$H(x_5) \ge G(x_4) + t = G(G^{-1}(y) - t/\eta) + t \ge y - t + t = y,$$

so $H^{\leftarrow}(y) \le x_5$. Or if $x_4 < a$, then $x_5 = a$ and $G^{-1}(y) < a + t/\eta$, so if $a + t/\eta \le b$, then $y \le G(a + t/\eta) \le G(a) + t \le H(a)$, so $H^{\leftarrow}(y) = a = x_5$. Or if $a + t/\eta > b$, then $t > \eta(b - a)$ and $H(a) \ge G(a) + \eta(b - a) \ge G(b) = d \ge y$, so $H^{\leftarrow}(y) = a$.

In any case, $x_5 \le b$, so $H^{\leftarrow}_{[A,B]}(y) \le H^{\leftarrow}(y) \le x_5$. $\square$

4.2 Lemma. Let G, H and h be functions from $[a, b]$ into $\mathbb{R}$ with $H \equiv G + h$, where for some $\gamma > 0$, $\beta > 0$ and $\eta < \infty$, $\mathrm{osc}_{[a,b]} h \le \gamma$ and $\beta(x - u) \le G(x) - G(u) \le \eta(x - u)$ for $a \le u < x \le b$. Let $c := G(a)$ and $d := G(b)$. Let $y := y_0$ be a point of $[c, d]$ and $\xi := G^{-1}(y)$. Let $s := h(\xi)$. If v is any number with $\beta \le v \le \eta$, and if $\xi - (s + \gamma)/\eta \le b$ and $\xi - (s - \gamma)/\eta \ge a$, then

$$|H^{\leftarrow}(y) - \xi + s/v| \le \gamma/\beta + |s| E_{\eta,\beta},$$

where $E_{\eta,\beta} := (\eta - \beta)/(\eta\beta)$. If $A \le a < b \le B$, the same holds for $H^{\leftarrow}_{[A,B]}(y)$ in place of $H^{\leftarrow}(y)$ if $H(x) < y$ for $A \le x < a$ and if $\xi + (\gamma - s)/\beta \le b$.

Proof. Here again $G^{\leftarrow} \equiv G^{-1}$ on $[c, d]$. We have $|h(x) - s| \le \gamma$, so that $H(x) \le G(x) + s + \gamma$ for $a \le x \le b$. If $s + \gamma \ge 0$, then, by Lemma 4.1(a1), $H^{\leftarrow}(y) \ge G^{-1}(y) - (s + \gamma)/\beta$. Since

$|s/\beta - s/v| \le |s|(\beta^{-1} - \eta^{-1}) = |s|E_{\eta,\beta}$, in this case

(4.3) $$H^{\leftarrow}(y) - G^{-1}(y) + s/v \ge -\gamma/\beta - |s|E_{\eta,\beta}.$$

If $s + \gamma < 0$, then by Lemma 4.1(b1), $H^{\leftarrow}(y) \ge x_2$, where $x_3 = x_2 \le b$ by hypothesis. Thus $H^{\leftarrow}(y) \ge G^{-1}(y) - (s+\gamma)/\eta$, and (4.3) again holds.

If (4.3) holds, and $H(x) < y$ for $A \le x < a$, then clearly (4.3) also holds for $H^{\leftarrow}_{[A,B]}(y)$.

For $a \le x \le b$, we have $H(x) \ge G(x) + s - \gamma$ since $|h(x) - s| \le \gamma$. So if $\gamma \ge s$, then by Lemma 4.1(a2), $H^{\leftarrow}(y) \le G^{-1}(y) + (\gamma - s)/\beta$, and

(4.4) $$H^{\leftarrow}(y) - G^{-1}(y) + s/v \le \gamma/\beta + |s|E_{\eta,\beta}.$$

The same holds for $H^{\leftarrow}_{[A,B]}(y)$ if $\xi + (\gamma - s)/\beta \le b$. Or if $\gamma < s$, then by Lemma 4.1(b2), $H^{\leftarrow}(y) \le x_4 := \xi - (s - \gamma)/\eta$, where $x_5 = x_4 \ge a$ by hypothesis. Thus (4.4) again holds, also for $H^{\leftarrow}_{[A,B]}(y)$. Combining (4.3) and (4.4), we get the conclusion.   □

In the following, note (again) that h is not restricted to functions for which $G + h$ is nondecreasing. For convenience, in the next fact the functions (G and h) are taken to be defined on $[0, 1]$.

A continuous real-valued function f on a closed interval $[a, b]$ will be called C^1 if it has a derivative f' on the open interval $]a, b[$ which extends to a continuous function on $[a, b]$. If f is also increasing from $[a, b]$ onto an interval $[c, d]$, it will be called an increasing **diffeomorphism** iff its inverse is also C^1. Recall that $R[0, 1]$ is the space of all Riemann functions on $[0, 1]$.

4.5 Theorem.   Let G be an increasing diffeomorphism from $[0, 1]$ onto itself. Then the map $h \mapsto (G+h)^{\leftarrow} \equiv (G+h)^{\leftarrow}_{[0,1]}$ is $\mathcal{C}$-differentiable at $h = 0$ from $R[0, 1]$ into L^p of Lebesgue measure λ on $[0, 1]$, where $1 \le p < \infty$ and $\mathcal{C}$ is the collection of all uniformly bounded and uniformly Riemann sets $\mathcal{F} \subset R[0, 1]$. Its derivative is $A(h) := -(h \circ G^{-1})/G' \circ G^{-1}$.

Proof.   Since G is an increasing diffeomorphism, there exist $0 < \beta < \eta < \infty$ such that $\beta \le G' \le \eta$ which is equivalent to the relation $\beta(x - y) \le G(x) - G(y) \le \eta(x - y)$ for $0 \le y < x \le 1$. Thus, G satisfies the conditions on it in Lemma 4.2 with $a = c = 0$, $b = d = 1$, and $0 < \beta < \eta < \infty$. Let $H := G + h$, so $H(x) = G(x) + h(x)$ for $0 \le x \le 1$. For any $h \in R[0, 1]$, also $h \circ G^{-1} \in R[0, 1]$ by Lemma A.4 of the Appendix and $A(h) \in R[0, 1]$, while $H^{\leftarrow}$ is nondecreasing, so all the functions arising are Lebesgue measurable by Prop. 3.4. For the supremum norm on $R[0, 1]$, sets in $\mathcal{C}$ are bounded, and A is a bounded linear operator into L^p since $G' \ge \beta > 0$. Let $\mathcal{F}$ be a uniformly bounded, uniformly Riemann set and $h := tg$ for $g \in \mathcal{F}$. Since $\mathcal{F}$ is bounded for the supremum norm, we can assume that $\|g\|_\infty \le 1$ for all $g \in \mathcal{F}$. We need to show that, uniformly in $g \in \mathcal{F}$, $\|H^{\leftarrow} - G^{-1} - (-h \circ G^{-1})/(G' \circ G^{-1})\|_p = o(t)$ as $t \to 0$ where $\|\cdot\|_p$ is the L^p norm. Suppose not. Then, for some $\alpha > 0$, there are $g_k \in \mathcal{F}$ and numbers $\varepsilon_k \to 0$ with $h_k := \varepsilon_k g_k$ such that $\|R_k\|_p \ge \alpha\varepsilon_k$ for $k = 1, 2, \ldots$, where

$$R_k := (G + h_k)^{\leftarrow} - G^{-1} + (h_k \circ G^{-1})/G' \circ G^{-1}.$$

For any $y \in [0, 1]$ let $\xi := G^{-1}(y) \in [0, 1]$ and $v := G'(\xi)$. Then $\beta \le v \le \eta$. Let $\gamma_k := 2\|h_k\|_\infty \le 2\varepsilon_k$. To check the hypotheses of Lemma 4.2 for $[0, 1]$, we have $\text{osc}_{[0,1]}h_k \le \gamma_k$, $\xi - (h_k(\xi) + \gamma_k)/\eta \le 1$ since $\xi \le 1$ and $h_k(\xi) \ge -\gamma_k$, and $\xi - (h_k(\xi) - \gamma_k)/\eta \ge 0$ since $\xi \ge 0$ and $h_k(\xi) \le \gamma_k$. So, Lemma 4.2 applies to give

$$|H^{\leftarrow}(y) - \xi + h_k(\xi)/v| \le D_{\eta,\beta}\gamma_k \,, \text{ where } D_{\eta,\beta} := \frac{2}{\beta} - \frac{1}{\eta}.$$

So $\|R_k\|_p \leq \|R_k\|_\infty \leq 2D_{\eta,\beta}\varepsilon_k$ for all $k,$.

Let C_k be the set where $|R_k| \geq \alpha\varepsilon_k/2$. Then

$$(\alpha\varepsilon_k)^p \leq \int_0^1 |R_k(x)|^p dx \leq (\alpha\varepsilon_k/2)^p + \lambda(C_k)(2D_{\eta,\beta}\varepsilon_k)^p.$$

Dividing by ε_k^p gives

$$\liminf_{k\to\infty} \lambda(C_k) \geq \alpha^p(1 - 2^{-p})/(2D_{\eta,\beta})^p > 0.$$

For any Borel set $B \subset [0,1]$, $\lambda(G^{-1}(B)) \geq \lambda(B)/\eta$: to see this, first suppose B is an interval, and note that $\inf(G^{-1})' \geq 1/\eta$. Then let B be a finite union of intervals, and lastly an arbitrary Borel set by the monotone class theorem (e.g., [20, Theorem 4.4.2]).

Let $D_k := G^{-1}(C_k)$. Then for some $c > 0$ not depending on k (though depending on α, p, β and η), we can assume that $\lambda(D_k) \geq c$ for all k. Let $0 < \varepsilon < \min(\alpha,\beta)$ and $0 < \varepsilon' < \min(c/2, \beta(\alpha-\varepsilon)/6)$. Apply Theorem 3.2(II), to $\mathcal{F}$, ε' and $H = [0,1]$ to get a $\delta > 0$. Since G' is continuous on $[0,1]$ there is a $\zeta > 0$ with $\zeta < \min(\delta, c/4)$ such that whenever $0 \leq x \leq u \leq 1$ and $u - x \leq \zeta$, we have $|G'(x) - G'(u)| < \beta\varepsilon^2/4$. Take k large enough so that $3\varepsilon_k/\beta < \zeta$. Let A_k be the set given in Theorem 3.2(II) for $f = g_k$. Then since $\lambda(A_k) > 1 - \varepsilon'$ and $\varepsilon' < c/2$, $\lambda(A_k \cap D_k) > c/2$, so $A_k \cap D_k \cap [c/4, 1 - c/4]$ is non-empty and contains a point x_0. Then $x_0 - \zeta > x_0 - c/4 \geq 0$ and likewise $x_0 + \zeta < 1$. Let $J_0 := [x_0 - \zeta, x_0 + \zeta]$. To apply Lemma 4.2 to the interval J_0, let $y := G(x_0)$, so $\xi = G^{-1}(y) = x_0$, $h := h_k$, and $v := G'(\xi)$, with $A := 0 < a := x_0 - \zeta < b := x_0 + \zeta < B := 1$. Let $\beta' := \inf\{G'(x): x \in J_0\}$ and $\eta' := \sup\{G'(x): x \in J_0\}$. Then $\eta' - \beta' < \beta\varepsilon^2/2 \leq \beta'\varepsilon^2/2$. Let $\gamma' := \mathrm{osc}_{J_0} h \leq \gamma_k \leq 2\varepsilon_k$. To check the hypotheses of Lemma 4.2, we have

$$x_0 - (h(\xi) + t)/\eta \leq x_0 + \varepsilon_k/\eta \leq x_0 + \zeta \quad \text{and}$$

$$x_0 - (h(\xi) - t)/\eta \geq x_0 - \varepsilon_k/\eta \geq x_0 - \zeta$$

for any $t \geq 0$ because $|h(\xi)| \leq \varepsilon_k$ and $\beta \leq \eta$. Specifically, take $t = \gamma'$. Also, $\xi + (\gamma' - s)/\beta \leq \xi + 3\varepsilon_k/\beta \leq b$, so Lemma 4.2 applies for $b < B = 1$. For $A = 0 \leq x < a$, we have $H(x) = G(x) + h(x) < G(x_0) - \beta(3\varepsilon_k)/\beta + \varepsilon_k < y$, so Lemma 4.2 applies for $A = 0 < a$. Then Lemma 4.2 yields, since $\varepsilon < \beta \leq \beta' \leq \eta'$,

$$|R_k(y)| \leq \gamma'/\beta' + |h_k(x_0)|(\eta' - \beta')/(\beta'\eta') \leq \gamma'/\beta + \varepsilon_k\varepsilon/2.$$

Then since $x_0 \in D_k$, we have $y \in C_k$, $|R_k(y)| \geq \alpha\varepsilon_k/2$ and $(\alpha-\varepsilon)\varepsilon_k/2 \leq \gamma'/\beta$, so $\mathrm{osc}_{J_0} g_k \geq \beta(\alpha-\varepsilon)/2$. On the other hand since $x_0 \in A_k$, and $\zeta < \delta$, we have $\mathrm{osc}_{J_0} g_k \leq 2\varepsilon' < \beta(\alpha-\varepsilon)/3$, a contradiction. $\qquad\square$

Let $DL[a, b]$ be the space of functions continuous from the left, with right limits, on $[a, b]$, and let $C_0(\mathbb{R})$ be the space of continuous real functions on $\mathbb{R}$ approaching 0 at $\pm\infty$, both with sup norm. Esty, Gillete, Hamilton and Taylor [25, p. 114] show that $h \mapsto (F + h)^{\leftarrow}$ restricted to $[a, b]$ is compactly differentiable at $h = 0$ from $C_0(\mathbb{R})$ into $DL[a, b]$ if F is a diffeomorphism from some interval onto an interval $[c, d]$ with $c < a < b < d$. Here, $h \mapsto (F + h)^{\leftarrow}$ is *not* $\mathcal{C}$-differentiable into $DL[a, b]$, even for the following small subclass $\mathcal{C}$ of those considered above: indeed, fix r, $1 \leq r < \infty$, let $\Phi(t) := t^r$, $0 \leq t < \infty$, and let $\mathcal{W}_r[a, b]$ be the space of all functions f of bounded Φ-variation, here called r-variation, with the norm $\|f\|_{[r]} := \sup_{[a,b]} |f| + v_\Phi(f)^{1/r}$. Let $\mathcal{C} = \{C_r\}$, $C_r := \{f \in \mathcal{W}_r: \|f\|_{[r]} \leq 3\}$, a single uniformly bounded and regulated set. Let $F(x) = x$, $0 \leq x \leq 1$, $a = 1/4$, $b = 3/4$, and consider $h_n(x) \equiv \max(0, n^{-1} - |x - 1/2|)$, $n \to \infty$, to see that $\mathcal{C}$-differentiability fails. On the other hand, limitation to the space $C_0(\mathbb{R})$ is rather restrictive, and it is perhaps not surprising if strong conditions are needed to get differentiability into a space with sup norm. Andersen, Borgan, Gill and Keiding [1, Proposition II.8.5] state a related compact differentiability property of the inverse operator.

Now, some facts on right and left limits will be useful. For any real-valued function f on an open interval J and $x \in J$, let $f^+(x-) := \limsup_{u \uparrow x} f(u)$, $f_-(x-) := \liminf_{u \uparrow x} f(u)$, $f^+(x+) := \limsup_{v \downarrow x} f(v)$, $f_-(x+) := \liminf_{v \downarrow x} f(v)$. Say f is **balanced** at x iff

$$-\infty \le f_-(x-) = f_-(x+) \le f(x) \le f^+(x-) = f^+(x+) \le +\infty.$$

In the following, f could be unbounded or nonmeasurable:

4.6 Theorem (W. H. Young).   For any open interval $J \subset \mathbb{R}$ and any real-valued function f on J, f is balanced at all points of J except at most for a countable set.

This theorem of Young ([59], [60]) is included (with proof) in Hobson [34, §228].

Now, for any real-valued function f on $[0, 1[$ the function $x \mapsto f^+(x+)$ is Borel measurable. Indeed, for any real t, let $B := \{x\colon 0 \le x < 1,\ f^+(x+) < t\}$. If $x_n \notin B$ and $x_n \downarrow x$, then for some $v_n \downarrow x$, $f(v_n) > t - 1/n$, so $x \notin B$. Thus, for every $x \in B$, there is an $\varepsilon > 0$ such that $[x, x+\varepsilon[\subset B$, and a maximal such $\varepsilon = \varepsilon_x$. Let V be the interior of B. For any $y \ne z$ in $B \backslash V$, the intervals $[y, y+\varepsilon_y[$ and $[z, z+\varepsilon_z[$ are disjoint: if not, we can assume $y < z$, and then $z < y + \varepsilon_y$, contradicting $z \notin V$. So, for $k = 1, 2, \ldots$, there are less than k values of $y \in B \backslash V$ with $\varepsilon_y > 1/k$, so $B \backslash V$ is countable and B is a Borel set, so $x \mapsto f^+(x+)$ is Borel measurable. Likewise, the other one-sided lim inf and lim sup functions mentioned in the definition of "balanced" are all Borel measurable.

A point x is called a **density point** of a set A if $\lim_{\delta \downarrow 0} \lambda(A \cap [x - \delta, x + \delta])/2\delta = 1$. Almost all points of a set of positive measure are density points of it (e.g. [20, Theorem 7.2.1 for $g = 1_A$], or [16, Theorem 6.3.6]).

The next fact will show that the Riemann condition on each function f in Theorem 4.5 is close to necessary. Let U be the identity function $U(x) \equiv x$, $0 \le x \le 1$. For a given function f, let $D_f(t, x) := [(U + tf)^{\leftarrow} - U](x)/t$, taking inverses with respect to $[0, 1]$.

In what follows, we consider the interval $[0, 1]$ with Lebesgue measure λ. Also, let $0 \le y \le 1$. For an arbitrary function H from $[0, 1]$ into $\mathbb{R}$, and $H^{\leftarrow}(y) := H^{\leftarrow}_{[0,1]}(y)$, $H^{\leftarrow}$ is non-decreasing and so Borel measurable. A function $V(t, y)$ will be said to converge in probability to $V(y)$ as $t \to a$ (resp. $t \uparrow a$, $t \downarrow a$) if for every sequence $t_n \to a$ (resp. $t_n \uparrow a$, $t_n \downarrow a$), $V(t_n, y) \to V(y)$ in probability (i.e., in measure for λ). Thus, measurability in t is not required.

4.7 Theorem.   Suppose that f is a function from $[0, 1]$ into $\mathbb{R}$ and $D_f(t, \cdot) \to h$ in probability on $[0, 1]$, where h is a finite valued measurable function, (a) as $t \downarrow 0$, (b) as $t \uparrow 0$, or (c) as $t \to 0$. Then on each interval $[0, c]$, $0 < c < 1$, f is bounded above in case (a), bounded below in case (b), and bounded in case (c). For almost all $y \in [0, 1]$, in case (a), $h(y) = -f^+(y-)$; in case (b), $h(y) = -f_-(y-)$; and in case (c), $h(y) = -f(y)$ and f is continuous at y.

Proof.   Case (a): if f is not bounded above on an interval $[0, c]$, $0 < c < 1$, then for any $t > 0$ and $y \in [0, 1]$, $(U + tf)^{\leftarrow}(y) \le c$, and for $c < y < 1$, $D_f(t, y) \to -\infty$ as $t \downarrow 0$, a contradiction.

Now given a y, $0 < y < 1$, take c with $y < c < 1$ and $M \ge 0$ such that $f(x) \le M < \infty$ for $0 \le x \le c$. Then for $t > 0$, $(U + tf)(x) < y$ if $0 \le x < y - tM$. By Theorem 4.6, $f^+(y-) = f^+(y+)$ for all but countably many y, so we can assume this holds for our y.

Thus, for any $\varepsilon > 0$ there is a $\delta > 0$ such that

(4.8) $$|x - y| < \delta \quad \text{implies} \quad f(x) < f^+(y-) + \varepsilon.$$

For k large enough so that $M/k < \delta$,

(4.9) $$(U + \tfrac{1}{k}f)(x) < y \quad \text{for} \quad 0 \le x \le y - \delta.$$

For $|x - y| < \delta$, $(U + \tfrac{1}{k}f)(x) < x + \tfrac{1}{k}(f^+(y-) + \varepsilon)$. Let $x_k := y - \tfrac{1}{k}(f^+(y-) + \varepsilon)$. Then $|x_k - y| < \delta$ for k large enough, say $k \ge k_0$, where we can take k_0 such that $M/k_0 < \delta$. Then for $y - \delta < x < x_k$, by (4.8)

$$(U + \tfrac{1}{k}f)(x) < x_k + \tfrac{1}{k}(f^+(y-) + \varepsilon) = y.$$

It follows from this and (4.9) that $(U + \tfrac{1}{k}f)^{\leftarrow}(y) \ge x_k$ for $k \ge k_0$. Then

(4.10) $$D_f(1/k, y) \ge \frac{x_k - y}{1/k} = -f^+(y-) - \varepsilon,$$

which holds for all but countably many $y \in (0, 1)$. There is a sequence $k = k_j \to \infty$ such that $D_f(1/k_j, y)$ converges for almost all $y \in [0, 1]$ to $h(y)$. Then $h(y) \ge -f^+(y-) - \varepsilon$. Letting $\varepsilon \downarrow 0$ gives

(4.11) $$h(y) \ge -f^+(y-) \quad \text{for almost all } y.$$

Now, for any $q \in \mathbb{Q}$ (the set of rational numbers), let A_q be the set of all accumulation points of $\{x \in [0, 1]: f(x) \ge q\}$. Then A_q is closed. Let D_q be the set of all density points of A_q. Then by facts noted above,

(4.12) $$\lambda \left(\bigcup_{q \in \mathbb{Q}} A_q \backslash D_q \right) = 0.$$

For a given $k = 2, 3, \ldots$, and $q \in \mathbb{Q}$, $q > 0$, let

(4.13) $$t_m := t_{m,k,q} := (1 - \tfrac{1}{k})^m / q.$$

Or, if $q < 0$, let $t_m := (1 - \tfrac{1}{k})^{m+1}/|q|$. Then by diagonalization, there exists a sequence $m = m(i)$ such that for each such k and q,

(4.14) $$D_f(t_{m(i),k,q}, y) \to h(y) \quad \text{for almost all } y \in [0, 1].$$

Take any y for which (4.14) holds and $y \notin \bigcup_{q \in \mathbb{Q}} A_q \backslash D_q$, which is almost all y by (4.12).

Then for any $q \in \mathbb{Q}$ with $q < f^+(y-)$, $\delta = 1/k$, and any m large enough, there exist $u_m = u_{m,k,q}$ and $v_m = v_{m,k,q}$ such that

$$y - (1 - \delta)^m < u_m < y - (1 - \delta)^{m+1}, \qquad y + (1 - \delta)^{m+1} < v_m < y + (1 - \delta)^m,$$

$f(u_m) \ge q$, and $f(v_m) \ge q$. Then if $f^+(y-) > 0$, take $q > 0$, and by (4.13)

$$(U + t_m f)(u_m) \ge u_m + t_m q = u_m + (1 - \delta)^m > y,$$

so $(U + t_m f)^{\leftarrow}(y) \le u_m < y - (1 - \delta)^{m+1}$, and $D_f(t_m, y) \le -(1 - \delta)q$. Thus by (4.14), $h(y) \le -(1 - \delta)q$. Letting $q \uparrow f^+(y-)$ and $\delta = 1/k \downarrow 0$ gives $h(y) \le -f^+(y-)$. Or, if $f^+(y-) \le 0$, take $q < 0$. Nearly the same argument with v_m in place of u_m again gives $h(y) \le -f^+(y-)$, which with (4.11) proves case (a).

Case (b) then follows from case (a) for $-f$.

Case (c) follows from cases (a) and (b) and Theorem 4.6.   □

An operator H will be called differentiable at F **along** f if $(H(F + tf) - H(F))/t$ converges as $t \to 0$.

4.15 Corollary. Let $1 \le p < \infty$. Let f be a real-valued function on $[0,1]$ and $g(x) :=$ $f(1 - x)$, $0 \le x \le 1$. Then $F \mapsto F^{\leftarrow}$ is differentiable into L^p at $F = U$ (the identity) along both f and g if and only if f is a Riemann function.

Proof. "If" follows from Theorem 4.5 and "only if" from Theorem 4.7.   □

5 The composition operator for real-valued functions.

Theorem 5.1, on joint differentiability of composition for some specific function spaces, at suitable F, G, extends the work of Reeds [41] and Fernholz [26] from sup-norm-compact differentiability to $\mathcal{C}$-differentiability in F. (With respect to G, one can either have Fréchet differentiability from L^s to L^q, $q < s$, as here, or compact differentiability from L^s to L^s as in Reeds and Fernholz.)

A function F from $\mathbb{R}$ into $\mathbb{R}$ is called **Lipschitz** if for some $K < \infty$, $|F(x) - F(y)| \le$ $K|x - y|$ for all x, y. Then for Lebesgue almost all x, the derivative $F'(x)$ exists, with $|F'(x)| \le K$. For any M-function Φ let $\mathcal{W}_\Phi(J)$ be the set of all real functions of bounded Φ-variation on the interval J. Let $\mathcal{W}_\Phi^*(J)$ be the set of all real functions f on J such that $\delta f \in \mathcal{W}_\Phi(J)$ for some $\delta > 0$, and $\|f\|_{(\Phi)} := \inf\{c > 0 : v_\Phi(f/c) \le 1\}$. Then $\mathcal{W}_\Phi(J)$ is included in $E(J)$ by Theorem 2.1, and hence so is $\mathcal{W}_\Phi^*(J)$. Musielak and Orlicz [38] proved that for $J = [a, b]$, $\{f \in \mathcal{W}_\Phi^*(J) : f(a) = 0\}$ is a Banach space with the norm $\| \cdot \|_{(\Phi)}$. Let $\|f\|_{[\Phi]} := \|f\|_\infty + \|f\|_{(\Phi)}$, where $\|f\|_\infty := \sup|f|$. Then $(\mathcal{W}_\Phi^*(J), \| \cdot \|_{[\Phi]})$ is a Banach space.

Given $h \colon \mathbb{R}^2 \mapsto \mathbb{R}$, a Nemytskii operator T is defined by $(Tg)(x) \equiv h(x, g(x))$. An assumption often made on such operators is that h is a **Caratheodory function** (e.g. [3, pp. 216, 219]), namely that $h(\cdot, u)$ is measurable for each u, and $h(x, \cdot)$ is continuous for almost all x. For T to take $L^s[0,1]$ continuously into $L^q[0,1]$ for $1 \le q < s$, a well known condition is the **growth condition**

$$|h(x, u)| \le a(x) + b|u|^{s/q} \quad \text{for some} \quad a(\cdot) \in L^q \quad \text{and} \quad b < \infty$$

(e.g. Zeidler [61, II/B p. 561]). In our case, h will be of the special form $h(x, u) \equiv H(G(x) + u)$ for some G and H. For $0 < \tau < \infty$ recall the space $R_\tau(\mathbb{R})$ of τ-Riemann functions from $\mathbb{R}$ into $\mathbb{R}$, defined in §1. If $h(x, u) \equiv H(G(x) + u)$, the growth condition follows from $H \in R_{s/q}(\mathbb{R})$, with $B := \sup_y |H(y)|/(1 + |y|^{s/q})$, and $G \in \mathcal{L}^s$, with $b := 2^{s/q}B$, since then $a := B(1 + |2G|^{s/q}) \in \mathcal{L}^q$.

In most of the literature on the Nemytskii operator, notably including Shragin [47, II, pp. 71-77], it is assumed that h is a Caratheodory function. The assumption is actually necessary for *global* continuity of the operator T, see Goldberg, Kampowsky and Tröltzsch [31, Section 2.4], Lucchetti and Patrone [36]. Under the assumptions of Theorem 5.1 just below, $(x, u) \mapsto F(G(x) + u)$ is indeed a Caratheodory function. If $H(G(x) + \cdot)$ is continuous, even for one x, then H is continuous (everywhere). In treating differentiability of the two-function composition operator $(f, g) \mapsto (F + f) \circ (G + g)$, or, more precisely, of $(t, g) \mapsto (F + tf) \circ (G + g)$ at $t = 0 \in \mathbb{R}$ and $g = 0$, for a fixed f, we need instead continuity of the Nemytskii operator $g \mapsto f \circ (G + g)$ from L^s to L^p at the point $g = 0$. To cover the case, of main interest e.g. in Part I, Section 6, $f = F_n - F$ where F is continuous and F_n is a step function (empirical distribution function), f must have jumps at some points. Then, $u \mapsto f(G(x) + u)$ is everywhere continuous in u for no x, so $h : (x, u) \mapsto f(G(x) + u)$ is not a Caratheodory function, while $F_n - F$ is not a "pathological counter-example" function.

Another point needing notice to obtain the full characterization of directions of differentiability f in Theorems 5.1 and 5.3 below is that if H is universally but not Borel measurable, then $h(x, u)$ is not Borel measurable in u for any x, so h is not a Shragin function as defined by Appell and Zabrejko [4, p. 17], and functions h need to be somewhat more general.

For the following main theorem, recall the definition of uniformly τ-Riemann set from §1.

5.1 Theorem. Let $(\Omega, \mathcal{A}, \mu)$ be a finite, complete measure space with no atoms. Suppose that F is a Lipschitz function from $\mathbb{R}$ into $\mathbb{R}$, that $\mu \circ G^{-1}$ is absolutely continuous with respect to Lebesgue measure λ on $\mathbb{R}$, and that $G \in \mathcal{L}^s := \mathcal{L}^s(\Omega, \mu)$. Then the composition operator $(f, g) \mapsto (F + f) \circ (G + g)$ is jointly differentiable at $f = 0$, $g = 0$ from $R_{s/q}(\mathbb{R}) \times L^s(\Omega, \mu)$ into $L^q(\Omega, \mu)$ whenever $1 \le q < s < \infty$, for C-differentiability on $R_{s/q}(\mathbb{R})$, with the norm $\|f\|_{\{s/q\}} := \sup_x |f(x)|/(1 + |x|^{s/q})$, where C is the collection of all uniformly (s/q)-Riemann sets $\mathcal{K} \subset R_{s/q}(\mathbb{R})$ and the usual norms on L^s and L^q. Here the joint differentiability means that for any such $\mathcal{K}$,

$$\|(F + tf) \circ (G + g) - F \circ G - tf \circ G - (F' \circ G) \cdot g\|_q = o(|t| + \|g\|_s)$$

as $|t| + \|g\|_s \to 0$, uniformly for $f \in \mathcal{K}$. In particular $\mathcal{K}$ can be any set of functions bounded for $\|\cdot\|_{\{s/q\}}$ and uniformly regulated on every bounded interval, or more specifically any set of uniformly bounded Φ-variation norm $\|\cdot\|_{[\Phi]}$ for an M-function Φ. Thus the composition operator is jointly Fréchet differentiable for any Φ-variation norm $\|\cdot\|_{[\Phi]}$ in f, from $\mathcal{W}_\Phi^* \times \mathcal{L}^s$ into $\mathcal{L}^q$, $1 \le q < s$, and

$$\|(F + f) \circ (G + g) - F \circ G - f \circ G - (F' \circ G) \cdot g\|_q = o(\|f\|_{(\Phi)} + \|g\|_s)$$

as $\|f\|_{(\Phi)} + \|g\|_s \to 0$.

Note. If μ had an atom, there is no G such that $\mu \circ G^{-1}$ is absolutely continuous with respect to λ.

Proof. Since F is Lipschitz, it has a derivative $F'(x)$ for $x \notin A$ for some set A with $\lambda(A) = 0$. Then $\mu(G^{-1}(A)) = 0$, so $F'(G(x))$ exists for μ-almost all x. For any continuous function F, the set of x for which $F'(x)$ exists is a Borel set, indeed an $F_{\sigma\delta}$ (countable intersection of countable unions of closed sets). On its Borel domain, F' is easily seen to be a Borel measurable function. This and universal measurability of f and F implies measurability of compositions, defined almost everywhere (see the Appendix, Theorem A.3).

The remainder in the (claimed) differentiation is a sum of two terms, $R_1 := t[f \circ (G + g) - f \circ G]$ and $R_2 := F \circ (G + g) - F \circ G - (F' \circ G) \cdot g$. For R_2 (which does not involve f) we need the following known fact:

5.2 Lemma. The map $g \mapsto F \circ (G + g)$ is Fréchet differentiable at $g = 0$ from L^s to L^q of any finite measure space if $1 \le q < s$ whenever F is a Lipschitz function, at any $G \in \mathcal{L}^s$ such that for almost all x, $F'(G(x))$ is defined. The derivative of the map is $g \mapsto (F' \circ G) \cdot g$.

Notes. Here $g \mapsto F \circ (G + g)$ is a Nemytskii operator. Fernholz [26, Prop. 6.1.2] shows under similar conditions that one can take $q = s$ if Fréchet is replaced by compact differentiability.

Proof.   This fact is apparently due to Wang Sheng-Wang [57], Cor. 2 of Theorem 2, who announced without proof a result that applies. Appell [2, Theorem 4]; [3, Theorem 2.10] characterized the F yielding differentiability. But checking Appell's hypotheses seems to take a little longer than the following direct proof, given for completeness, essentially from Krasnosel'skiĭ, Zabreĭko, Pustyl'nik and Sobolevskiĭ [35, §20.2, where $L_\alpha := L^p$ for $p = 1/\alpha$]. We want to show that

$$\|F(G + g) - F(G) - F'(G)g\|_q = o(\|g\|_s) \quad \text{as} \quad \|g\|_s \to 0.$$

Let

$$R(x, u) := u^{-1}[F(G(x) + u) - F(G(x))] - F'(G(x)), \quad u \neq 0$$

$$:= 0 \quad \text{if } u = 0,$$

defined for almost all x and all u. Let $K := \sup_y |F'(y)|$ (supremum over all y for which $F'(y)$ is defined). Then $|R(x, u)| \leq 2K$ and $R(x, u) \to 0$ as $|u| \to 0$ for almost all x. Let $R_g(x) := R(x, g(x))$. Then the remainder in the differentiation is $R_g g$ and by the Hölder (or Rogers-Hölder: [20, p. 140]) inequality,

$$\|R_g g\|_q \leq \|g\|_s \|R_g\|_{qs/(s-q)}.$$

As $\|g\|_s \to 0$, $R_g \to 0$ in measure: if not, there are g_k with $\|g_k\|_s \to 0$ and R_{g_k} not going to 0 in measure. Taking subsequences twice, we can assume that no subsequence of R_{g_k} converges to 0 in measure, while $g_k \to 0$ a.e., but then $R_{g_k} \to 0$ a.e., a contradiction. So by dominated convergence, $\|R_g\|_{qs/(s-q)} \to 0$, and the remainder is $o(\|g\|_s)$.   □

Now to finish the proof of Theorem 5.1, consider the other remainder term $R_1 := tf(G + g) - tf(G)$. The operator $f \mapsto f \circ G$ is easily seen to be bounded and linear from $(R_{s/q}, \| \cdot \|_{\{s/q\}})$ into $L^q(\Omega)$ since $G \in \mathcal{L}^s(\Omega)$. Then it is enough to show that $\|R_1\|_q = o(|t|)$ as $t \to 0$ and $\|g\|_s \to 0$, or equivalently that $\|f(G + g) - f(G)\|_q \to 0$ as $\|g\|_s \to 0$, uniformly for $f \in \mathcal{K}$. If not, then for some $\varepsilon > 0$ and sequences $f_k \in \mathcal{K}$ and g_k with $\|g_k\|_s \to 0$,

$$(*) \qquad \|f_k(G + g_k) - f_k(G)\|_q > 2\varepsilon \quad \text{for all} \quad k.$$

Let H_k be any sequence of measurable sets in Ω with $\mu(H_k) \to 0$ as $k \to \infty$. Let $B := \sup\{\|f\|_{\{s/q\}}: f \in \mathcal{K}\} < \infty$ by assumption. Then

$$\zeta_k := \|(f_k(G + g_k) - f_k(G))1_{H_k}\|_q \leq 2B\mu(H_k)^{1/q} + B[\|(G + g_k)1_{H_k}\|_s^{s/q} + \|G1_{H_k}\|_s^{s/q}],$$

and $\|(G + g_k)1_{H_k}\|_s \leq \|G1_{H_k}\|_s + \|g_k\|_s$. Now $\|G1_{H_k}\|_s \to 0$ as $k \to \infty$ by dominated convergence, so $\zeta_k \to 0$. Thus for some k_0 and $\gamma > 0$, we have $\|(f_k(G+g_k) - f_k(G))1_H\|_q < \varepsilon$ whenever $k \geq k_0$ and $\mu(H) < 3\gamma$.

Let $H_k := \{x : |f_k(G(x) + g_k(x)) - f_k(G(x))| > \varepsilon\}$. Then by $(*)$, $\mu(H_k) \geq 3\gamma$. Take N large enough so that $\mu(\Omega \backslash G^{-1}([-N, N])) < 2\gamma$. By absolute continuity there is a $\zeta > 0$ with $\zeta \leq \varepsilon$ such that if $J \subset [-N, N]$ and $\lambda(J) < \zeta$ then $\mu(G^{-1}(J)) < \gamma/3$. If $A_k := H_k \cap G^{-1}([-N, N])$ then $\mu(A_k) > \gamma$. By Theorem 3.2(II) for $H = [-N, N]$ and ζ, there is a $\delta > 0$ such that there is a Borel set $C_k \subset [-N, N]$ with $\lambda(C_k) > 2N - \zeta$ and $|f_k(x) - f_k(y)| < \varepsilon$ whenever $|x - y| < \delta$, $x \in C_k$, and $y \in \mathbb{R}$. Letting $J = [-N, N]\backslash C_k$, we have $\mu(G^{-1}(J)) < \gamma/3$.

For k large enough, $|g_k| < \delta$ except on a set S in Ω with $\mu(S) \leq \gamma/2$. Then there is a set $B_k \subset A_k$ with $\mu(B_k) \geq \gamma/2$ on which $|g_k| < \delta$, and $D_k := B_k \cap G^{-1}(C_k)$ is non-empty. But for $x \in D_k$, since $x \in A_k \subset H_k$, we have

$$|f_k(G(x) + g_k(x)) - f_k(G(x))| > \varepsilon.$$

Since $x \in B_k$, we have $|g_k(x)| < \delta$. Then since $G(x) \in C_k$, we have

$$|f_k(G(x) + g_k(x)) - f_k(G(x))| < \varepsilon,$$

a contradiction. So the first statement in Theorem 5.1 is proved. The second statement then follows by Prop. 3.3. For the third and last statement, let $\mathcal{F}$ be a set of functions on $\mathbb{R}$ such that for some $M < \infty$, $\|f\|_{[\Phi]} \leq M$ for all $f \in \mathcal{F}$. Then by definition of $\|\cdot\|_{(\Phi)}$, if $\mathcal{G} := \{f/(M+1): f \in \mathcal{F}\}$, then $v_\Phi(g) \leq 1$ for all $g \in \mathcal{G}$. So $\mathcal{G}$ is uniformly regulated on any bounded interval by Theorem 2.2. It follows that $\mathcal{F}$ is uniformly regulated on any bounded interval by Theorem 2.2(c). Also, $\mathcal{F}$ is uniformly bounded, thus bounded in $\|\cdot\|_{\{s/q\}}$. $\square$

With respect to the directions of differentiability f at F, Theorem 5.1 as just proved is an extension of the Reeds-Fernholz results from sup norm-compact sets to much larger sets. It is not possible to take $q = s$ and have Fréchet differentiability in Lemma 5.2 (or, *a fortiori*, in Theorem 5.1) unless F is an affine function, $F(t) \equiv c + dt$ for some constants c, d, as shown in [35, Theorem 20.1], see also [3, Theorem 2.11]; both mention Vaĭnberg [54].

A finite measure space $(\Omega, \mathcal{S}, \mu)$ is called **perfect** if for every measurable real-valued function η on Ω and every set $E \subset \mathbb{R}$ such that $\eta^{-1}(E)$ is measurable for the completion of μ, E is measurable for the completion of the restriction of $\mu \circ \eta^{-1}$ to the Borel σ-algebra in $\mathbb{R}$. If Ω is a complete separable metric space with Borel σ-algebra $\mathcal{S}$, then, for any finite measure μ on $\mathcal{S}$, $(\Omega, \mathcal{S}, \mu)$ is perfect (e.g., Gnedenko and Kolmogorov [29, Sec. 3 Theorem 1] in light of [20, Theorem 7.1.3]).

The following shows that if (Ω, μ) is perfect, the space $R_{s/q}(\mathbb{R})$ in Theorem 5.1 is as large as possible for the theorem to hold, even if $\mathcal{C}$ is replaced by the class of all singletons $\{f\}$. Thus the a.e. continuity, universal measurability and the growth condition are all needed. Necessity of the conditions on sets $\mathcal{K}$ for uniformity is not addressed here.

5.3 Theorem. Let (Ω, μ) be a perfect finite measure space, $1 \leq q < s$, and $G \in \mathcal{L}^s(\Omega, \mu)$. Suppose Lebesgue measure λ on the Borel σ-algebra in $\mathbb{R}$ is absolutely continuous with respect to $\mu \circ G^{-1}$, and for some F, composition $(h, g) \mapsto (F + h) \circ (G + g)$ is defined and jointly Fréchet differentiable into $L^q(\Omega, \mu)$ at $h = g = 0$ for $g \in L^s(\Omega, \mu)$ and h in the one-dimensional subspace spanned by a function f. Then $f \in R_{s/q}(\mathbb{R})$.

Proof. The remainder for given t and g is

$$(F + tf) \circ (G + g) - tf \circ (G + g) - (F' \circ G)g.$$

Subtracting the remainder for $t = 0$ from that for $t \neq 0$, we can assume $F \equiv 0$. Then $\|tf(G+g) - tf(G)\|_q = o(|t| + \|g\|_s)$ as $|t| + \|g\|_s \to 0$, or equivalently $\|f(G+g) - f(G)\|_q \to 0$ as $\|g\|_s \to 0$. This reduces to the continuity of the Nemytskii operator $T : (Tg)(x) := f(G(x) + g(x))$ from L^s into L^q at $g = 0$. For that, $|f(y)|/(1 + |y|^{s/q})$ must be bounded, as shown e.g. in [3, Theorem 2.1] under unnecessary, slightly stronger assumptions. Here is a proof. Suppose not. Then for some y_n, $|f(y_n)|/(1 + |y_n|^{s/q}) \to \infty$ as $n \to \infty$. We can assume that $|f(y_n)| \geq n^2(1 + |y_n|^{s/q})$ for all n. Let $p_n := C/(n^2(1 + |y_n|^s))$, choosing C so that $\sum_n p_n = 1$. Then $\sum_n p_n |y_n|^s < \infty$ while $\sum_n p_n |f(y_n)|^q = +\infty$.

Since μ is finite, it has at most countably many atoms. Since λ is absolutely continuous with respect to $\mu \circ G^{-1}$, it is easily seen that there is a measurable set $A \subset \Omega$ with $\mu(A) > 0$ on which μ is nonatomic (cf. Shilov and Gurevich [45, Sec. 9.1 Theorem 1]). Thus there exist disjoint measurable sets $D_n \subset A$ with $\mu(D_n) = p_n\mu(A)$ for all n (e.g. Mukherjea and Pothoven [37, Lemma 3.9 p. 163]). Let $g_k(x) := y_n - G(x)$ for $x \in D_n$ and $n \geq k$, with $g_k(x) = 0$ otherwise. Then $f \circ (G + g_k)$ is not even in $\mathcal{L}^q$ for any k although $\|g_k\|_s \to 0$, a contradiction. So $|f(y)|/(1 + |y|^{s/q})$ is bounded.

Next, since $(F+f) \circ (G+g)$ and $F \circ (G+g)$ are μ-measurable, $f \circ (G+g)$ is μ-measurable for every $g \in \mathcal{L}^s(\Omega, \mu)$. Thus, $f \circ \psi$ is μ-measurable for every $\psi \in \mathcal{L}^\infty(\Omega, \mu)$ (let $g = \psi - G$). In proving that f is u.m., we can assume that $\mu(A) = 1$ for a measurable set A as above on which μ is nonatomic. Let ν be μ restricted to A. There exists a measurable function ψ_1 on A such that $\nu \circ \psi_1^{-1}$ is the uniform distribution $U[0,1]$ on $[0,1]$: decompose A into disjoint measurable sets A_0, A_1 with $\nu(A_0) = \nu(A_1) = 1/2$, let ψ_1 take A_0 into $[0, 1/2]$ and A_1 into $[1/2, 1]$, and continue recursively. Now let $0 < M < \infty$ and let Q be any probability measure on the Borel sets of $[-M, M]$ with a distribution function F_Q. Then $U[0,1] \circ (F_Q^{\leftarrow})^{-1} = Q$ (e.g. [20, Proposition 9.1.2]), so $\nu \circ \eta^{-1} = Q$, where $\eta := F_Q^{\leftarrow} \circ \psi_1$. Let B be a Borel set in $\mathbb{R}$. Then $(f \circ \eta)^{-1}(B) = \eta^{-1}(f^{-1}(B))$ is μ- and so ν-measurable (we can let $\eta(x) := 0$ for $x \in \Omega \backslash A$). So since (A, ν) is perfect, $f^{-1}(B)$ is measurable for the completion of Q. Thus $f^{-1}(B) \cap [-M, M]$ is a u.m. set. Letting $M \to \infty$, we get that $f^{-1}(B)$ is a u.m. set in $\mathbb{R}$ and f is a u.m. function.

Next it will be shown that f is continuous almost everywhere on $\mathbb{R}$. (Krasnosel'skii et al. [35, p. 372] prove continuity in measure.) Suppose not. Then on some interval $K := [-M, M]$, $0 < M < \infty$, f is not continuous a.e. on K. So, for some $\varepsilon > 0$, the condition in Theorem 3.1(c) fails for $\delta = 1/n$ for all $n = 1, 2, \ldots$, and there is no measurable set $A \subset K$ with $\lambda(A) > 2M - \varepsilon$ such that for all $x \in A$ and all $y \in \mathbb{R}$ with $|x - y| < 1/n$, we have $|f(x) - f(y)| \leq \varepsilon$. Let

$$B_n := \left\{ (x, y) \in \mathbb{R}^2 : |x| \leq M, \ y \in \mathbb{R}, \ |x - y| < 1/n, \ |f(x) - f(y)| > \varepsilon \right\}.$$

Let $\mathbb{Q}$ be the set of rational numbers. Then

$$\left\{ (x, y) \in \mathbb{R}^2 : |x - y| < 1/n \right\} = \bigcup_{q \in \mathbb{Q}} \left] q, q + \frac{1}{n} \right[\times \left] q, q + \frac{1}{n} \right[$$

and

$$\left\{ (x, y) \in \mathbb{R}^2 : |f(x) - f(y)| > \varepsilon \right\} =$$

$$\bigcup_{s \in \mathbb{Q}, s > \varepsilon} \bigcup_{r \in \mathbb{Q}} (\{x : \ f(x) < r\} \times \{y : \ f(y) > r + s\}) \cup (\{x : \ f(x) > r + s\} \times \{y : \ f(y) < r\}).$$

Since f is u.m., and by the distributive laws for intersections of countable unions, we get $B_n = \cup_{k=1}^\infty S_{nk} \times T_{nk}$ for some u.m. sets S_{nk}, T_{nk}, which we can assume are all non-empty (B_n is non-empty, or we could take $A = K$). Then

$$\{x : \ (x, y) \in B_n \text{ for some } y\} = \cup_{k=1}^\infty S_{nk} =: C_n,$$

a u.m. set. Let

$$A_n := \{x \in K : |f(x) - f(y)| \leq \varepsilon \text{ for all } y \in \mathbb{R} \text{ with } |x - y| < 1/n\}.$$

Then $A_n = K \backslash C_n$, so A_n is Lebesgue measurable and $\lambda(A_n) \leq 2M - \varepsilon$. For each $k = 1, 2, \ldots,$, let $t_{nk} \in T_{nk}$. For $x \in C_n$, let $h_n(x) := t_{nk}$ for the least k such that $x \in S_{nk}$. Then h_n is a u.m. function, and $(x, h_n(x)) \in B_n$ for all $x \in C_n$.

Let λ and $\mu \circ G^{-1}$ both be defined on the respective completions of the Borel σ-algebra. Any set of $\mu \circ G^{-1}$ measure 0 is included in a Borel set E with $(\mu \circ G^{-1})(E) = 0$, so $\lambda(E) = 0$. So any $\mu \circ G^{-1}$ measurable set is λ measurable. Each C_n is $\mu \circ G^{-1}$ measurable.

We have $\lambda(C_n) \geq \varepsilon$ for all n. Since λ on K is absolutely continuous with respect to $\mu \circ G^{-1}$, there is a $\delta > 0$ such that for any $\mu \circ G^{-1}$ measurable set $C \subset K$ with $\mu \circ G^{-1}(C) < \delta$, we have $\lambda(C) < \varepsilon$ (e.g. Cohn [16, Lemma 4.2.1]). Thus $\mu(G^{-1}(C_n)) \geq \delta$ for all n. For $\omega \in G^{-1}(C_n)$ let $g_n(\omega) := h_n(G(\omega)) - G(\omega)$, and let $g_n(\omega) := 0$ otherwise. Then $|g_n(\omega)| < 1/n$

for all ω and g_n is μ-measurable, so $g_n \to 0$ in $L^s(\Omega, \mu)$, while $|f \circ (G + g_n) - f \circ G|(\omega) > \epsilon$ for $\omega \in G^{-1}(C_n)$, so $\|f \circ (G + g_n) - f \circ G\|_q > \epsilon \delta^{1/q}$ for all n. This contradicts continuity of the Nemytskii operator T. So f is continuous a.e., which completes the proof. $\qquad\square$

Example. For a class of cases where Theorem 5.3 applies, with its somewhat unusual hypothesis on $\mu \circ G^{-1}$, let $\Omega = [0, 1]$, $\mu = U[0, 1]$, let H be the distribution function of a random variable X having a strictly positive density on $\mathbb{R}$ and $E|X|^s < \infty$, and let $G = H^{-1}$.

To treat the operator $(f, g) \mapsto (F + f) \circ (G + g)^\leftarrow$, combining Theorems 4.5 and 5.1, it will be useful to have two chain rules for composition of Fréchet differentiable with C-differentiable functions. The first was announced in [44] (but not included in the short section on chain rules in [6, pp. 216–217]). We do not have references for the proofs of the next two theorems, but the proofs are omitted because they are straightforward extensions of proofs of chain rules for Fréchet differentiability.

5.4 Theorem (Sebastião e Silva). Let $(X, \|\cdot\|_X)$, $(Y, \|\cdot\|)$ and $(Z, |\cdot|)$ be normed spaces. Let C be a collection of bounded subsets of X. Suppose T is C-differentiable at a point $x_0 \in X$ into Y. Let $y_0 := T(x_0)$ and let U be a function from a neighborhood of y_0 in Y into Z, Fréchet differentiable at y_0. Then $U \circ T$ is C-differentiable at x_0 from X into Z, with derivative $U'(y_0) \circ T'(x_0)$.

Theorem 5.4 extends to a case where the function depends on another variable. Let X, Y and Z be normed spaces, and let $\|\cdot\|_A$ denote the norm on A for $A = X$, Y or Z. Let C be a class of subsets of X and D a class of subsets of Y. Let T be a function from a subset of $X \times Y$ into Z. Then T will be called (C, D)-**differentiable** at (x_0, y_0) into Z if it is $\mathcal{E}$-differentiable at (x_0, y_0) into Z where $\mathcal{E} := \{C \times D : C \in C, D \in D\}$ in the normed space $X \times Y$ with norm $\|(x, y)\| := \|x\|_X + \|y\|_Y$.

5.5 Theorem. Let W, X, Y, Z be normed spaces. Let C and D be collections of bounded subsets of X and W, respectively. Let B be the class of all bounded subsets of Y. Let T be a function from an open subset of $X \times Y$ into Z, (C, B)-differentiable at a point (x_0, y_0). Let H be a function from an open subset of W into Y which is D-differentiable at a point w_0 with $H(w_0) = y_0$. Then the function $(x, w) \mapsto T(x, H(w))$ is (C, D)-differentiable at (x_0, w_0) into Z.

If F and G are two probability distribution functions on $\mathbb{R}$ then $F \circ G^\leftarrow$ from $]0, 1[$ into $]0, 1[$ is called a **procentile-procentile** function or $P - P$ plot. Reeds [41] treated composition as a step toward treating $F \circ G^\leftarrow$. We have

5.6 Theorem. Let F be Lipschitz and G a diffeomorphism from $[0, 1]$ onto itself. Then the operator $(f, g) \mapsto (F + f) \circ (G + g)^\leftarrow$ is (C, D)-differentiable at $f = g = 0$ from $R_\tau(\mathbb{R}) \times R[0, 1]$ into $L^q[0, 1]$ whenever $1 \leq q < \infty$ and $1 < \tau < \infty$, for the collection C of all uniformly τ-Riemann sets for f, and the collection D of all uniformly bounded and uniformly Riemann sets for g. So, C can be the class of $\|\cdot\|_{[\Phi]}$-bounded sets for any M-function Φ, and D the class of uniformly bounded and regulated sets. Thus the same map is jointly Fréchet differentiable from $(\mathcal{W}_\Phi^*(\mathbb{R}), \|\cdot\|_{[\Phi]}) \times (W_\Psi^*[0, 1], \|\cdot\|_{[\Psi]})$ into L^q for any two M-functions Φ, Ψ. Specifically, we have

$$\|(F + f) \circ (G + g)^\leftarrow - F \circ G^{-1} - f \circ G^{-1} + (F' \circ G^{-1}) \cdot (g \circ G^{-1})/(G' \circ G^{-1})\|_q$$

$$= \quad o(\|f\|_{(\Phi)} + \|g\|_{[\Psi]}) \quad \text{as} \quad \|f\|_{(\Phi)} + \|g\|_{[\Psi]} \to 0.$$

Proof.   Let $s := \tau q$. The map $g \mapsto (G+g)^{\leftarrow}$ is $\mathcal{D}$-differentiable from $R[0,1]$ into L^s at $g = 0$ by Theorem 4.5. Next, the diffeomorphism G^{-1} satisfies the hypotheses on G in Theorem 5.1, by Lemma A.4 of the Appendix. So $(f,h) \mapsto (F+f) \circ (G^{-1}+h)$ is $(\mathcal{C}, \mathcal{B})$-differentiable, where $\mathcal{B}$ is the class of bounded sets in L^s, at $f = h = 0$ from $R_{s/q}(\mathbb{R}) \times L^s[0,1]$ into $L^q[0,1]$ by Theorem 5.1. Composing the two maps, Theorem 5.5 gives the first conclusion. The others follow as in the proof of Theorem 5.1, and $\{f : \|f\|_{[\Phi]} \leq 1\}$ is clearly uniformly bounded and is uniformly regulated on any bounded interval as shown in the proof of Theorem 5.1, so Proposition 3.3 and Theorem 4.5 apply. $\qquad\square$

Reeds proved sup norm-compact differentiability of $(F,G) \mapsto F \circ G^{\leftarrow}$ in [41, Theorem 6.4.3] at (F, U), where F is C^1 with a bounded, uniformly continuous derivative.

6   The composition operator for Banach-valued functions.

Let X be any set. Let $(Y, |\cdot|)$ and $(Z, |\cdot|)$ be two normed vector spaces. For a bounded linear operator T from Y into Z, let $|T| := \sup\{|Ty| : |y| \leq 1\}$ be the usual operator norm. Let $B(X, Y)$ be the Banach space of all bounded functions from X into Y with supremum norm $\|g\| := \sup\{|g(x)| : x \in X\}$. Let U be an open set in Y. For $\delta > 0$, let $U_\delta := \{y \in U : \text{if } v \in Y \text{ and } |v| < \delta, \text{ then } y + v \in U\}$.

6.1 Theorem.   Let $G \in B(X, Y)$ take values in U_δ for some $\delta > 0$. Let $F \in B(U, Z)$ be Fréchet differentiable at each $u \in U$. Suppose the derivative $(DF)(u)(\cdot)$ is bounded on U_δ, that is $\sup\{|(DF)(u)(\cdot)| : u \in U_\delta\} < \infty$, and continuous uniformly on U_δ, in the sense that

$$\sup\{|(DF)(u)(\cdot) - (DF)(v)(\cdot)| : u \in U, \ v \in U_\delta, \ |u - v| < \alpha\} \to 0$$

as $\alpha \downarrow 0$. Let $\mathcal{C}$ be the class of all sets $A \subset B(U, Z)$, equicontinuous uniformly on U_δ in the sense that

$$\sup\{|f(u) - f(v)| : u \in U, \ v \in U_\delta, \ |u - v| < \alpha, \ f \in A\} \downarrow 0 \text{ as } \alpha \downarrow 0.$$

Then $(f, g) \mapsto (F + f) \circ (G + g)$ is jointly differentiable at $f = g = 0$ from $B(U, Z) \times B(X, Y)$ into $B(X, Z)$ for $\mathcal{C}$-differentiability in f and Fréchet differentiability in g for $\|\cdot\|$. The derivative is $(f, g) \mapsto f \circ G + (DF)(G)(g)$, where $(DF)(G)(g)$ is the function $x \mapsto (DF)(G(x))(g(x))$.

Proof.   As in §5 we have

$$(F + f) \circ (G + g) - F \circ G = f \circ G + (DF)(G)(g) + R_1 + R_2,$$

where $R_1 := f \circ (G + g) - f \circ G$ and $R_2 := F \circ (G + g) - F \circ G - (DF)(G)(g)$. Note that $(DF)(G)(g)$ is linear in g. Let A be a set of functions from U into Z, equicontinuous uniformly on U_δ, and $f \in A$. Then as $g \to 0$ in sup norm and $t \to 0$,

$$\|R_1\| = \|tf \circ (G + g) - tf \circ G\| = |t| \, \|f \circ (G + g) - f \circ G\| = o(|t|)$$

by the uniform equicontinuity. Next, boundedness of DF on U_δ implies that $(DF)(G)(g) \in B(X, Z)$. Then $\|R_2\| = o(\|g\|)$ by the continuity of DF uniformly on U_δ and the mean value theorem [19, 8.6.2]. □

Note that Theorem 6.1, since the functions G and g are defined on an arbitrary set, has nothing to do with possible derivatives of G or g. The theorem was suggested by the main theorem of Gray [32] on differentiation with respect to a parameter ξ of $\xi \mapsto F(\xi) \circ G(\xi)$. Neither theorem implies the other.

Appendix: Universally measurable sets and functions; preservation of Lebesgue measurability.

Let X be a set and $\mathcal{B}$ a σ-algebra of subsets of X. A set $A \subset X$ is called **universally measurable** (**u.m.**) iff it is measurable for the completion of every finite, countably additive measure on $(X, \mathcal{B})$. The u.m. sets form a σ-algebra $\mathcal{U}$. A function f from X to Y, where $\mathcal{C}$ is a σ-algebra of subsets of Y, will be called **u.m.** iff $f^{-1}(C) \in \mathcal{U}$ for every $C \in \mathcal{C}$.

Let $X = Y = \mathbb{R}$, $\mathcal{B} = \mathcal{C} =$ Borel σ-algebra, $\mathcal{L} = \sigma$-algebra of Lebesgue measurable sets, $\mathcal{L}^0 =$ the set of all Lebesgue measurable functions.

A.1 Theorem.   A set $E \subset \mathbb{R}$ is u.m. if and only if $f^{-1}(E) \in \mathcal{L}$ for every $f \in \mathcal{L}^0$.

A.2 Theorem.   A function $g: \mathbb{R} \mapsto \mathbb{R}$ is such that $g \circ f \in \mathcal{L}^0$ for every $f \in \mathcal{L}^0$ if and only if g is a u.m. function.

References for these facts are Darst [17] for Theorem A.1 and Shragin [46, Remark 1] for Theorem A.2, which also follows easily from Theorem A.1.

It follows directly from Theorem A.2 that a real-valued function g defined on an interval J is u.m. if and only if $g \circ f \in \mathcal{L}^0$ for all functions $f \in \mathcal{L}^0$ taking values in J. Another fact is:

A.3 Theorem.   If $g: \mathbb{R} \mapsto \mathbb{R}$ is u.m. and f is measurable from $(\Omega, \mathcal{A}, \mu)$ to $(\mathbb{R}, \mathcal{B})$, then $g \circ f$ is measurable for the μ-completion of $\mathcal{A}$.

Proof.   For any Borel set $B \subset \mathbb{R}$, $(g \circ f)^{-1}(B) = f^{-1}(g^{-1}(B))$, and $g^{-1}(B)$ is a u.m. set in $\mathbb{R}$. Thus $g^{-1}(B)$ is measurable for the completion of $\mu \circ f^{-1}$. If follows that $f^{-1}(g^{-1}(B))$ is μ-completion measurable. $\square$

Next, here is a fact about preservation of Lebesgue measurability:

A.4 Theorem.   Let G be an increasing function on an interval $J \subset \mathbb{R}$ such that for some $\beta > 0$, $G(y) - G(x) \geq \beta(y - x)$ for any $x \leq y$ in J. Let λ_J be Lebesgue measure on J and let λ be Lebesgue measure on $\mathbb{R}$. Then

(a) For any Lebesgue measurable set $B \subset \mathbb{R}$, $G^{-1}(B)$ is Lebesgue measurable and

(A.5) $\lambda_J(G^{-1}(B)) \leq \lambda(B)/\beta$.

(b) If f is any Lebesgue measurable function, then $f \circ G$ is Lebesgue measurable on J.

Proof.   For (a), first suppose B is an interval $[c, d]$. Let H be the range of G. If $H \cap [c, d]$ is empty or contains only one point, then (A.5) holds since G is strictly increasing. Otherwise, let

$$c_1 := \inf\{H \cap [c, d]\}, \ d_1 := \sup\{H \cap [c, d]\}, \ a := \inf\{G^{-1}([c, d])\}, \ b := \sup\{G^{-1}([c, d])\}.$$

Then $c_1 < d_1$ and $a < b$. Let $a \leq u < v \leq b$. Then $G(v) - G(u) \geq \beta(v - u)$. If $G(a) \geq c$ let $u = a$. Then $G(u) = c_1$. Otherwise let $u \downarrow a$ and $G(u) \downarrow c_1$. Likewise we have either $v = b$ and $G(v) = d_1$ or $v \uparrow b$ and $G(v) \uparrow d_1$. Thus

$$d - c \geq d_1 - c_1 \geq \beta(b - a) \geq \beta\lambda(G^{-1}([c, d])).$$

So (A.5) is proved for closed intervals. It is also easily seen to hold for intervals which may be open or closed at either end, and for finite, disjoint unions of such intervals. Then by the monotone class theorem (e.g. [20, Theorem 4.4.2]) it holds for any Borel set B. Then, if B is any set with $\lambda(B) = 0$, we have $\lambda_J(G^{-1}(B)) = 0$. It follows that (A.5) holds for any Lebesgue measurable set B, so (a) holds. In particular, for any Lebesgue measurable set L, $G^{-1}(L)$ is λ_J measurable, and (b) follows. □

Acknowledgments. We are thankful to Richard Gill for several conversations and for pointing out the book of Andersen et al. [1], and to Vladimir Koltchinskii for pointing out the paper of Billingsley and Topsøe.

References

[1] P. K. ANDERSEN - Ø. BORGAN - R. D. GILL - N. KEIDING, *Statistical Models Based on Counting Processes*, Springer-Verlag, Berlin, 1993.

[2] J. APPELL, *Upper estimates for superposition operators and some applications*, Ann. Acad. Sci. Fenn. (= Suomalaisen Tiedeakatemian Helsingfors Toimitsukia) Ser. A I. Math., **8** (1983), pp. 149–159.

[3] J. APPELL, *The superposition operator in function spaces – A survey*, Expositiones Math., **6** (1988), pp. 209–270.

[4] J. APPELL - P. P. ZABREJKO, *Nonlinear superposition operators*, Cambridge University Press, 1990.

[5] V. O. ASATIANI - Z. A. CHANTURIA, *The modulus of variation of a function and the Banach indicatrix*, Acta Sci. Math., **45** (1983), pp. 51–66.

[6] V. I. AVERBUKH - O. G. SMOLYANOV, *The theory of differentiation in linear topological spaces*, Russian Math. Surveys, **22** (1967), no. 6, pp. 201–258 = Uspekhi Mat. Nauk, **22** (1967), no. 6, pp. 201–260.

[7] V. I. AVERBUKH - O. G. SMOLYANOV, *The various definitions of the derivative in linear topological spaces*, Russian Math. Surveys, **23** (1968), no. 4, pp. 67–113 = Uspekhi Mat. Nauk, **23** (1968), no. 4, pp. 67–116.

[8] S. K. BERBERIAN, *The character space of the algebra of regulated functions*, Pacific J. Math., **74** (1978), pp. 15–36.

[9] P. BILLINGSLEY, *Convergence of Probability Measures*, Wiley, New York, 1968.

[10] P. BILLINGSLEY - F. TOPSØE, *Uniformity in weak convergence*, Z. Wahrscheinlichkeitsth. verw. Geb., **7** (1967), pp. 1–16.

[11] N. H. BINGHAM, *Fluctuation theory in continuous time*, Adv. Appl. Prob. **7** (1975), pp. 705–766.

[12] N. BOURBAKI, *Fonctions d'une variable réelle*, Hermann, Paris, 1976.

[13] M. BROKATE - F. COLONIUS, *Linearizing equations with state-dependent delays*, Appl. Math. Optimiz., **21** (1990), pp. 45–52.

[14] W. Bücher, *Differentiabilité de la composition et complétitude de certains espaces fonctionnels*, Comm. Math. Helv., **43** (1968), pp. 256–288.

[15] Z. A. Chanturia [Čanturija], *The modulus of variation of a function and its application in the theory of Fourier series*, Dokl. Akad. Nauk SSSR, **214** (1974), pp. 63–66 = Soviet Math. Dokl., **15** (1974), pp. 67–71.

[16] D. L. Cohn, *Measure Theory*, Birkhäuser, Boston, 1980.

[17] R. B. Darst, *A characterization of universally measurable sets*, Proc. Camb. Philos. Soc., **65** (1969), pp. 617–618.

[18] C. Dellacherie - P.-A. Meyer, *Probabilities and Potential*, Hermann, Paris, 1975; English transl. North-Holland, Amsterdam, 1978.

[19] J. Dieudonné, *Foundations of Modern Analysis*, Academic Press, New York, 1960; *Fondements de l'analyse moderne*, **1**, Gauthier-Villars, Paris, 1963.

[20] R. M. Dudley, *Real Analysis and Probability* (2d printing, corrected), Chapman and Hall, New York and London, 1993.

[21] R. M. Dudley, *Fréchet differentiability, p-variation and uniform Donsker classes*, Ann. Probab., **20** (1992), pp. 1968–1982.

[22] R. M. Dudley, *The order of the remainder in derivatives of composition and inverse operators for p-variation norms*, Ann. Statist., **22** (1994), pp. 1–20.

[23] R. M. Dudley, *Empirical processes and p-variation*, in *Festschrift for Lucien Le Cam*, Eds. D. Pollard, E. Torgersen, G. L. Yang, Springer-Verlag, New York, 1997, pp. 219–233.

[24] N. Dunford - J. T. Schwartz, *Linear Operators, Part I*, Interscience, New York, 1958.

[25] W. Esty - R. Gillette - M. Hamilton - D. Taylor, *Asymptotic distribution theory of statistical functionals: the compact derivative approach for robust estimators*, Ann. Inst. Statist. Math., **37** (1985), pp. 109–129.

[26] L. T. Fernholz, *von Mises calculus for statistical functionals*, Lect. Notes in Statist. (Springer-Verlag), **19**, 1983.

[27] A. Filippova, *Mises' theorem on the asymptotic behavior of functionals of empirical distribution functions and its statistical applications*, Theory Probab. Appl., **7** (1961), pp. 24–57.

[28] M. Fréchet, *La notion de différentielle dans l'analyse générale*, Ann. Sci. Ecole Norm. Sup. (Sér. 3), **42** (1925), pp. 293–323.

[29] B. V. Gnedenko - A. N. Kolmogorov, *Limit Distributions for Sums of Independent Random Variables*, 2d ed. Transl. and Ed. by K. L. Chung, Addison-Wesley, Reading, Mass, 1968.

[30] C. Goffman - G. Moran - D. Waterman, *The structure of regulated functions*, Proc. Amer. Math. Soc., **57** (1976), pp. 61–65.

[31] H. Goldberg - W. Kampowsky - F. Tröltzsch, *On Nemytskij operators in L^p-spaces of abstract functions*, Math. Nachr., **155**, pp. 127–140.

[32] A. GRAY, *Differentiation of composites with respect to a parameter*, J. Austral. Math. Soc. (Ser. A), **19** (1975), pp. 121–128.

[33] T. H. HILDEBRANDT, *Introduction to the Theory of Integration*, Academic Press, New York, 1963.

[34] E. W. HOBSON, *The Theory of Functions of a Real Variable and the Theory of Fourier's Series*, 1, 3d ed. (1927), repr. Dover, New York, 1957.

[35] M. A. KRASNOSEL'SKIĬ - P. P. ZABREĬKO - E. I. PUSTYL'NIK - P. SOBOLEVSKIĬ, *Integral operators in spaces of summable functions*, Nauka, Moscow, 1966; transl. by T. ANDO, Noordhoff, Leyden, 1976.

[36] R. LUCCHETTI - F. PATRONE, *On Nemytskii's operator and its application to the lower semicontinuity of integral functionals*, Indiana Univ. Math. J., **29**, pp. 703-713.

[37] A. MUKHERJEA - K. POTHOVEN, *Real and Functional Analysis*, Plenum, New York and London, 1978.

[38] J. MUSIELAK - W. ORLICZ, *On generalized variations (I)*, Studia Math., **18** (1959), pp. 11–41.

[39] E. NELSON, *Regular probability measures on function space*, Ann. Math., **69** (1959), pp. 630–643.

[40] S. PERLMAN, *Functions of generalized variation*, Fund. Math., **105** (1980), pp. 199–211.

[41] J. A. REEDS III, *On the definition of von Mises functionals*, Ph. D. Dissertation, Harvard University, 1976.

[42] B. RIEMANN, *Über die Darstellbarkeit einer Funktion durch eine trigonometrische Reihe*, Abh. Gesell. Wiss. Göttingen Math. Kl. **13**, pp. 87-132; repr. in Bernhard Riemann: Gesammelte mathematische Werke und wissenschaftlicher Nachlass, with commentaries, 2d. ed., ed. Raghavan Narasimhan, Springer-Verlag (Heidelberg) and Teubner (Leipzig), 1990.

[43] F. RIESZ - B. SZ.-NAGY, *Leçons d'analyse fonctionelle*, 3d ed., Gauthier-Villars, Paris, 1955; *Functional Analysis* (transl. by L. F. BORON), Ungar, New York, 1955.

[44] J. SEBASTIÃO E SILVA, *Le calcul différentiel et intégral dans les espaces localement convexes, réels ou complexes I, II*, Rend. Accad. Lincei Sci. Fis. Mat. Nat., (Ser. 8) **20** (1956), pp. 743-750, **21** (1956), pp. 40-46.

[45] G. E. SHILOV - B. L. GUREVICH, *Integral, Measure and Derivative: A Unified Approach*, Transl. and Ed. by R. A. SILVERMAN, Prentice-Hall, Englewood Cliffs, N.J., 1966.

[46] I. V. SHRAGIN, *Superposition measurability*, Sov. Math. (Iz. Vuz.), **19** (1975), pp. 69–76 = Izv. Vyssh. Uch. Zaved., **1975**, no. 1, pp. 82–92.

[47] I. V. SHRAGIN, *Classes of measurable vector functions and Nemytskii's operators I, II*, Russian Math. (Iz. Vuz.), **38** (1994), no. 4, pp. 45–55, no. 5, pp. 70–79, = Izv. Vyssh. Uch. Zaved., **1994**, no. 4, pp. 48-58, no. 5, pp. 70-79.

[48] *W. SIERPIŃSKI, *Sur une propriété des fonctions qui n'ont que des discontinuités de première espèce*, Bull. Sect. Scient. Acad. Roumaine, **16** (1933), no. 1/3, pp. 1–4.

[49] A. V. SKOROHOD, *Limit theorems for stochastic processes with independent increments*, Theory Prob. Appl. **2** (1957), pp. 138-171.

[50] R. TABERSKI, *On the power variations and pseudovariations of positive integer orders*, Demonstratio Math., **19** (1986), pp. 881-893.

[51] A. E. TAYLOR, *The differential: nineteenth and twentieth century developments*, Arch. Hist. Exact Sci., **12** (1974), pp. 355–383.

[52] *O. D. TSERETELI (CERETELI), *The metric properties of a function of bounded variation*, (in Russian), Akad. Nauk Gruzin. SSR Trudy Tbiliss. Mat. Inst. Razmadze, **26** (1959), pp. 23–64.

[53] *O. D. TSERETELI (CERETELI), *On the Banach indicatrix and some of its applications*, (in Russian), Soobshch. Akad. Gruzin. SSR **25** (1960), pp. 129–136.

[54] M. M. VAĬNBERG, *Variational methods in the study of nonlinear operators*, Gostekhizdat, Moscow, 1956; English transl. Holden-Day, San Francisco, 1964.

[55] R. VON MISES, *Les lois de probabilité pour les fonctions statistiques*, Ann. Inst. H. Poincaré, **6** (1936), pp. 185–212.

[56] R. VON MISES, *On the asymptotic behavior of differentiable statistical functions*, Ann. Math. Statist., **18** (1947), pp. 309–348.

[57] WANG SHENG-WANG, *Differentiability of the Nemyckii operator*, Doklady Akad. Nauk SSSR, **150** (1963), pp. 1198–1201 (Russian); Sov. Math. Doklady, **4** (1963), pp. 834–837.

[58] L. C. YOUNG, *General inequalities for Stieltjes integrals and the convergence of Fourier series*, Math. Ann., **115** (1938), pp. 581–612.

[59] W. H. YOUNG, *On the distinction of right and left at points of discontinuity*, Quarterly J. Pure and Applied Math., **39** (1908), pp. 67–83.

[60] W. H. YOUNG, *On the discontinuities of a function of one or more real variables*, Proc. London Math. Soc., (Ser. 2) **8** (1909), pp. 117–124.

[61] E. ZEIDLER, *Nonlinear Functional Analysis and its Applications*, Vols. I, II/B, Springer-Verlag, Berlin, 1985, 1990.

---

*We found these references from secondary sources but have not seen them in the original.

PART IV

BIBLIOGRAPHIES ON p-VARIATION AND ϕ-VARIATION

BY R. M. DUDLEY, R. NORVAIŠA AND JINGHUA QIAN

ABSTRACT. This is a pair of annotated reference lists, including all items the authors could find, on

(1) p-variation of real-valued functions f as defined by Wiener in 1924 and developed by L. C. Young and E. R. Love in the late 1930's and others since then. Usually f is defined on an interval, but some papers give extensions to multidimensional domains;

(2) ϕ-variation, namely the supremum of all sums $\sum_i \phi(|\Delta_i f|)$, where $\Delta_i f :=$ $f(x_i) - f(x_{i-1})$, ϕ is a continuous, increasing function, 0 at 0, and $x_0 < x_1 < \cdots < x_n$, $n = 1, 2, \ldots$. Thus $\phi(y) = y^p$ gives p-variation.

Not included, however, are works on: (a) "quadratic variation" as studied in probability theory and defined as a limit along a sequence of partitions $\{x_j\}$ with mesh $\max_j(x_j - x_{j-1}) \to 0$, at some rate, or where the sums converge only in probability; (b) the special case $p = 1$ of ordinary bounded variation; or (c) sequence spaces, called James spaces.

CONTENTS

1. INTRODUCTION

This is not "the" reference list for the entire volume. Each of Parts I, II and III has its own reference list which includes some items, not on p-variation or ϕ-variation, not included here. On the other hand, the following lists contain many items not cited in any of Parts I, II or III.

An effort has been made to compile reference lists as complete as possible on p-variation, as defined by Wiener (1924) and developed by L. C. Young (1936) and others, and the related notion of ϕ-variation. There is much literature on a kind of quadratic variation for stochastic processes where there are some restrictions on the partitions. The lists cover only the notions of p-variation and ϕ-variation, for functions of either one or several real variables, where the partitions are unrestricted. The authors have compiled a separate, less complete list, not included here, for the case where partitions are restricted. The theory of functions with bounded variation (p-variation for $p = 1$) is another related large area of research not covered by the present list. Finally, sequence spaces with bounded 2-variation, known as James spaces, have been studied in Banach space theory, but works on James spaces are not included here except for a few which also treat functions of a continuous variable, indicated by the code *J.

One- or two-sentence summaries are given after most entries. In some cases the title of the paper seemed to provide a sufficient summary. Announcements are listed before the corresponding longer papers, with one summary for both.

We will be glad to learn of any papers we have overlooked or other corrections to the list.

Notation and historical comments. Given a real-valued function f on an interval $[a, b]$, a partition $\kappa := \{a = x_0 < \cdots < x_n = b\}$ and a number $p \in [1, \infty)$, let $v_p(f; \kappa) := \sum_{i=1}^{n} |f(x_i) - f(x_{i-1})|^p$. Denoting the mesh of κ by $|\kappa|$, for each $\epsilon > 0$, let $v_p(f; \epsilon) := \sup\{v_p(f; \kappa) : |\kappa| \le \epsilon\}$ and $\overline{V}_p(f; [a, b]) := \lim_{\epsilon \downarrow 0} v_p(f; \epsilon)^{1/p}$. A function f is of bounded p-variation if $V_p(f; [a, b]) := \sup\{v_p(f; \kappa)^{1/p} : \kappa\} < \infty$. The class of all such functions will be denoted by $\mathcal{W}_p$. If $f \in \mathcal{W}_p$ then f is regulated and $\overline{\mathfrak{S}}_p(f; [a, b])^p := \sum_{x \in (a, b]} |f(x+) - f(x-)|^p < \infty$. The sum $\overline{\mathfrak{S}}_p$ is applied in some listed papers either: (a) when f is periodic of period $b - a$, so that the sum is over a complete cycle, e.g. $(a, b] := (0, 2\pi]$; or (b) when f is right-continuous, with $f(b+) := f(b)$ if $f(b+)$ is not defined otherwise. Then $f(a+) = f(a)$, so there is no term at the left endpoint a.

For a function of bounded variation, in his 1924 paper, N. Wiener gave a necessary and sufficient condition for its continuity in terms of Fourier series coefficients. For the proof, he defined and used the p-variation $\overline{V}_p(f; [a, b])$. In addition to establishing certain properties of the class of functions f with $\overline{V}_p(f; [a, b]) < \infty$, Wiener proved that their Fourier series converge (to $(f(x+) + f(x-))/2$) everywhere when $p = 2$ and almost everywhere if $2 < p < \infty$. Marcinkiewicz (1934) proved the convergence everywhere for $2 < p < \infty$. L. C. Young (1936) found a different proof of this fact. His main observation was that the indefinite integral of the Dirichlet kernel in the representation of the n-th partial sum of the Fourier series has bounded p-variation with $p > 1$, uniformly in n. Another very useful discovery of Young in this paper was a new set of conditions for the existence of the (extended) Riemann-Stieltjes integral expressed in terms of the p-variations of both the integrand and the integrator (Part II, Theorem 3.27).

What Young and we call p-variation has also been called by some authors α-variation (e.g. Gehring, 1954) or β-variation (e.g. Blumenthal and Getoor, 1960). Also, "p-variation" has been used in the literature for notions different from and in some cases unrelated to Wiener-Young p-variation. We tried to exclude unrelated work from the list. The star * on a publication year indicates works we learned about from secondary sources but have not seen in the original.

The following codes appear after some entries:

*E = ergodic theory
*F = Fourier series
*J = James spaces
*L = local times
*I = integration
*M = multidimensional extensions
*P = probability

BIBLIOGRAPHY ON p-VARIATION

Abundo, M. (1987)* An almost sure invariance principle for the maps $S_a(x) = ax(1-x)$ in the interval $[0,1]$. *Matematiche* (Catania), **42**, 11–18. MR 91f:58049 *E *P

An invariance principle for the composition $\{f \circ S_a\}$ with f a function of bounded p-variation.

Avdispahić, M. (1985). An Izumi's theorem on the absolute convergence of Fourier series and its generalizations. *Radovi Matematički* (Sarajevo), **1**, 231–240. MR 87f:42020 *F

Absolute convergence of Fourier series for functions of bounded p-variation with a bound on integrals of increments.

Avdispahić, M. (1986). Fejér's theorem for the class W_p. *Rend. Circ. Mat. Palermo* (2), **35**, 90–101. MR 88m:42012 *F

The derived Fourier series of a function with bounded p-variation, $p > 1$, is Cesàro summable of any order $\alpha > 1 - 1/p$.

Babaev, A. H. (1978)* The asymptotic behavior of the best approximation in the metric $L_q[-1,1]$ of a function of bounded p-variation. (Russian) *Izv. Akad. Nauk Azerbaĭdžan. SSR Ser. Fiz.-Tehn. Mat. Nauk*, **5**, 32–37. MR 80j:41039

An asymptotic formula for the best polynomial approximation of a function whose rth successive integral is a function of bounded p-variation.

Bary, N. K. (1961). *Trigonometric series.* I, II. (Russian) Fizmatgiz, Moscow; translation by M. F. Mullins. *A treatise on trigonometric series.* I, II. MacMillan, New York, 1964. MR 23 #A3411, MR 30#1347 *F

A proof (on p. 310) of uniform convergence of Fourier series for continuous functions with bounded p-variation, using the Salem test.

Bellenot, S. F. (1982). Transfinite duals of quasireflexive Banach spaces. *Trans. Amer. Math. Soc.*, **273**, 551–577. *J

Certain transfinite duals of the James sequence space and the James function space are isomorphic to subspaces of one another.

Bergh, J. and Peetre, J. (1974). On the spaces V_p $(0 < p \leq \infty)$. *Boll. Un. Mat. Ital.*, Ser. 4, **10**, 632–648. Zbl 308.46032

Spline (piecewise polynomial) approximation using a kind of p-variation.

Berezhnoĭ, E. I. (1988)*. On a set of correctable functions of bounded *p*-variation. (Russian) *Studies in the theory of functions of several real variables* (Russian), 34–38; Matematika, Yaroslav. Gos. Univ., Yaroslavl'. R.Zh. 1B89 1989

A function f of bounded p-variation is correctable if it can be related to a function $\bar{f}$ with $\overline{V}_p(\bar{f}) = 0$ in a certain way defined by the author. The set of such functions has first category in $\mathcal{W}_p$.

Berkson, E. and Gillespie, T. A. (1994). La *q*-variation des fonctions et l'intégration spectrale des multiplicateurs de Fourier. (French) [The *q*-variation of functions and the spectral integration of Fourier multipliers] *C. R. Acad. Sci. Paris Sér. I Math.*, **318**, 817–820. MR 95a:43005 *F

An extension of the classical Marcinkiewicz multiplier theorem to functions of bounded p-variation.

Berkson, E. and Gillespie, T. A. (1997). The *q*-variation of functions and spectral integration of Fourier multipliers. *Duke Math. J.*, **88**, 103–132. *I

Let $1 < p < \infty$, $1 \leq q < \infty$ and $|(1/p)-(1/2)| < 1/q$. Functions of bounded q-variation are Right Cauchy integrable with respect to the Stone-type spectral family of a strongly continuous one-parameter group of operators on a closed subspace of L^p. See also Krabbe (1961b).

Berman, R. and Singman, D. (1994). Intermittent oscillation and tangential growth of functions with respect to Nagel-Stein regions on a half-space. *Illinois J. Math.*, **38**, 19–46. MR 94i:31002 *M

The p-variation for functions $u : R^{n+1} \to R$ is defined and used to bound the Hausdorff measure of the tail of a maximal function.

Berman, S. M. (1969). Harmonic analysis of local times and sample functions of Gaussian processes. *Trans. Amer. Math. Soc.*, **143**, 269–281. MR 40 #2155 *L

If $X(t)$, $0 \leq t \leq 1$, is a Gaussian process with mean zero satisfying $E(X(t) - X(s))^2 \geq c|t - s|^\beta$ for some β, $0 < \beta < 2$, then almost all sample functions of X are of unbounded p-variation for $p < 2/\beta$.

Bertoin, J. (1988)*. Une extension d'une inégalité de Burkholder, Davis, Gundy pour les processus à α-variation bornée et applications. (French) [An extension of a Burkholder-Davis-Gundy inequality for processes with bounded α-variation; applications] *Stochastics*, **24**, 75–86. MR 89m:60095

Bertoin, J. (1989). Sur une intégrale pour les processus à α-variation bornée. *Ann. Probab.*, **17**, 1521–1535. MR 91g:60065 *P

Given a stochastic process X, α-variation is defined by $\sup\{EV_\alpha(X; \kappa): \kappa\}$ where κ is a sequence of stopping times. Conditions for the existence of a modified Left Cauchy integral are given in terms of α-variation.

Bertoin, J. (1990). Complements on the Hilbert transform and the fractional derivative of Brownian local times. *J. Math. Kyoto Univ.*, **30** 651–670. MR 92f:60139 *L *P

A Tanaka-type formula and Ray-Knight-type results for the Hilbert transform and the one-sided fractional derivative of a local time. The results are based on the α-variation notion of Bertoin (1989).

Blei, R. C. (1995). An extension theorem concerning Fréchet measures. *Canad. Math. Bull.*, **38**, 278–285. MR 96k:28007 *M

The p-variation of bimeasures and polymeasures.

Blumenthal, R. M. and Getoor, R. K. (1960). Some theorems on stable processes. *Trans. Amer. Math. Soc.*, **95**, 263–273. MR 22 #10013 *P

A symmetric stable process of index α has sample paths of bounded p-variation if and only if p > α.

Blumenthal, R. M. and Getoor, R. K. (1961). Sample functions of stochastic processes with stationary independent increments. *J. Math. Mech.*, **10**, 493–516. MR 23 #A689 *P

Boundedness of p-variation of sample functions under conditions on the Lévy measure.

Bretagnolle, J. (1972). p-variation de fonctions aléatoires. I. Séries de Rademacher. (French) *Séminaire de Probabilités, VI* (Univ. Strasbourg, 1970-1971), pp. 51–63; *Lect. Notes in Math.* (Springer-Verlag), **258**, Berlin. MR 51 #9221 *P

An inequality for the expectation of the p-variation of Rademacher series with p ∈ (0, 2).

Bretagnolle, J. (1972). p-variation de fonctions aléatoires. II. Processus à accroissements indépendants. (French) *Séminaire de Probabilités, VI* (Univ. Strasbourg, 1970-1971), pp. 64–71; *Lect. Notes in Math.* (Springer-Verlag), **258**, Berlin. MR 51 #9222 *P

Extension of the first part to stochastic processes with independent increments.

Brudnyi, Ju. A. (1974). Spline approximation and functions of bounded variation. *Dokl. Akad. Nauk SSSR*, **215**; transl. in: *Soviet Math. Dokl.*, **15**, 518–521. Zbl 299.41007

Rates of approximation by piecewise polynomial functions follow from boundedness of p-variation.

Bruneau, M. (1967). Calcul de la p-variation d'une fonction. (French) *C. R. Acad. Sci. Paris Sér. A-B*, **265**, A173–A176. MR 37 #4218

A factorization of a function of bounded p-variation.

Bruneau, M. (1970). Fonctions d'une variable réelle; p-variation fine d'une fonction à p-variation bornée. (French) *C. R. Acad. Sci. Paris Sér. A-B*, **270**, A585–A588. MR 41 #3686

A measure and a density corresponding to a function of bounded p-variation.

Bruneau, M. (1972). Fonctions d'une variable réelle; caractérisation des points extrémaux de la boule unité de l'espace des fonctions à p-variation bornée. (French) C. R. Acad. Sci. Paris Sér. A-B, **274**, A51–A54. MR 44 #7270

Bruneau, M. (1974). Variation totale d'une fonction. (French) Lect. Notes in Math. (Springer-Verlag), **413**, Berlin. MR 52 #14190

A systematic study of p-variation which includes earlier results of the author.

Bruneau, M. (1979). Sur la p-variation d'une surmartingale continue. (French) Séminaire de Probabilités, XIII (Univ. Strasbourg, 1977/78), 227–232; Lect. Notes in Math. (Springer-Verlag), **721**, Berlin. MR 82c:60084 *P

The p-variation of a continuous function is bounded by the number of crossings of strips.

Burkill, J. C. and Gehring, F. W. (1953). A scale of integrals from Lebesgue's to Denjoy's. Quart. J. Math., Oxford, Ser. 2, 4, 210–220. MR 15, 204i

A class D_p, with $1 \leq p < \infty$, of continuous functions such that their indefinite Perron integrals have bounded p-variation.

Ciemnoczołowski, J. and Orlicz, W. (1985). Variation and compactness. Comment. Math. Prace Mat., **25**, 201–214. MR 87h:46037

Boundedness and convergence of series in Banach spaces with applications to functions with bounded p-variation.

Coifman, R., Rubio de Francia, J. L. and Semmes, S. (1988). Multiplicateurs de Fourier de $L^p(R)$ et estimations quadratiques. C. R. Acad. Sci. Paris, Série I, **306**, 351–354. MR 89e:42009 *F

A bounded function is a Fourier multiplier for $L^q(R)$ if a certain condition on its p-variation holds. See also Krée (1966).

Denker, M. (1989).* The central limit theorem for dynamical systems. Dynamical systems and ergodic theory (Warsaw, 1986), 33–62, Banach Center Publ., 23, PWN, Warsaw. MR 92d:28007 *E *P

Dickmeis, G. and Dickmeis, W. (1977).* Beste Approximation in Räumen von beschränkter p-Variation. (German) Forschungsbericht des Landes Nordrhein Westfalen, No. 2697: Fachgruppe Mathematik-Informatik. Westdeutscher Verlag, Opladen, 36 pp. ISBN: 3-531-02697-6 MR 57 #6981

Dobrescu, E. and Masca, G. (1960).* On the Stieltjes integral. (Romanian) An. Şti. Univ. "Al. I. Cuza" Iaşi secţ. I (Ser. 2), **6**, 541–546. MR 25 #1265

An inequality analogous to one proved by Young (1936) is used to define the Stieltjes integral for a pair of functions subject to certain modified p- and q-variation conditions.

Dudley, R. M. (1992). Fréchet differentiability, p-variation and uniform Donsker classes. Ann. Probab., **20**, 1968–1982. MR 94a:60004 *P

A ball in $\mathcal{W}_p$ with $0 < p < 2$ is a universal Donsker class. Duality inequalities for p-variation spaces.

Dudley, R. M. (1992). Nonlinear functionals of empirical measures. Probability in Banach spaces, 8 (Brunswick, ME, 1991), 403–410; Progr. Probab., 30, Birkhäuser Boston, Boston, MA. MR 94f:60011 *P

A comparison of differentiability of statistical functionals with respect to the supremum norm and p-variation norms.

Dudley, R. M. (1994). The order of the remainder in derivatives of composition and inverse operators for p-variation norms. *Ann. Statist.*, **22**, 1–20. MR 96a:62038 *P

Fréchet differentiability of the two operators acting on W_p with $p \geq 1$.

Dudley, R. M. (1997). Empirical processes and p-variation. In: *Festschrift for Lucien Le Cam*, Eds. D. Pollard, E. Torgersen, G. L. Yang. Springer-Verlag, New York, pp. 219–233.

The order of the p-variation norm, $1 \leq p \leq 2$, of the empirical process with applications to remainder bounds in Fréchet differentiability of statistical functionals.

D' yachkov, A. M. (1980). On the existence of the Stieltjes integral of the class V_p. (Russian) *Soobshch. Akad. Nauk Gruzin. SSR*, **99**, 353–536. MR 83h:26007 *I

Given $\alpha \in (0,1)$, there are functions f and g satisfying Hölder conditions of order α and $1 - \alpha$, respectively, and such that the Riemann-Stieltjes integral $\int f \, dg$ does not exist. See also Kondurar (1937), L. C. Young (1936).

Freedman, M. A. (1983). Operators of p-variation and the evolution representation problem. *Trans. Amer. Math. Soc.*, **279**, 95–112. MR 85b:47042

A representation of an evolution of bounded p-variation, $1 \leq p < 2$, via the product integral.

Garsia, A. M. and Sawyer, S. (1964). On some classes of continuous functions with convergent Fourier series. *J. Math. Mech.*, **13**, 589–601. MR 33 #7777 *F

Instead of boundedness of the p-variation of a continuous periodic function f, two new conditions on the number of open intervals comprising the set $\{x: f(x) > y\}$ are introduced.

Gehring, F. W. (1954). A study of α-variation. I. *Trans. Amer. Math. Soc.*, **76**, 420–443. MR 16, 346a

Functions with bounded p-variation are characterized in terms of moment constants. A Stieltjes convolution problem raised by J. E. Littlewood and L. C. Young is solved.

Gehring, F. W. (1955). A note on a paper by L. C. Young. *Pacific J. Math.*, **5**, 67–72. MR 16, 910g

A relation between an integrated Lipschitz condition and p-variation.

Golubov, B. I. (1967). Continuous functions of bounded p-variation. (Russian) *Mat. Zametki*, **1**, 305-312; transl. in: *Math. Notes*, 1967, 1, 203–207. MR 35 #2063 *F

Conditions for continuity in terms of Fourier coefficients.

Golubov, B. I. (1968a). Functions of bounded p-variation. (Russian) *Uspehi Mat. Nauk*, **23**, no. 1 (139) 219–220. MR 36 #6859 *F

A characterization of continuous functions with bounded p-variation. Properties of Fourier coefficients of functions with bounded p-variation.

Golubov, B. I. (1968b). Functions of bounded p-variation. (Russian) *Izv. Akad. Nauk SSSR Ser. Mat.*, **32**, 837–858; transl. in: *Math. USSR Izv.*, **2**, 799–819. MR 38 #3392 *F

For periodic functions whose r-th fractional derivatives for $r > 0$ are of bounded p-variation, estimates of the degree of approximation by partial sums of the corresponding Fourier series are given in the sup and L^q norms, $q \geq p$.

Golubov, B. I. (1968c). The Fourier integral and the continuity of functions of bounded p-variation. (Russian) *Izv. Vysš. Učebn. Zaved. Mat.*, **78**(11), 83–92. MR 40 #3173

Continuity conditions for square integrable functions of bounded p-variation via their Fourier transforms.

Golubov, B. I. (1969). The p-variation of functions. (Russian) *Mat. Zametki*, **5**, 195–204; transl. in: *Math. Notes*, **5-6**, 119–124. MR 39 #1608

The L_q-norm of $[f(\cdot + h) - f(\cdot)]/h^{1/q}$ tends to $\overline{\mathfrak{S}}_q(f)$ as $h \downarrow 0$ whenever periodic function $f \in \mathcal{W}_p$ with $1 \leq p < q < \infty$.

Golubov, B. I. (1970). The analogues of Wiener's and Lozinskiĭ's theorems for functions of several variables. (Russian) *Sakharth. SSR Mecn. Akad. Math. Inst. Šrom.*, **38**, 31–34. MR 44 #7220 *M

Golubov, B. I. (1971). The p-variation of functions of two variables. (Russian) *Izv. Vysš. Učebn. Zaved. Mat.*, **9** (112) 40–49. MR 44 #6921 *M

Expressions of p-variation via limits of certain integrals.

Golubov, B. I. (1972a). Determination of the jump of a function of bounded p-variation from its Fourier series. (Russian) *Mat. Zametki*, **12**, 19–28; transl. in: *Math. Notes* **12**, 444–454. MR 47 #8788 *F

A formula for the jump of a function of bounded p-variation at a given point in terms of derivatives of partial sums of its Fourier series.

Golubov, B. I. (1972b). Tests of the continuity of functions of bounded p-variation. (Russian) *Sibirsk. Mat. Ž.*, **13**, 1002–1015, 1197; transl. in: *Siberian Math. J.*, **13**, 693–702. MR 48 #11413

New conditions for the continuity of functions of bounded p-variation.

Golubov, B. I. (1974). An analogue of a theorem of Wiener. (Russian) *Sakharth. SSR Mecn. Akad. Moambe,* **74**, 297–300. MR 50 #2422 *M

Continuity of a function of two variables and of bounded p-variation in terms of Fourier coefficients.

Góra, P. (1989)* Countably piecewise expanding transformations without absolutely continuous invariant measure. *Dynamical systems and ergodic theory* (Warsaw, 1986), 113–117, Banach Center Publ., 23, PWN, Warsaw. MR 92c:58063 *E

For these transformations f the function $g_f(x) := |f'(x)|^{-1}$ for $x \in [a, b] \setminus A$ and $g_f(x) := 0$ for $x \in A$ (A is the countable discontinuity set) has finite p-variation for some $p > 1$.

Greenwood, P. E. (1969). The variation of a stable path is stable. *Z. Wahrsch. Verw. Gebiete*, **14**, 140–148. MR 41 #4653 *P

If X is a symmetric α-stable process, $0 < \alpha < 2$ and $\alpha < p$ then the limit of $v_p(X; n^{-1})$ as $n \to \infty$ exists almost surely and is an (α/p)-stable random variable Y. Moreover, the p-variation $v_p(X; [0, 1])$ is in the domain of normal attraction of Y.

Hambly, B. M. and Lyons, T. J. (1998). Stochastic area for Brownian motion on the Sierpinski gasket. *Ann. Probab.*, **26**, 132–148. *P

Sample functions of a Brownian motion and the associated area process on the Sierpinski gasket have bounded p-variation for each $p > \log 5/\log 2$.

Herda, H. H. (1968). Uniformly convex and reflexive modulared variation spaces. *Studia Math.*, **30**, 373–381. MR 38 #2587

The p-variation of a function over $[a, b]$ with $p = p(s, t)$ for $s, t \in [a, b]$.

Hirschman, I. I., Jr. (1959). On multiplier transformations. *Duke Math. J.*, **26**, 221–242. MR 21 #3721

The p-variation property and boundedness of multiplier transformations on L^r, $r \neq 2$.

Hirschman, I. I., Jr. (1962). Multiplier transformations. III. *Proc. Amer. Math. Soc.*, **13**, 851–857. MR 26 #581

Multiplier transformations generated by functions of bounded p-variation.

Hofbauer, F. and Raith, P. (1992). The Hausdorff dimension of an ergodic invariant measure for a piecewise monotonic map of the interval. *Canad. Math. Bull.*, **35**, 84–98. MR 93c:58113 *E

The derivative of such a map is assumed to have bounded p-variation.

Hofbauer, F. (1995). Local dimension for piecewise monotonic maps on the interval. *Ergodic Theory Dynam. Systems*, **15**, 1119–1142. MR 97c:28042 *E

Maps with derivatives of bounded p-variation

Huang, Yen-Chin (1994). *Empirical distribution function statistics, speed of convergence, and p-variation*. Ph. D. thesis, Massachusetts Institute of Technology. *P

There exists a sequence of Brownian bridges such that, for $2 < p < \infty$, the expected p-variation norm of the difference between the empirical process and the Brownian bridge is of order $n^{(1/p)-(1/2)}$.

Iosifescu, M. (1987)." Mixing properties for f-expansions: the bounded p-variation case. *Mathematical statistics and probability theory*, Vol. A (Bad Tatzmannsdorf, 1986), 195–199, Reidel, Dordrecht-Boston, Mass.-London. MR 89b:11069 *P

Izumi, M. and Izumi, S. (1967). On absolute convergence of Fourier series. *Ark. Mat.*, **7**, 177–184. MR 36 #4247 *F

Under a condition on the integrated modulus of continuity of a function of bounded p-variation.

Izumi, S. (1950). Notes on Fourier analysis (XVI). *Tôhoku Math. J. Ser. 2*, **1**, 144–166. MR 11, 656e *I *F

A new sufficient condition for existence of the Stieltjes integral defined by Izumi and Kawata (1938).

Izumi, S. and Kawata, S. (1938). Notes on Fourier series (IV): Stieltjes integrals. *Tôhoku Math. J.*, **44**, 410–420. *I *F

A new definition of Stieltjes integral for functions with convergent Fourier series is based on an extension of the Parseval formula for functions with bounded p-variation due to Young (1936).

Jain, N. C. and Monrad, D. (1981). Gaussian measures in certain function spaces. *Probability in Banach spaces, III* (Medford, Mass., 1980), 246–256; *Lect. Notes in Math.* (Springer-Verlag), **860**, Berlin. MR 84i:60052 *P

Conditions for a Gaussian process X to have sample paths of bounded p-variation are given in terms of the function $\sigma(s,t) = E|X(t) - X(s)|$.

Jain, N. C. and Monrad, D. (1983). Gaussian measures in B_p. *Ann. Probab.*, **11**, 46–57. MR 84c:60060 *P

Conditions for tightness of the induced Gaussian probability measure on $\mathcal{W}_p$.

Jedryka, T. M. (1982).* An estimation of the solution of Volterra's integral equation for vector-valued functions with values in some function spaces. *Problemy Mat.* (Poland), No. 3, 71–76. MR 84m:45011

The author considers operators acting on spaces of continuous functions of bounded p-variation.

Kalinauskaitė, N. (1980). Strong γ-variation and random measures with values in R^s. (Russian) *Litovsk. Mat. Sb.*, **20**, 55–60. MR 82d:60094 *P *M

An extension of Blumenthal and Getoor (1961) to random measures.

Keller, G. (1985). Generalized bounded variation and applications to piecewise monotonic transformations. *Z. Wahrsch. Verw. Gebiete*, **69**, 461–478. MR 86i:28024 *E

Quasi-compactness of the Perron-Frobenius operator of a piecewise monotonic transformation when the inverse of the derivative is of bounded p-variation.

Kel'zon, A. A. (1978). Interpolation of functions with bounded p-variation. (Russian) *Izv. Vysš. Učebn. Zaved. Mat.*, **5**, 131–134. MR 58 #23251

A convergence result of Névai (1974) for functions with bounded ϕ-variation holds under less stringent assumptions for functions with bounded p-variation.

Kel'zon, A. A. (1984). Interpolation of continuous functions of bounded p-variation. (Russian) *Izv. Vyssh. Uchebn. Zaved. Mat.*, **8**, 14–20. MR 86b:41005

Uniform convergence of the Lagrange interpolation process.

Kisliakov, S. V. (1984). A remark on the space of functions of bounded p-variation. *Math. Nachr.*, **119**, 137–140. MR 86m:46026

For $p > 1$, the first dual of the separable Banach space of continuous functions $f \in \mathcal{W}_p$ with $f(0) = \overline{V}_p(f) = 0$ is nonseparable and the second dual can be identified with $\mathcal{W}_p$.

Kolyada, V. I. (1983). On the metric Darboux property. (Russian) *Anal. Math.* (Budapest), **9**, 291–312. MR 85g:26002

A decomposition of certain L_p-functions into a function having the metric Darboux property and a function of bounded p-variation.

Kondurar, V. (1937). Sur l'intégrale de Stieltjes. (French) *Recueil Math.* (*Mat. Sbornik* (N.S.)), **2**, 361–366. *I

The integral $\int_a^b f\, dg$ exists in the Riemann-Stieltjes sense if f and g satisfy Hölder conditions of order α and β, respectively, with $\alpha + \beta > 1$. See also L. C. Young (1936).

Kováčik, O. (1993).* On the embedding $H^\omega \subset V_p$. *Math. Slovaca,* **43**, 573–578. MR 95e:46031

A characterization of the embedding $H^w \subset \mathcal{W}_p$, where H^w is a class of functions with modulus of continuity majorized by w.

Krabbe, G. L. (1961a). Integration with respect to operator-valued functions. *Bull. Amer. Math. Soc.,* **67**, 214–218. MR 23 #A1235

The space of functions with bounded p-variation is a Banach algebra.

Krabbe, G. L. (1961b). Integration with respect to operator-valued functions. *Acta Sci. Math.* (Szeged), **22**, 301–319. MR 24 #A2650 *I

Given a measure space (R, μ), suppose that E is a resolution of the identity in $L_2(R, \mu)$ and $E_r(\lambda)$, $\lambda \in [a, b]$, is a uniformly bounded family of continuous extensions of $E(\lambda)$ on $L_r(R, \mu)$. If a continuous function $f \in \mathcal{W}_p([a, b])$ with $1 \leq p < \infty$ then an interval $I(p)$ exists such that the Riemann-Stieltjes integral $\int_a^b f(\lambda)\, dE_r(\lambda)$ converges in the Banach space of endomorphisms of $L_r(R, \mu)$ for each $r \in I(p)$.

Krabbe, G. L. (1961c). Réfractions non-hilbertiennes d'une transformation symétrique bornée. (French) *Studia Math.,* **20**, 347–357. MR 25 #438 *I

For $0 < r < 1$, let $\mathcal{F}_r := \bigcup\{\mathcal{W}_{1/p}\colon 2|r - (1/2)| < p \leq 1\}$ with the inductive limit topology and let $\mathcal{E}_r$ be the Banach space of endomorphisms of $L_{1/r}$. An integral of Stieltjes type provides a homomorphism between the algebras $\mathcal{F}_r$ and $\mathcal{E}_r$.

Krabbe, G. L. (1962). Generalized measures whose values are operators into intermediate space. *Bull. Amer. Math. Soc.,* **68**, 42–46. MR 25 #439 *I

Krabbe, G. (1963a). Generalized measures whose values are operators into intermediate space. *Math. Ann.,* **151**, 219–238. MR 30 #450 *I

A Riemann-Stieltjes integral representation of a bounded linear operator acting from a subspace of $\mathcal{W}_p$ into an interpolation space.

Krabbe, G. L. (1963b). Spectral permanence of scalar operators. *Pacific J. Math.,* **13**, 1289–1303. MR 28 #2443

The result of Krabbe (1963a) is applied to study the spectrum of a bounded linear operator on an interpolation space.

Krée, P. (1966). Sur les multiplicateurs dans $\mathcal{F}L^p$ avec poids. *Ann. Inst. Fourier* (Grenoble), **16** no. 2, 91–121. MR 35 #7080 *M

Functions of bounded n-dimensional p-variation provide such multipliers (Sec. 5), where $\mathcal{F}$ = Fourier transform.

Krotov, V. G. (1981). On differentiability of functions from L^p and H^p for $0 < p < 1$. (Russian) *Dokl. Akad. Nauk SSSR*, **256**, 1311–1314. MR 82f:26007

Krotov, V. G. (1982). On the differentiability of functions from L^p, $0 < p < 1$. (Russian) *Mat. Sbornik*, **117**, 95–113; transl. in: *Math. USSR Sbornik*, **45** (1983), 101-119. MR 83c:26004

Comparison of Hardy-Littlewood classes of functions satisfying integrated Lipschitz conditions with the space of functions with bounded p-variation.

Kruglov, A. A. and Solomjak, M. Z. (1971). Interpolation of operators in the spaces V_p. *Vestnik Leningrad Univ.*, **13** (Mat. Meh. Astron. Nr. 3) 54–60; transl. in: *Vestnik Leningrad Univ. Math.*, **4** (1977), 209–216. MR 45 #2464 Zbl 224.46039

Relations between spaces of functions of bounded p-variation and certain interpolation spaces. An example concerning the non-existence of the Riemann-Stieltjes integral for functions with bounded p- and q-variation, respectively, such that $1/p + 1/q = 1$.

Laczkovich, M. and Preiss, D. (1985). α-variation and transformation into C^n functions. *Indiana Univ. Math. J.*, **34**, 405–424. MR 86g:26010

Connections between boundedness of the p-variation of a given function and Lipschitz or differentiability conditions for a transformation of the function.

Lal, S. N. and Ram, S. (1977). The absolute summability of Fourier series of a function of Wiener's class by Nörlund means. *Proc. Amer. Math. Soc.*, **67**, 87–94. MR 56 #12753 *F

Generalization of Izumi and Izumi (1967) to Nörlund summability.

Lépingle, D. (1975). Sur la variation d'ordre p des martingales locales. (French) *C. R. Acad. Sci. Paris Sér. A-B*, **281**, 917–919. MR 54 #6266 *P

If $p > 2$ then, with probability 1, the p-variation $v_p(M;\kappa)$ over a partition κ approaches the jump p-variation $\overline{\mathfrak{S}}_p(M;[0,1])^p$ as the mesh $|\kappa| \to 0$ for any local martingale M.

Lépingle, D. (1976). La variation d'ordre p des semi-martingales. (French) *Z. Wahrsch. Verw. Gebiete*, **36**, 295–316. MR 54 #8849 *P

p-variation of sample paths of semimartingales.

Lévy, P. (1941). Intégrales stochastiques. (French) *C. R. Acad. Sci. Paris*, **212**, 1066–1068. MR 5, 126a *I *P

If continuous functions $f \in \mathcal{W}_p$, $g \in \mathcal{W}_q$ and $1/p + 1/q > 1/2$ then certain Riemann-Stieltjes sums are close to their averages as $n \to \infty$ when partitions are formed from ordered samples of n independent uniformly distributed random variables.

Lindenstrauss, J. and Stegall, C. (1975). Examples of separable spaces which do not contain ℓ_1 and whose duals are non-separable. *Studia Math.*, **54**, 81–105. MR 52 #11543 *J

The James function space is identified with the set $\mathcal{C}W_2^*$ of continuous functions f such that $V_2^*(f) = 0$ (cf. Section 2.5 in Part II).

Love, E. R. (1951). A generalization of absolute continuity. *J. London Math. Soc.*, **26**, 1–13. MR 12, 599a

The class of functions f such that $v_p(f(\cdot + h) - f; [a, b - h]) \to 0$ as $h \downarrow 0$ is characterized for $1 < p < \infty$.

Love, E. R. (1993). The refinement-Ross-Riemann-Stieltjes (R^3S) integral. In: *Analysis, geometry and groups: a Riemann legacy volume*, Eds. H. M. Srivastava, Th. M. Rassias, Part I. Hadronic Press, Palm Harbor, FL, pp. 289–233. MR 95k: 26011

A simpler variant of Theorem 5.1 of L. C. Young (1938) concerning existence of the W. H. Young integral is proved using the approach of L. C. Young (1936).

Love, E. R. and Young, L. C. (1937). Sur une classe de fonctionelles linéaires. *Fund. Math.*, **28**, 243–257.

A representation of bounded linear functionals over the space of continuous functions with bounded p-variation.

Love, E. R. and Young, L. C. (1938). On fractional integration by parts. *Proc. London Math. Soc.*, **44**, 1–35.

A representation of bounded linear functionals on $\mathcal{W}_p^*$ (cf. Section 2.5 in Part II).

Lyons, T. (1994). Differential equations driven by rough signals (I): An extension of an inequality of L. C. Young. *Math. Res. Lett.*, **1**, 451–464. MR 96b:60150

A non-linear Stieltjes integral equation driven by a continuous R^d-valued function with bounded p-variation, $1 \le p < 2$.

Lyons, T. J. and Qian, Z. M. (1996). Calculus for multiplicative functionals, Itô's formula and differential equations. In: *Itô's Stochastic Calculus and Probability Theory*, Eds. N. Ikeda, S. Watanabe, M. Fukushima, H. Kunita. Springer-Verlag, Tokyo, pp. 233–250. *I

The calculus is based on a pathwise treatment of stochastic integrals obtained using p-variation.

Lyons, T. J. and Qian, Z. M. (1997). Flow equations on spaces of rough paths. *J. Functional Analysis* **149**, 135–159.

For a suitable nonlinear first order differential equation in a function space, the solution is shown to be continuous with respect to the p-variation of the initial function at time $t = 0$.

Macak, I. K. (1976a)* The β-variation of Gaussian processes with stationary increments. (Russian) *Studies in the theory of random processes*, 106–114; Izdanie Inst. Mat. Akad. Nauk Ukrain. SSR, Kiev. MR 56 #16762 *P

The p-variation of such processes.

Macak, I. K. (1976b). On the β-variation of a random process. (Russian) *Teor. Verojatnost. i Mat. Statist.*, **14**, 104–114, 157; transl. in: *Theor. Probab. Math. Statist.*, **14** (1977), 113–122. MR 54 #1366 *P

Conditions for a stochastic process to have almost all sample paths of bounded p-variation.

Marcinkiewicz, J. (1934). On a class of functions and their Fourier series. *Comptes Rendus Soc. Sci., Varsovie*, **26**, 71-77. Reprinted in: J. Marcinkiewicz. *Collected Papers*, pp. 36–41; Ed. A. Zygmund. Polish Scientific Publishers, Warsaw, 1964. Zbl 9. 160 *F

See the "historical comments" above.

Marcus, M. B. and Rosen, J. (1992). *p*-variation of the local times of symmetric stable processes and of Gaussian processes with stationary increments. *Ann. Probab.*, **20**, 1685–1713. MR 94c:60124 *L *P

The ϕ-variation for certain classes of local times and functions ϕ is calculated. The main results concern the p-variation under restricted partitions.

Marcus, M. B. and Rosen, J. (1993). Φ-variation of the local times of symmetric Lévy processes and stationary Gaussian processes. *Seminar on Stochastic Processes*, (Seatle, WA, 1992), 209–220. *Progr. Probab.*, **33**, Birkhäuser, Boston, MA. MR 95g:60102

An extension of Marcus and Rosen (1992) to a class of Lévy processes.

McLaughlin, J. R. (1973). Absolute convergence of series of Fourier coefficients. *Trans. Amer. Math. Soc.*, **184**, 291–316. MR 49 #979*F

Unification and generalization of earlier results including some for functions with bounded p-variation.

Millar, P. W. (1971). Path behaviour of processes with stationary independent increments. *Z. Wahrsch. Verw. Gebiete*, **17**, 53–73. MR 48 #3130 *P

The p-variation under restricted partitions with a discussion of the unrestricted case.

Monrad, D. (1983) On the *p*-variation of Gaussian random fields with separable increments. *Seminar on stochastic processes*, 1982 (Evanston, Ill., 1982), 203–218; *Progr. Probab. Statist.*, 5, Birkhäuser Boston, Boston, Mass. MR 86m:60133 *M *P

The p-variation of a Gaussian random field with separable stationary increments over rectangles.

Monroe, I. (1972). On the γ-variation of processes with stationary independent increments. *Ann. Math. Statist.*, **43**, 1213–1220. MR 47 #1135 *P

An extension of Blumenthal and Getoor (1961).

Oehring, Ch. (1990). Asymptotics of singular numbers of smooth kernels via trigonometric transforms. *J. Math. Analysis Appl.*, **145**, 573–605. MR 91h:47052

Oehring, Ch. (1992). Asymptotics of rearranged trigonometric and Walsh-Fourier coefficients of smooth functions. *J. Math. Analysis Appl.*, **164**, 422–446. MR 93d:42006 *F

The Wiener-Lozinski characterization of continuous functions with bounded p-variation, $p < 2$, in terms of their Fourier coefficients is extended for both trigonometric and Walsh-Fourier series.

Olevskii, V. (1994). A note on multiplier transformations. *Internat. Math. Res. Notices*, No. 1, 13–17. Addendum: *ibid.*, No. 7, 311. MR 95a:42012b *F

The main result is a discrete analogue of Lemma 3 of Coifman, Rubio de Francia and Semmes (1988).

Orlicz, W. (1946). Une généralisation d'un théoréme de MM. S. Banach et S. Mazur. *Ann. Soc. Polon. Math.*, **19**, 62–65. MR 9, 17

An almost sure bound for the p-variation of a linear combination of functions with bounded p-variation and Rademacher functions.

Postolică, V. (1978).* On set functions of bounded p-variation. *An. Ştiinţ. Univ. "Al. I. Cuza" Iaşi Secţ. I a Mat.* (N.S.), **24**, 251–258. MR 80m:28002

Postolică, V. (1981). On the primitives of functions with bounded p-variation. (Romanian) *Stud. Cerc. Mat.*, **33**, 141–146. MR 83g:26016

A characterization of indefinite Lebesgue-Stieltjes integrals of functions of bounded p-variation.

Postolică, V. (1982).* On set functions of bounded slope p-variation. *An. Ştiinţ. Univ. "Al. I. Cuza" Iaşi Secţ. I a Mat.* (N.S.), **28**, 5-9. MR 85c:28003

An extension of Postolică (1981) to set functions.

Pisier, G. (1988). The dual J^* of the James space has cotype 2 and the Gordon-Lewis property. *Math. Proc. Cambridge Philos. Soc.*, **103**, 323–331. MR 89a:46041 *J

The space W_p, $1 < p < \infty$, has the properties stated in the title.

Pisier, G. and Xu, Q. H. (1987). Random series in the real interpolation spaces between the spaces v_p. *Geometrical Aspects of Functional Analysis*, Eds. J. Lindenstrauss and V. Milman; *Lect. Notes in Math.* (Springer-Verlag), **1267**, 185–209. MR 89d:46011 *J

Spaces v_p of sequences of bounded p-variation, $1 < p < \infty$, and interpolation spaces between v_1 and ℓ_∞.

Pisier, G. and Xu, Q. H. (1988). The strong p-variation of martingales and orthogonal series. *Probab. Theory Related Fields*, **77**, 497–514. MR 89d:60086 *P

Boundedness of p-variation with $1 \leq p < 2$ for a martingale sequence in L_p.

Pych-Taberska, P. (1986). Pointwise approximation by partial sums of Fourier series and conjugate series. *Funct. Approx. Comment. Math.* (Poznań), **15**, 231–244. MR 85m: 42005

The rate of pointwise convergence of Fourier series is studied for functions with bounded p-variation (cf. also the review MR 85m: 42005 by E. R. Love).

Qian, Jinghua (1998). The p-variation of partial sum processes and the empirical process. *Ann. Probab.* **26**, 1370-1383. *P

The exact orders of the p-variation norm of the empirical process in probability
for $1 \leq p \leq 2$ and almost surely for $1 \leq p < 2$.

Riazanov, B. W. (1968). On functions of the class V. (Russian) Vestnik Moskov.
Univ., Ser. I Math. Mekh., **6**, 36–39.

Any function of bounded p-variation, $p \geq 1$, can be changed on a set of
arbitrarily small measure so as to give a function of class Lip$(1/p)$.

Siddiqi, R. N. MR: 48 #783, 48 #9224, 54 #779, 56 #3553, 56 #16240, 58 #17676,
81d:42010, 82d:42007, 82f:42006, 83d:42014, 84d:42011, 87i:42012, 88h:42007,
89h:42017, 90b:42011, 90f:42006, 90h:42010, 90m:42010, 91a:42003, 91e:42006,
93h:42003, 95b:42004, 96m:42007. *F

A series of papers on Fourier series for functions of bounded p-variation, many
on Nörlund summability.

Singman, D. (1993). Lower derivatives of functions of finite variation and generalized
BCH sets. J. Math. Anal. Appl., **173**, 483–496. MR 94c:26018 *M

Given a function $u: R^n \times R_+ \to [0, \infty)$ with bounded p-variation, the set
where the generalized lower derivative of u is positive is studied.

Skvortsov (Skvorcov), P. G. (1964)* Fourier coefficients of functions with bounded
γ-variation. (Russian) Kabardino-Balkarsk. Gos. Univ. Učen. Zap., **22**, 146–
148. MR 36#4242 *F

Convergence rates for sums of powers of Fourier coefficients.

Skvortsov (Skvorcov), P. G. (1965)* Unconditional convergence of Fourier series for
functions of bounded γ-variation. (Russian) Kabardino-Balkarsk. Gos. Univ.
Učen. Zap., **24**, 349–352. MR 36#4243

Unconditional convergence almost everywhere of a series expansion with respect to certain orthonormal systems.

Stoiński, S. (1996)* Real-valued functions almost periodic in p-variation. Fasciculi
Mathematici (Poznań), No. **26**, 155–162. MR 98b:42018

A class of almost periodic functions defined in terms of p-variation is linear
and is a proper subset of the class of Bohr almost periodic functions.

Stricker, C. (1979). Sur la p-variation des surmartingales. (French) Séminaire de
Probabilités, XIII (Univ. Strasbourg, 1977/78), 233–237; Lect. Notes in Math.
(Springer-Verlag), **721**, Berlin. MR 82a:60066 *P

A result of Bruneau (1979) is used to prove that the sample paths of a continuous semimartingale have bounded p-variation with $p > 2$.

Stricker, C. (1979). Une remarque sur l'exposé précédent. (French) Séminaire de
Probabilités, XIII (Univ. Strasbourg, 1977/78), 238–239; Lect. Notes in Math.
(Springer-Verlag), **721**, Berlin. MR 82a:60067

Száz, Á. (1993)* The fundamental theorem of calculus in an abstract setting. Tatra
Mt. Math. Publ., **2**, 167–174. MR 94m:26017

Set functions of bounded p-variation are used as integrators.

Taberski, R. (1967). On Dini series. II. Bull. Acad. Polon. Sci. Sèr. Sci. Math.
Astronom. Phys., **15**, 703–710. MR 40 #1736

The convergence of Fourier-Dini series for a continuous function of bounded p-variation.

Taberski, R (1969). Convergence criteria for Hankel's repeated integrals. *Bull. Acad. Polon. Sci. Sér. Sci. Math. Astronom. Phys.*, **17**, 1–10. MR 40 #3208

A function is of bounded p-variation in a definition of Hankel's integral.

Taberski, R. (1969). Trigonometric interpolation. I. *Colloq. Math.*, **20**, 287–294. MR 39 #7341 *F

An extension of several tests of convergence of Fourier series and bounds of Fourier coefficients of functions of bounded p-variation.

Taberski, R. (1970). Trigonometric interpolation. II. *Colloq. Math.*, **21**, 111–126. MR 41 #5861 *M *F

Review of definitions of p-variation for a function on a rectangle and an extension of the inequality of Hölder type due to Young (1936).

Taberski, R. (1984). Trigonometric approximation in the norms and seminorms. *Studia Math.*, **80**, 197–217. MR 86j: 42004 *F

Approximation of functions having Weyl derivative of bounded p-variation.

Taberski, R. (1988). Quantitative versions of L. C. Young criteria for Fourier-Bessel series. *Funct. Approx. Comment. Math.* (Poznań), **16**, 81–106. MR 89j: 42021

Rates of convergence by means of p-variation.

Terehin, A. P. (1965). The approximation of functions of bounded p-variation. (Russian) *Izv. Vysš. Učebn. Zaved. Mat.*, **2** (45), 171–187. MR 32 #2814

Generalization of the theorem of F. Riesz about integrability of the p-th power of a function with bounded variation. Generalization of inequalities of Jackson and of S. N. Bernstein. The dual space of a Banach space of continuous functions with bounded p-variation.

Terehin, A. P. (1967). Integral smoothness properties of periodic functions of bounded p-variation. (Russian) *Mat. Zametki*, **2**, 289–300; transl. in: *Math. Notes*, **1–2**, 659–665. MR 36 #6560

Relations between p-variation and integrated Lipschitz conditions.

Terehin, A. P. (1967). The Lebesgue constant for the space of functions of bounded p-variation. (Russian) *Mat. Zametki*, **2**, 505–512; transl. in: *Math. Notes*, **2**, 798–802. MR 36 #1890 *F

The order of the L_p-norm of the difference between the indefinite integral of the Dirichlet kernel and a function with values 0 and 1 on $(-\infty, 0]$ and $(0, \infty)$, respectively.

Terehin, A. P. (1969)* Integral constructive properties of periodic functions of bounded p-variation. (Russian) Works of young scientists: mathematics and mechanics, No. 2 , 131–135. Izdat. Saratov. Univ., Saratov. MR 50 #10659

Connections between best approximations using norms of L_p and $\mathcal{W}_p$ spaces.

Terehin, A. P. (1972). Functions of bounded *p*-variation with given order of modulus of *p*-continuity. (Russian) *Mat. Zametki*, **12**, 523–530; transl. in: *Math. Notes*, **11-12**, 751-755. MR 47 #6963

  A construction of a continuous function f with a given order of $v_p(f; \epsilon)$.

Terehin, A. P. (1972). Multidimensional *q*-integral *p*-variation, and generalized Sobolev differentiability in L_p of functions from L_q. (Russian) *Sibirsk. Mat. Ž.*, **13**, 1358–1373, 1421. MR 48 #11414 *M

  The increments of a function in the definition of p-variation are replaced by L_q-norms.

Terehin, A. P. (1979). A mixed *q*-integral *p*-variation, and theorems of equivalence and embedding of classes of functions with a mixed modulus of smoothness. (Russian) *Studies in the theory of differentiable functions of several variables and its applications*, VII. Trudy Mat. Inst. Steklov. **150**, 306–319, 324. MR 81f:46050 *M

Terekhin, A. P. (1982). Mixed *q*-integral *p*-variation and mixed differentiability in L_p of functions from L_q. (Russian) *Mat. Zametki*, **32**, 151–167, 269; transl. in: *Math. Notes*, 32(1982), 556-565. MR 84c:46035 *M

  Relations between generalized differentiability, integrated Lipschitz conditions and p-variation.

Tietz, H. (1973). Permanenz- und Taubersätze bei *pV*-Summierung. (German) *J. Reine und Angew. Math.*, **260**, 151–177. MR 48 #754

  Methods of summability.

Towghi, N. (1995). Stochastic integration of processes with finite generalized variations. I. *Ann. Probab.*, **23**, 629–667. MR 96g:60071 *M *P

  A kind of p-variation condition on a Fréchet pseudomeasure on an n-fold product of measurable sets is defined and used to prove the existence of a stochastic integral.

Trofimov, E. I. (1992). Sobolev topologies in semimartingale theory. *Séminaire de Probabilités*, XXVI, 596–607; *Lect. Notes in Math.* (Springer-Verlag), **1526**, Berlin. MR 94g:60013 *P

  Moment inequalities for the p-variation of martingales due to Lépingle (1976) are used to prove tightness of the corresponding distributions in Sobolev spaces.

Upton, C. J. F. (1977). On classes of continuous almost periodic functions. *Proc. London Math. Soc.*, **35**, 159–179. MR 56 #6277

  Almost periodic functions are defined using p-variation.

Upton, C. J. F. (1989). Some new classes of almost periodic functions. *Chinese Ann. Math. Ser. B*, **10**, 200–213. MR 90m:42018

  Almost periodicity of a function is defined using the p-variation of the indefinite Denjoy integral.

Volosivets, S. S. (1992). On the ϵ-entropy of some sets of functions of bounded *p*-variation. (Russian) *Izv. Vysš. Učebn. Zaved. Mat.*, no. 2, 83–85. Russian Math. (Iz. Vuz.) **36**, 83–85. MR 93m:26003

Bounds for the metric entropy of a set of functions with a given order of modulus of continuity related to a p-variation norm.

Volosivets, S. S. (1992). On the ϵ-entropy and widths of a compact set of smooth functions in the space of functions of bounded p-variation. (Russian) *Vestnik Moskov. Univ. Ser. I Mat. Mekh.*,no. 5, 81–84; transl. in: *Moscow Univ. Math. Bull.*, 47 (1992), no. 5, 56–58. MR 95f:41045

Volosivets, S. S. (1993). Approximation of functions of bounded p-variation by polynomials in Haar and Walsh systems. (Russian) *Mat. Zametki*, **53**, no. 6, 11–21, 157; transl. in: *Math. Notes* 53 (1993), 569–575. MR 96a:41023 *F

Conditions are given on approximation in L^p norm for a function to be equal a.e. to some function in W_p.

Volosivets, S. S. (1995a). Asymptotic characteristics of a compact set of smooth functions in a space of functions of bounded p-variation. (Russian) *Mat. Zametki*, **57**, 214–227, 318; transl. in *Math. Notes*, 57 (1995), 148– 157. MR 96h:26014

Metric entropy of a class of functions of bounded p-variation which can be expressed via fractional integrals.

Volosivets, S. S. (1995b)? Approximation of functions of bounded p-variation by polynomials in multiplicative systems. (Russian) *Anal. Math.* (Budapest), **21**, 61–77. MR 96m:41013 *F

A new discrete modulus of continuity is introduced for functions of bounded p-variation.

Volosivets, S. S. (1996). Polynomials of best approximation and relations between moduli of continuity in spaces of functions of bounded p-variation. (Russian) *Izv. Vyssh. Uchebn. Zaved. Mat.*, No 9, 21–26; transl. in: *Russian Math. (Iz. VUZ)*, **40**, 1996, 18–23. MR 97j:41010

Several results about approximation of continuous functions are extended to functions of bounded p-variation.

Wang, X. H. (1982). On an inequality of the integral $\int_a^b f(x)g(x)dx$ and its application to the numerical approximation of functions with low order smoothness. *Scientia Sinica*, **25**, 1241–1249. MR 85b:26022 *I

The absolute value of the integral is bounded in terms of the p-variation of f and the q-variation of the indefinite integral G of g whenever $G(b) = 0$ and $1/p + 1/q = 1$.

Wiener, N. (1924). The quadratic variation of a function and its Fourier coefficients. *J. Math. and Phys.* (MIT, Cambridge, Mass), **3**, 73–94. Reprinted in: *Norbert Wiener: Collected Works.* Ed. P. Masani, Vol. II, 36–58; MIT Press, Cambridge, 1979. *F

Wong, S. (1980). Hölder continuous derivatives and ergodic theory. *J. London Math. Soc.*, (2) **22**, 506–520. MR 82e:58061 *E

The density of an invariant measure is of bounded p-variation for certain p.

Xu, Q. H. (1988). Espaces d'interpolation réels entre les espaces V_p: propriétés géométriques et applications probabilistes. (French) [Real interpolation spaces between the spaces V_p: geometric properties and probabilistic applications] *Séminaire d'Analyse Fonctionnelle* 1985 / 1986 / 1987, Paris VI–VII, 77–123; Publ. Math. Univ. Paris VII, 28, Paris. MR 89m:46059 *P

A comparison between the space of functions of bounded p-variation and certain interpolation spaces with applications to stable processes.

Xu, Q. H. (1996). Fourier multipliers for $L_p(R^n)$ via q-variation. *Pacific J. Math.*, **176**, 287–296. MR 98b:42024 *M *F

An extension of the result of Coifman et al. (1988) to multivariate functions. See also Krée (1966).

Yoder, L. (1974). Variation of multiparameter Brownian motion. *Proc. Amer. Math. Soc.*, **46**, 302–309. MR 54 #6301 *M

The results of Goffman and Loughlin (1972/73) are extended to the multivariate case.

Young, L. C. (1936). An inequality of the Hölder type, connected with Stieltjes integration. *Acta Math.*, **67**, 251–282. *I *F

Existence of a new integral of Stieltjes type and convergence of Fourier series for functions with bounded p-variation.

Young, L. C. (1937b). Inequalities connected with bounded p-th power variation in the Wiener sense and with integrated Lipschitz conditions. *Proc. London Math. Soc.* (Ser. 2), **43**, 449–467.

A p-variation inequality for Stieltjes Faltung (convolution).

Young, L. C. (1938a). Inequalities connected with binary derivation, integrated Lipschitz conditions and Fourier series. *Math. Z.*, **43**, 255–270. *F

A further development of Young (1937b).

Young, L. C. (1938b). A convergence criterion for Fourier series. *Quart. J. Math.*, Oxford, **9**, 115–120.

An extension of W. H. Young's criterion, with certain refinements of Hardy and Littlewood, to functions of bounded p-variation.

Young, L. C. (1938c). Intégrales généralisées de Stieltjes et convergence des séries de Fourier. Société Math. France, Gauthier-Villars, Paris.

An overview of the author's results.

Young, L. C. (1942). On the convergence of Fourier-Bessel series. *Proc. London Math. Soc.*, **47**, 290–307. MR 4, 39

For functions of bounded p- or ϕ-variation.

Zerekidze, T. (1974)* The functions of classes V_p and the Banach indicatrix. *Conf. of young scient. and post-graduate students*, Inst. Appl. Math. Tbilisi.

A function $f = F \circ \chi$, with χ increasing and F continuous, has bounded p-variation, $p > 1$, if $N(\cdot, F)^{1/p}$ is Lebesgue integrable, where $N(y, F) := card\{x : F(x) = y\}$, called the Banach indicatrix.

Ziemian, K. (1985). Almost sure invariance principle for some maps of an interval. *Ergodic Theory Dynamical Systems*, **5**, 625–640. MR 87g:58076 *E *P

For the process $(F \circ f^n)_{n \geq 0}$, where F is a function of bounded p-variation.

Zinčenko, N. M. (1978). On the p-variation of Gaussian random fields. (Russian) *Teor. Veroyatnost. i Mat. Statist.*, **19**, 72–76. MR 80a:60044 *M *P

Strong and weak p-variations are calculated for a stationary, isotropic Gaussian field with the variance of increments decreasing like a power function.

BIBLIOGRAPHY ON ϕ-VARIATION

Notation. Let ϕ be a continuous, unbounded, nondecreasing function on $[0, \infty)$ with $\phi(u) = 0$ if and only if $u = 0$. Let $v_\phi(f; \kappa) := \sum_{i=1}^{n} \phi(|f(x_i) - f(x_{i-1})|)$ for a partition $\kappa = \{a = x_0 < \cdots < x_n = b\}$. A function f has bounded ϕ-variation if $v_\phi(f; [a, b]) := \sup\{v_\phi(f; \kappa): \kappa\} < \infty$. The class of all such functions will be denoted by $\mathcal{W}_\phi$. The notion of ϕ-variation was introduced by Young (1937a).

Adell, J. A. and de la Cal, J. (1996). Bernstein-type operators diminish the ϕ-variation. *Constr. Approx.*, **12**, 489–507. MR 98e:41029

  Given ϕ, the relations $v_\phi(L_t f) \le v_\phi(f)$, $t \in T$, $v_\phi(L_t f - f) \to 0$ as $t \to \infty$, and their local ϕ-variation analogues are proved under certain conditions for several families of linear positive operators $\{L_t: t \in T\}$.

Akhobadze, T. I. (1976). The convergence of Cesàro means of negative order for functions of bounded generalized second variation. (Russian) *Mat. Zametki*, **20**, 631–644. MR 56 #3550

Akhobadze, T. I. (1979). Functions of a bounded generalized second variation. (Russian) *Math. Sb. (N.S.)*, **109(151)**, 291–326. MR 80j:42028. *F *M

  ϕ-variation of a function of two variables.

Akhobadze, T. I. (1981). Classes of functions and convergence of Fourier series. (Russian) *Acta Math. Acad. Sci. Hungar.*, **37**, 95–119. MR 83b:42005 *F

  Relations between p-variation and integrated Lipschitz conditions.

Akhobadze, T. I. (1986).* Continuity of functions of generalized bounded variation. (Russian) *Soobshch. Akad. Nauk Gruzin. SSR*, **121**, 17–20. MR 87j:26013

Akhobadze, T. I. (1986). Continuity of a function of several variables in a class of generalized bounded variation. (Russian) *Acta Math. Hungar.*, **48**, 317–341. MR 88f:42019 *M *F

  A generalization of ϕ-variation to functions of several variables.

Albrycht, J. and Musielak, J. (1970). On a class of functions with finite generalized variation. *Ganita* (Lucknow, India), **21**, 49–57. MR 45 #8787

  Relations between the ϕ-variation of a function f and the least upper bound with respect to partitions of sums of terms $\phi(|f(y+) - f(x-)|)$ taken over intervals $[x, y]$ in partitions of $[a, b]$.

Asatiani, V. O. and Čanturija, Z. A. (1977).* The modulus of the variation of a function, and Banach's indicatrix. (Russian) *Sakharth. SSR Mecn. Akad. Moambe*, **86**, 549–552. MR 56 #15850

Asatiani, V. O. and Chanturia, Z. A. (1983). The modulus of variation of a function and the Banach indicatrix. *Acta Sci. Math. (Szeged)*, **45**, 51–66. MR 85c:26006

  The relationship between the degree of summability of the Banach indicatrix and a modulus of variation related to ϕ-variation.

Avdispahić, M. (1987). Concepts of generalized variation on Vilenkin groups and convergence of Fourier-Vilenkin series. *A. Haar memorial conference*, Vol. I, Budapest, 1985, 145–163; North-Holland, Amsterdam. MR 88j:43003

Several notions of ϕ-variation of functions on bounded Vilenkin groups are discussed.

Baernstein, A. (1972). On the Fourier series of functions of bounded Φ-variation. *Studia Math.*, **42**, 243–248. MR 46 #4085 *F

Necessity of the condition of Salem (1940).

Bardaro, C. and Vinti, G. (1991a). Some estimates of integral operators with respect to the multidimensional Vitali Φ-variation, and applications in fractional calculus. *Rend. Mat. Appl.*, **11**, 405–416. MR 92k:26050

Inequalities for integral operators with homogeneous kernels.

Bardaro, C. and Vinti, G. (1991b). On convergence of moment operators with respect to ϕ-variation. *Appl. Anal.*, **41**, 247–256. MR 92h:26013

For a ϕ-absolutely continuous function f, $v_\phi(T_n f - f) \to 0$ as $n \to \infty$, where $T_n f(t) = \int w_n(s) f(st) \, ds$.

Bardaro, C. and Vinti, G. (1991c)* Modular estimates of integral operators with homogeneous kernels in Orlicz type spaces. *Resultate Math.*, **19**, 46–53. MR 92c:47057

Estimates are also given with respect to the ϕ-variation.

Berezhnoĭ, E. I. (1994). A sharp correctability theorem for spaces of functions of generalized bounded variation. (Russian) *Mat. Zametki*, **56**, 10–21; transl. in: *Math. Notes*, **56**, 1105–1112. MR 97c:46035

An extension of the Wiener-Young notion.

Berman, R. (1988). Generalized variation and functions of slow growth. *Canad. J. Math.*, **40**, 55–85. MR 89c:30088

Cauchy-Stieltjes and Poisson-Stieltjes representations known in H^p theory are generalized and studied replacing integrators of bounded variation by integrators with bounded ϕ-variation.

Beurling, A. (1964). Analyse harmonique de pseudomesures. Intégration par rapport aux pseudomesures. *C. R. Acad. Sc. Paris*, **258**, 1984-1987. MR 29 #5062d *I

Let ρ, χ be moduli of continuity of continuous functions f, g, respectively, such that the integral $\int_0^1 \rho(t) t^{-1} d\chi(t)$ exists. Then there exists an integral $\int f \, d\mu$ with respect to a pseudomeasure μ defined by g.

Bruneau, M. (1975). Mouvement brownien et ϕ-variation. *J. Math. Pures Appl.*, **54**, 11–25. MR 55 #9299 *P

A set $\{t \in [0,1]: v_\phi(B; [0,1]) = v_\phi(B; [0,t]) + v_\phi(B; [t,1])\}$ is studied, where B is a standard Brownian motion and ϕ is a convex function such that $\phi^{-1}(u) \sim \sqrt{2u \log \log 1/u}$ as $u \downarrow 0$.

Burkill, J. C. (1948). Differential properties of Young-Stieltjes integrals. *J. London Math. Soc.*, **23**, 22-28. MR 10, 185 *I

Let ρ, χ be moduli of continuity of continuous functions f, g, respectively, such that $\rho(t)\chi(t)t^{-2}$ is integrable in the neighborhood of 0. Then the indefinite Riemann-Stieltjes integral F of f with respect to g exists and satisfies the relation $F(x+h) - F(x) = f(x)[g(x+h) - g(x)] + o(h)$ as $h \downarrow 0$.

Čanturija, Z. A. (1974). The modulus of variation of a function and its application in the theory of Fourier series. (Russian) *Doklady Akad. Nauk SSSR*, **214**, 63-66; transl. in: *Soviet Math. Dokl.*, **15**, 67-71. MR49 #7682 *F

The modulus of variation of a function is defined and used to generalize ϕ-variation conditions for the uniform convergence of Fourier series.

Chanturiya, Z. A. (1987). On the continuity of functions of the class V_Φ. (Russian) *Mat. Zametki*, **41**, 191-201. MR 88e:42010 *F

A characterization of continuity via Fourier coefficients for certain classes of functions with bounded ϕ-variation.

Chanturiya, Z. A. (1989). Conditions for continuity of functions of generalized bounded variation. (Russian) *Dokl. Akad. Nauk SSSR*, **307**, 1329-1331; transl. in: *Soviet Math. Dokl.*, **40** (1990), 263-265. *F

Continuation of Chanturiya (1987).

Chang, M. H. (1983). On a limit theorem for variation of Brownian motion in Banach spaces. *Stochastic Anal. and Appl.*, **1**, 341-351. MR 85m:60134 *P

An extension of the main result of Taylor (1972) to a Banach space valued Brownian motion.

Ciemnoczołowski, J. and Orlicz, W. (1984). Inclusion theorems for classes of functions of generalized bounded variations. *Comment. Math. Prace Mat.*, **24**, 181-194. MR 86g:26009

Relation between several notions of generalized variation.

Ciemnoczołowski, J. and Orlicz, W. (1985a). Functions of bounded ϕ-variation and some related operators. *Demonstratio Math.*, **18**, 231-251. MR 87g:26016

An extension of Riazanov (1968) to functions with bounded ϕ-variation.

Ciemnoczołowski, J. and Orlicz, W. (1985b). Subseries convergence in the space of functions of bounded ϕ-variation. *Resultate Math.*, **8**, 1-8. MR 88a:46027

Let $\{f_i\} \subset \mathcal{W}_\phi$ and $\phi(u)/\psi(u) \to 0$ as $u \downarrow 0$. If $v_\psi(\sum_1^n \epsilon_i f_i) \leq K$ for $n \geq 1$ and any $\epsilon_i = 0, 1$, then, for some f_ϵ, $v_\phi(\sum_1^n \epsilon_i f_i - f_\epsilon) \to 0$ uniformly on the set of $\epsilon = \{\epsilon_i\}$.

Ciemnoczołowski, J. and Orlicz, W. (1986a). A Lindenstrauss-Stegall theorem for ϕ-variation. *Bull. Polish Acad. Sci. Math.*, **34**, 173-180. MR 88a:26018

An extension of Kisliakov (1984) to functions with bounded ϕ-variation.

Ciemnoczołowski, J. and Orlicz, W. (1986b). Composing functions of bounded φ-variation. *Proc. Amer. Math. Soc.*, **96**, 431-436. MR 87k:26012

Necessary and sufficient conditions on F for the composite function Fof to have bounded ψ-variation, for all f having bounded ϕ-variation, are given.

Ciemnoczołowski, J., Matuszewska, W. and Orlicz, W. (1987). Some properties of functions of bounded ϕ-variation and of bounded ϕ-variation in the sense of Wiener. *Bull. Polish Acad. Sci. Math.*, **35**, 185–194. MR 88i:26020

Necessary and sufficient conditions for additivity and subadditivity with respect to intervals of the ϕ-variation of a continuous function.

Cohen, E. (1978). On the degree of approximation of a function by the partial sums of its Fourier series. *Trans. Amer. Math. Soc.*, **235**, 35–74. MR 57 #992 *F

The result of Golubov (1968b) is extended to functions whose fractional derivatives are of bounded ϕ-variation and with estimates taken in an Orlicz space norm.

Cohen, E. (1979). On the Fourier coefficients and continuity of functions of class V_ϕ^*. *Rocky Mountain J. Math.*, **9**, 227–237. MR 80m:42006 *F

Continuity of fractional derivatives and integrals of functions with bounded ϕ-variation.

Cybertowicz, Z. and Matuszewska, W. (1977).* Functions of bounded generalized variations. *Comment. Math. Prace Mat.*, **20**, 19–52. MR 57 #3333

ϕ-variation over subsets $A \subset [0,1]$ of measure 1.

Darmawijaya, S. (1988).* An Orlicz scale of integrals. *Southeast Asian Bull. Math.*, **12**, 123–133. MR 90c:26029 *I

An extension of Burkill and Gehring (1953) to ϕ-variation.

D'yačkov, A. M. (1988).* Conditions for the existence of Stieltjes integral of functions of bounded generalized variation. *Anal. Math.* (Budapest), **14**, 295–313. MR 90h:26011 *I

A generalization of Leśniewicz and Orlicz (1973).

D'yachkov, A. M. (1996). On the existence of the Stieltjes integral. *Doklady Math.*, **54**, 676–679. *I

Given a suitable function ϕ, the functions f such that the Riemann-Stieltjes integral $\int f\, dg$ exists for all functions g of bounded ϕ-variation are characterized.

Fristedt, B. E. (1967). Sample function behaviour of increasing processes with stationary, independent increments. *Pacific J. Math.*, **21**, 21–33. MR 35 #108 *P

Theorem 3 is related to ϕ-variation.

Fristedt, B. (1974). Sample functions of stochastic processes with stationary, independent increments. In: *Advances in Probability and Related Topics*, Eds. P. Ney and S. Port, **3**, 241–396. Dekker, New York. MR 53 #4240 *P

A survey of ϕ-variation for sample paths of stable processes.

Fristedt, B. and Taylor S. J. (1973). Strong variation for the sample functions of a stable process. *Duke Math. J.*, **40**, 259–278. MR 47 #5968

$\overline{V}_\phi(X) = \overline{\mathfrak{S}}_\phi(X)$ *for sample paths of a stable Lévy process* X.

Gniłka, S. (1978). Modular spaces of functions of bounded M-variation. *Funct. Approx. Comment. Math.*, **6**, 3–24. MR 81j:46035

A generalization of ϕ-variation.

Gniłka, S. (1992)* $\mathcal{V}_\phi$ spaces from another point of view. *Funct. Approx. Comment. Math.* (Poznań), **21**, 141–147. MR 95f:46041

A new topological description of $\mathcal{W}_\phi$.

Goffman, C. and Loughlin, J. J. (1972/73). Strong and weak ϕ-variation of Brownian motion. *Indiana Univ. Math. J.*, **22**, 135–138. MR 45 #5288 *P

Weak ϕ-variation is defined as the $\liminf$ *of* $\sum \phi(Osc(f; [x_{i-1}, x_i]))$ *as* $|\kappa| \to$ 0.

Goffman, C. and Waterman, D. (1970). Some aspects of Fourier series. *Amer. Math. Monthly*, **77**, 119–133. MR 40 #6155 *F

A survey listing several problems concerning the convergence of the Fourier series of a function with bounded ϕ-variation. One of them was solved by Oskolkov (1972).

Golubov, B. I. (1968)* Criteria for the compactness of sets in spaces of functions of bounded generalized variation. (Russian) *Izv. Akad. Nauk Armjan. SSR Ser. Mat.*, **3**, 409–416. MR 46 #2415

Compact sets of absolutely continuous functions in the sense of Musielak and Orlicz (1959).

Golubov, B. I. (1972a). The convergence of spherical Riesz means of multiple series and of Fourier integrals of functions of bounded generalized variation. (Russian) *Mat. Sb.*, **89**, 630–653. MR 48 #2658 *F *M

A new definition of multivariate ϕ-variation. Tests of R. Salem, de la Vallée Poussin and W. H. Young are generalized to summability of multiple Fourier series by spherical means.

Golubov, B. I. (1972b). Functions of generalized bounded variation, the convergence of their Fourier series and of their conjugate trigonometric series. (Russian) *Doklady Akad. Nauk SSSR*, **205**, 1277–1280. MR 48 #4635 *F

Several results on convergence of multiple Fourier series.

Golubov, B. I. (1974). The convergence of the double Fourier series of functions of bounded generalized variation. I. (Russian) *Sibirsk. Mat. Ž.*, **15**, 262–291. MR 54 #5737

Golubov, B. I. (1974). The convergence of the double Fourier series of functions of bounded generalized variation. II. (Russian) *Sibirsk. Mat. Ž.*, **15**, 767–783. MR 54 #5738

Two dimensional analogues of the results of Salem (1940), Oskolkov (1972), and Baernstein (1972).

Golubov, B. I. (1975). The convergence of spherical Riesz means of multiple Fourier series. *Mat. Sb. (N.S.)*, **96**, 189–211. MR 52 #3875 *M *F

Golubov, B. I. (1977). The summability of Fourier integrals by spherical Riesz means. (Russian) *Mat. Sb. (N.S.)*, **104**, 577–596, 663; transl. in: *Math. USSR Sbornik*, **33**, 501-518. MR 57 #3739 *M *F

Extension of the results of Golubov (1975) to integral spherical Riesz means.

Herda, H.-H. (1968). Modular spaces of generalized variation. *Studia Math.* **30**, 21–42. MR 37 #6742

Continuation of Musielak and Orlicz (1959). Comparison of function spaces of bounded φ-variation with different φ.

Kasperski, A. (1995). Notes on the space of multifunctions of finite generalized variation. *Math. Japon.*, **41**, 399–404. MR 96a:46075

A study of φ-variation of multifunctions by extending chapter II section 10 of Musielak (1983).

Kawada, T. and Kôno N. (1973). On the variation of Gaussian processes. In: *Proc. of the Second Japan-USSR Symposium on Probability Theory, Lect. Notes in Math.* (Springer-Verlag), **330**, 176–192. MR 55 #11365 *P

Boundedness of φ-variation of sample paths for certain continuous Gaussian process is proved.

Kluvánek, I. (1987). Additive set functions of bounded Φ-variation. *Miniconference on harmonic analysis and operator algebras* (Canberra, 1987), 89–100; Proc. Centre Math. Anal. Austral. Nat. Univ., **15**. Austral. Nat. Univ., Canberra. MR 89f:28026 *I

The φ-variation of additive set functions on a semialgebra and applications to integration theory.

Leśniewicz, M. and Leśniewicz, R. (1985). On double Riemann-Stieltjes integrals for functions of finite generalized variations. *Fasc. Math.* (Poznań), **15**, 37–61. MR 87d:26014 *M *I

Conditions on (φ, ψ)-variation, as defined by the authors, of a function on a rectangle.

Leśniewicz, R. and Orlicz, W. (1973). On generalized variations (II). *Studia Math.*, **45**, 71–109. MR 49 #11234

Two other proofs of the main result of Young (1938c).

Lévy, P. (1941). Propriétés intrinsèques des fonctions, et intégrales de Stieltjes. (French) *Bull. Soc. Math. France*, **69**, 5–9. MR 7, 198f *I

Continuous functions of bounded φ-variation are the smallest class of functions invariant under one-to-one monotone transformation of the argument containing all functions for which a φ-Hölder condition holds. Necessity of the φ-variation condition for the existence of a Stieltjes integral due to Young (1938c).

Lévy, P. (1953). Random functions: general theory with special reference to Laplacian random functions. In *University of California Publications in Statistics*, Eds. J. Neyman et. al., **1**, No 12, 331–390. University of California Press, Berkeley. MR 14, 1099f *I *P φ-variation conditions for Stieltjes integrability with integrator or integrand being a sample path of Brownian motion (pp. 353–357).

Littlewood, J. E. (1938). Mathematical notes (14): on a theorem of Hardy and Littlewood. J. London Math. Soc., **13**, 194–195.

A theorem of Hardy and Littlewood (1926) states that, if a function f and its conjugate g are both of bounded variation, then the Fourier series is absolutely convergent. It is shown that if instead g has bounded φ-variation with $\phi(u)/u \downarrow 0$ as $u \downarrow 0$ then the theorem no longer holds.

Maligranda, L. and Orlicz, W. (1987)* On some properties of functions of generalized variation. Monatsh. Math., **104**, 53–65. MR 88i:46072.

A class of functions F is introduced such that $F \circ f \in \mathcal{W}_\phi$ if $f \in \mathcal{W}_\phi$ and ϕ is convex.

Mantellini, I. and Vinti, G. (1998). Φ-variation and nonlinear integral operators. Atti Sem. Mat. Fis. Univ. Modena, Supplemento **46**, 847-862.

Given a nonlinear operator $(Tf)(s) = \int K(t, f(ts))dt$ and functions ϕ, ψ, the inequality $v_\phi(c_1 Tf) \leq v_\psi(c_2 f)$ is proved under certain conditions, for suitable constants c_1, c_2.

Matuszewska, W. and Orlicz, W. (1987). On property B_1 for functions of bounded φ-variation. Bull. Polish Acad. Sci. Math., **35**, 57–69. MR 88i:26023

Property B_1 means that $v_\phi(\lambda_n f_n) \to 0$ for any sequence f_n of functions of uniformly bounded φ-variation and for any sequence $\lambda_n \to 0$.

Moricz, F (1992). On the uniform convergence and L^1-convergence of double Walsh-Fourier series. Studia Math., **102**, 225–237. MR 93e:42040 *M *F

Extension of the condition of Salem (1940).

Musielak, J. (1983). Orlicz spaces and modular spaces. Lect. Notes in Math. (Springer-Verlag), **1034**. MR 85m:46028

Several properties of the space of functions with bounded φ-variation.

Musielak, J. and Orlicz, W. (1957). On spaces of functions of finite generalized variation. Bull. Polish Acad. Sci. Cl. III., **5**, 389–392. MR 19, 638d

Relations between spaces $\mathcal{W}_\phi$ for different ϕ.

Musielak, J. and Orlicz, W. (1959). On generalized variations (I). Studia Math., **18**, 11–41. MR 21 #3524

Develops the theory of Banach spaces of functions of bounded φ-variation. Moreover, proves a Helly extraction theorem and studies an extension of absolute continuity as defined by Love (1951).

Musielak, J. and Waszak, A. (1988). Generalized variation and translation operator in some sequence spaces. Hokkaido Math. J., **17**, 345–353. MR 90h:26011

Spaces of sequences of bounded φ-variation.

Névai, G. P. (1974). Remarks on interpolation. (Russian) Acta Math. Hungar., **25**, 123–144. MR 52 #3799

Uniform convergence of the Lagrange interpolation process for functions with bounded φ-variation and bounds of accuracy of this approximation.

Onneweer, C. W. (1970). On uniform convergence of Walsh-Fourier series. *Pacific J. Math.*, **34**, 117–122. MR 43 #806 *F

An extension of the condition of Salem (1940).

Oskolkov, K. I. (1972). Generalized variation, the Banach indicatrix, and the uniform convergence of Fourier series. *Mat. Zametki*, **12**, 313–324; transl. in: *Math. Notes*, **12**, 619-625. MR 47 #5507 *F

The deviation of a function f from the partial sum of its Fourier series is estimated whenever f is continuous and has bounded φ-variation. The necessity of the condition of Salem (1940) is proved for continuous functions.

Prus-Wiśniowski, F. (1989). On superposition of functions of bounded φ-variation. *Proc. Amer. Math. Soc.*, **107**, 361–366. MR 90i:26008

The Δ_2-condition in the main result of Ciemnoczołowski and Orlicz (1986b) is shown to be unnecessary.

Prus-Wiśniowski, F. (1990). Some remarks on functions of bounded φ-variation. *Comment. Math. Prace Mat.*, **30**, 147–166. MR 92g:26012

New proofs of known results concerning characterization of regulated functions and functions of bounded φ-variation.

Prus-Wiśniowski, F. (1991). Continuous functions of bounded φ-variation. *Comment. Math. Prace Mat.*, **31**, 127–146. MR 92m:26010

For a certain class of functions φ there exists a superadditive function $φ^v$ such that φ-variation and $φ^v$-variation are the same for continuous functions.

Prus-Wiśniowski, F. (1994). A continuous function of bounded φ-variation that is not φ-absolutely continuous. *Comment. Math. Prace Mat.*, **34**, 165–172. MR 1 325 083

For each convex φ-function φ with $φ(u)/u → 0$ as $u ↓ 0$, there exists a continuous function in $W_φ$ which is not φ-absolutely continuous defined by extending the definition of Love (1951).

Pych-Taberska, P. (1985). Approximation of periodic functions by the Euler and Borel means of Fourier series. *Demonstratio Math.*, **18**, 115–129. MR 87c: 42008 *F

Pych-Taberska, P. (1985). On the rate of convergence of Fourier-Legendre series. *Bull. Polish Acad. Sci. Math.*, **33**, 267–275. MR 87j:42071

Pych-Taberska, P. (1987)* On the convergence of triangular matrix means of Fourier series. *Funct. Approx. Comment. Math.* (Poznań), **17**, 121–130. MR 88g:42009

Pych-Taberska, P. (1988)* On the rate of convergence of Sturm-Liouville expansions. *Bull. Polish Acad. Sci. Math.*, **36**, 289–297 (1989). MR 92e:34118

Eigenfunction expansions of functions with bounded φ-variation with respect to eigenfunctions of Sturm-Liouville problems.

Pych-Taberska, P. (1989)* On the rate of pointwise convergence of Szász-Mirakyan operators. *Approximation and function spaces* (Warsawa, 1986), 323–330; Banach Center Publ., 22, PWN, Warsaw. MR 92d:41038

Pych-Taberska, P. (1990)* On the rate of pointwise convergence of double Fourier-Legendre series. *Fasc. Math.*, **19**, 207–213. MR 92c:42029

  Convergence of Fourier-Legendre series for functions with bounded ϕ-variation of two variables.

Pych-Taberska, P. (1990). Approximation properties of the partial sums of Fourier series of some almost periodic functions. *Studia Math.*, **96**, 91–103. MR 91k:42020

Ramazanov, A.-R. K. (1981). Uniform rational approximations of functions with derivatives of finite Φ-variation. (Russian) *Vestnik Moskov. Univ. Ser. I Mat. Mekh.* no. 5, 15–19; transl. in: *Moscow Univ. Math. Bull.*, **36**, 17–21. MR 84h:41033

  The order of the best approximation of a function by rational functions of order n.

Ramazanov, A.-R.K. (1994)* On approximation of functions in terms of Φ-variation. *Anal. Math.* (Budapest), **20**, 263–281. MR 95k:41017

  An analogue of Jackson's theorem on the rate of convergence to zero of the best polynomial approximation.

Salem, R. (1940). Essais sur les séries trigonométriques. *Actualités Scientifiques et Industrielles*, 862, Paris. Reprinted in: Œuvres Mathématiques, pp. 111–195. Hermann, Paris 1967. MR 2, 93c *F

  Uniform convergence of the Fourier series of a function of bounded ϕ-variation.

Schembari, N. P. and Schramm, M. (1990). $\Phi V[h]$ and Riemann-Stieltjes integration. *Colloq. Math.*, **60/61**, 421–441. MR 92g:26013

  Several modifications of ϕ-variation are compared.

Schramm, M. (1985). Functions of Φ-bounded variation and Riemann-Stieltjes integration. *Trans. Amer. Math. Soc.*, **287**, 49–63. MR 86d:26018 *I *F

  A generalization and comparison of several notions of bounded ϕ-variation with applications to Fourier series and existence of Riemann-Stieltjes integral.

Sevast'janov, E. A. (1974). Uniform approximations by piecewise monotone functions, and some applications of them to Φ-variations and Fourier series. (Russian) *Dokl. Akad. Nauk SSSR*, **217**, 27–30. MR 50#2767

  Bounds of the ϕ-variation of a function f by a series containing the best uniform approximation $M_n(f)$ by means of piecewise monotone functions having n intervals of monotonocity.

Sevast'janov, E. A. (1975)* Piecewise monotone approximation, and Φ-variations. (Russian) *Anal. Math.* , **1**, 141–164. MR 52#8747

  Continuation of Sevast'janov (1974).

Sevast'yanov, E. A. (1983)* Piecewise-monotone approximations. (Russian) *Constructive function theory '81* (Varna, 1981), 150–156; Bulgar. Acad. Sci. Sofia. MR 85b:41029

Relations between smallest piecewise-monotone deviations and ϕ-variation of the function being approximated.

Szelmeczka, J. (1986)* On convergence of singular integrals in the generalized variation metric. *Funct. Approx. Comment. Math.* (Poznań), **15**, 53–58. MR 88e:41053

An approximation of certain functions of bounded ϕ-variation by convolution operators.

Taberski, R. (1971). Trigonometric interpolation. III. *Colloq. Math.*, **23**, 145–156. MR 46 #2335 *F

An extension of some Fourier series convergence results of Young (1938c).

Taberski, R. (1971). Some properties of M-variation. *Comment. Math. Prace Mat.*, **15**, 141–146. MR 47 #412

Several theorems from Marcinkiewicz (1934) are extended to functions with bounded ϕ-variation.

Taberski, R. (1985). On variation of the Wiener type. *Demonstratio Math.*, **18**, 161–175. MR 87d:26012 *F

Properties of the ϕ-variation and its modifications with applications to Fourier series and Stieltjes integrals.

Taylor, S. J. (1972). Exact asymptotic estimates of Brownian path variation. *Duke Math. J.*, **39**, 219–241. MR 45 #4500 *P

It is proved that $V_\psi(B;[0,1]) < \infty$ almost surely, where B is a Brownian motion and $\psi(u) = u^2/(2\log_e |\log_e u|)$ for $u > 0$, while if $\psi(u) = o(\eta(u))$ as $u \downarrow 0$ then $V_\eta(B,[0,1]) = +\infty$ almost surely.

Trofimov, E. I. (1990). Standardized supports and limit theorems for semimartingales. (Russian) *Teor. Veroyatnost. i Primenen.* **35** (1990), no. 3, 515–530; transl. in: *Theory Probab. Appl.* **35**, (1990), no. 3, 523–538 (1991). MR 92f:60052 *P

Relations between integrated Lipschitz conditions and ϕ-variation are explored.

Vaĭnerman, Ju. R. (1972). Sufficient conditions for the existence of the Stieltjes integral. (Russian) *Vestnik Leningrad. Univ.*, No. **13**, Mat. Meh. Astronom. Vyp. 3, 15–20. MR 49 #3058 *I

Let ρ, χ be moduli of continuity of continuous functions f, g, respectively, such that the integral $\Phi(s) := \int_0^s \rho(t)t^{-1}d\chi(t)$ exists for $s = 1$. Then the Riemann-Stieltjes integral I of f with respect to g exists, and, for any Riemann-Stieltjes sum S based on a partition with the mesh less than s, the inequality $|I - S| \leq c\Phi(s)$ holds.

Vinti, G. (1994). Generalized ϕ-variation in the sense of Vitali: estimates for integral operators and applications in fractional calculus. *Comment. Math. Prace Mat.*, **34**, 199–213. Errata in: *ibid.*, **35**, 301. MR 96c:47073

An extension of Bardaro and Vinti (1991a) to the case when the function ϕ may depend on a parameter in $\mathbb{R}^n$.

Young, L. C. (1937a). Sur une généralisation de la notion de variation de puissance p-ième bornée au sens de N. Wiener et sur la convergence des séries de Fourier. *C. R. Acad. Sci. Paris Sér. A*, **204**, 470–472. *F

    An extension of Young (1936) to functions of bounded φ-variation.

Young, L. C. (1938d). General inequalities for Stieltjes integrals and the convergence of Fourier series. *Math. Ann.*, **115**, 581–612. *F

    An existence of Stieltjes integrals for functions with bounded φ-variation.

Young, L. C. (1943). A further inequality for Stieltjes integrals. *J. London Math. Soc.*, **18**, 78–82. MR 5, 202d *I *F

    An extension of the inequality for Stieltjes integrals in terms of φ-variation given by Young (1938b).

Young, L. C. (1970). Some new stochastic integrals and Stieltjes integrals. I. Analogues of Hardy-Littlewood classes. *Advances in Probability and Related Topics*, Vol. II, Ed. P. Ney, Dekker, New York, 161–240. MR 44 #4177 *I *P

    A definition of a stochastic integral of a deterministic function with respect to a stochastic process possibly having non-orthogonal increments over non-overlapping time intervals.

Young, L. C. (1974). Some new stochastic integrals and Stieltjes integrals, Part II. *Advances in Probability and Related Topics*, Vol. III, Ed. P. Ney, Dekker, New York, 101–178. MR 54 #8863 *I *P

    An extension of the first part to the case when the integrand is a stochastic process.

Young, L. C. (1976). A converse to some inequalities and approximations in the theory of Stieltjes and stochastic integrals, and for nth derivatives. *Studia Math.*, **55**, 215–223. MR 54 #3849 *I

    Optimality of the φ-variation conditions for the existence of integrals established by L. C. Young (1938c, 1943, 1970).

Zerekidze, T. Š. (1980)* Uniform convergence of Fourier series. (Russian) *Boundary properties of analytic functions, singular integral equations and some questions of harmonic analysis.* Akad. Nauk Gruzin. SSR Trudy Tbiliss. Mat. Inst. Razmadze, **65**, 85–98. MR 82a:42011

    The equivalence of two tests for the uniform convergence of Fourier series for continuous functions of bounded φ-variation in two variables.

Zinčenko, N. M. (1974)* The ψ-variation of a Wiener random field. (Russian) *Theory of random processes. Questions of statistics and control*, pp. 47–56; Izdanie Inst. Mat. Akad. Nauk Ukrain. SSR, Kiev. MR 58 #3007 *P *M

    Conditions on φ for boundedness of φ-variation of the n-parameter Brownian motion.

SUBJECT INDEX

The reference code is *Part.(sub)section*; page(s).

The author index is compiled only for Parts I, II and III, and not for their reference lists (each reference is cited in the text). The reference code is *Part.(sub)section*; page(s).

[1] L. C. Young's paper listed as "1938b" in Part I is his only 1938 paper cited in Parts II and III.

Druck: Strauss Offsetdruck, Mörlenbach
Verarbeitung: Schäffer, Grünstadt

Vol. 1653: R. Benedetti, C. Petronio, Branched Standard Spines of 3-manifolds. VIII, 132 pages. 1997.

Vol. 1654: R. W. Ghrist, P. J. Holmes, M. C. Sullivan, Knots and Links in Three-Dimensional Flows. X, 208 pages. 1997.

Vol. 1655: J. Azéma, M. Emery, M. Yor (Eds.), Séminaire de Probabilités XXXI. VIII, 329 pages. 1997.

Vol. 1656: B. Biais, T. Björk, J. Cvitanic, N. El Karoui, E. Jouini, J. C. Rochet, Financial Mathematics. Bressanone, 1996. Editor: W. J. Runggaldier. VII, 316 pages. 1997.

Vol. 1657: H. Reimann, The semi-simple zeta function of quaternionic Shimura varieties. IX, 143 pages. 1997.

Vol. 1658: A. Pumarino, J. A. Rodrıguez, Coexistence and Persistence of Strange Attractors. VIII, 195 pages. 1997.

Vol. 1659: V, Kozlov, V. Maz'ya, Theory of a Higher-Order Sturm-Liouville Equation. XI, 140 pages. 1997.

Vol. 1660: M. Bardi, M. G. Crandall, L. C. Evans, H. M. Soner, P. E. Souganidis, Viscosity Solutions and Applications. Montecatini Terme, 1995. Editors: I. Capuzzo Dolcetta, P. L. Lions. IX, 259 pages. 1997.

Vol. 1661: A. Tralle, J. Oprea, Symplectic Manifolds with no Kähler Structure. VIII, 207 pages. 1997.

Vol. 1662: J. W. Rutter, Spaces of Homotopy Self-Equivalences – A Survey. IX, 170 pages. 1997.

Vol. 1663: Y. E. Karpeshina; Perturbation Theory for the Schrödinger Operator with a Periodic Potential. VII, 352 pages. 1997.

Vol. 1664: M. Väth, Ideal Spaces. V, 146 pages. 1997.

Vol. 1665: E. Giné, G. R. Grimmett, L. Saloff-Coste, Lectures on Probability Theory and Statistics 1996. Editor: P. Bernard. X, 424 pages, 1997.

Vol. 1666: M. van der Put, M. F. Singer, Galois Theory of Difference Equations. VII, 179 pages. 1997.

Vol. 1667: J. M. F. Castillo, M. González, Three-space Problems in Banach Space Theory. XII, 267 pages. 1997.

Vol. 1668: D. B. Dix, Large-Time Behavior of Solutions of Linear Dispersive Equations. XIV, 203 pages. 1997.

Vol. 1669: U. Kaiser, Link Theory in Manifolds. XIV, 167 pages. 1997.

Vol. 1670: J. W. Neuberger, Sobolev Gradients and Differential Equations. VIII, 150 pages. 1997.

Vol. 1671: S. Bouc, Green Functors and G-sets. VII, 342 pages. 1997.

Vol. 1672: S. Mandal, Projective Modules and Complete Intersections. VIII, 114 pages. 1997.

Vol. 1673: F. D. Grosshans, Algebraic Homogeneous Spaces and Invariant Theory. VI, 148 pages. 1997.

Vol. 1674: G. Klaas, C. R. Leedham-Green, W. Plesken, Linear Pro-p-Groups of Finite Width. VIII, 115 pages. 1997.

Vol. 1675: J. E. Yukich, Probability Theory of Classical Euclidean Optimization Problems. X, 152 pages. 1998.

Vol. 1676: P. Cembranos, J. Mendoza, Banach Spaces of Vector-Valued Functions. VIII, 118 pages. 1997.

Vol. 1677: N. Proskurin, Cubic Metaplectic Forms and Theta Functions. VIII, 196 pages. 1998.

Vol. 1678: O. Krupková, The Geometry of Ordinary Variational Equations. X, 251 pages. 1997.

Vol. 1679: K.-G. Grosse-Erdmann, The Blocking Technique. Weighted Mean Operators and Hardy's Inequality. IX, 114 pages. 1998.

Vol. 1680: K.-Z. Li, F. Oort, Moduli of Supersingular Abelian Varieties. V, 116 pages. 1998.

Vol. 1681: G. J. Wirsching, The Dynamical System Generated by the 3n+1 Function. VII, 158 pages. 1998.

Vol. 1682: H.-D. Alber, Materials with Memory. X, 166 pages. 1998.

Vol. 1683: A. Pomp, The Boundary-Domain Integral Method for Elliptic Systems. XVI, 163 pages. 1998.

Vol. 1684: C. A. Berenstein, P. F. Ebenfelt, S. G. Gindikin, S. Helgason, A. E. Tumanov, Integral Geometry, Radon Transforms and Complex Analysis. Firenze, 1996. Editors: E. Casadio Tarabusi, M. A. Picardello, G. Zampieri. VII, 160 pages. 1998.

Vol. 1685: S. König, A. Zimmermann, Derived Equivalences for Group Rings. X, 146 pages. 1998.

Vol. 1686: J. Azéma, M. Émery, M. Ledoux, M. Yor (Eds.), Séminaire de Probabilités XXXII. VI, 440 pages. 1998.

Vol. 1687: F. Bornemann, Homogenization in Time of Singularly Perturbed Mechanical Systems. XII, 156 pages. 1998.

Vol. 1688: S. Assing, W. Schmidt, Continuous Strong Markov Processes in Dimension One. XII, 137 page. 1998.

Vol. 1689: W. Fulton, P. Pragacz, Schubert Varieties and Degeneracy Loci. XI, 148 pages. 1998.

Vol. 1690: M. T. Barlow, D. Nualart, Lectures on Probability Theory and Statistics. Editor: P. Bernard. VIII, 237 pages. 1998.

Vol. 1691: R. Bezrukavnikov, M. Finkelberg, V. Schechtman, Factorizable Sheaves and Quantum Groups. X, 282 pages. 1998.

Vol. 1692: T. M. W. Eyre, Quantum Stochastic Calculus and Representations of Lie Superalgebras. IX, 138 pages. 1998.

Vol. 1694: A. Braides, Approximation of Free-Discontinuity Problems. XI, 149 pages. 1998.

Vol. 1695: D. J. Hartfiel, Markov Set-Chains. VIII, 131 pages. 1998.

Vol. 1696: E. Bouscaren (Ed.): Model Theory and Algebraic Geometry. XV, 211 pages. 1998.

Vol. 1697: B. Cockburn, C. Johnson, C.-W. Shu, E. Tadmor, Advanced Numerical Approximation of Nonlinear Hyperbolic Equations. Cetraro, Italy, 1997. Editor: A. Quarteroni. VII, 390 pages. 1998.

Vol. 1698: M. Bhattacharjee, D. Macpherson, R. G. Möller, P. Neumann, Notes on Infinite Permutation Groups. XI, 202 pages. 1998.

Vol. 1699: A. Inoue, Tomita-Takesaki Theory in Algebras of Unbounded Operators. VIII, 241 pages. 1998.

Vol. 1700: W. A. Woyczyński, Burgers-KPZ Turbulence, XI, 318 pages. 1998.

Vol. 1701: Ti-Jun Xiao, J. Liang, The Cauchy Problem of Higher Order Abstract Differential Equations, XII, 302 pages. 1998.

Vol. 1702: J. Ma, J. Yong, Forward-Backward Stochastic Differential Equations and Their Applications. XIII, 270 pages. 1999.

Vol. 1703: R. M. Dudley, R. Norvaiša, Differentiability of Six Operators on Nonsmooth Functions and p-Variation. VIII, 272 pages. 1999.

4. Lecture Notes are printed by photo-offset from the master-copy delivered in camera-ready form by the authors. Springer-Verlag provides technical instructions for the preparation of manuscripts. Macro packages in T_EX, L^AT_EX2e, $L^AT_EX2.09$ are available from Springer's web-pages at http://www.springer.de/math/authors. Careful preparation of the manuscripts will help keep production time short and ensure satisfactory appearance of the finished book.

The actual production of a Lecture Notes volume takes approximately 12 weeks.

5. Authors receive a total of 50 free copies of their volume, but no royalties. They are entitled to a discount of 33.3 % on the price of Springer books purchase for their personal use, if ordering directly from Springer-Verlag.

Commitment to publish is made by letter of intent rather than by signing a formal contract. Springer-Verlag secures the copyright for each volume. Authors are free to reuse material contained in their LNM volumes in later publications: A brief written (or e-mail) request for formal permission is sufficient.

Addresses:

Professor F. Takens, Mathematisch Instituut,
Rijksuniversiteit Groningen, Postbus 800,
9700 AV Groningen, The Netherlands
E-mail: F.Takens@math.rug.nl

Professor B. Teissier, DMI, École Normale Supérieure
45, rue d'Ulm,
F-7500 Paris, France
E-mail: Teissier@ens.fr

Springer-Verlag, Mathematics Editorial, Tiergartenstr. 17,
D-69121 Heidelberg, Germany,
Tel.: *49 (6221) 487-701
Fax: *49 (6221) 487-355
E-mail: C.Byrne@Springer.de

GPSR Compliance

*The European Union's (EU) General Product Safety Regulation (GPSR)
is a set of rules that requires consumer products to be safe and our
obligations to ensure this.*

*If you have any concerns about our products, you can contact us on
ProductSafety@springernature.com*

In case Publisher is established outside the EU, the EU authorized
representative is:

Springer Nature Customer Service Center GmbH
Europaplatz 3
69115 Heidelberg, Germany

Batch number: 09624486

Printed by Printforce, the Netherlands